Pesticides: Advances in Chemical and Botanical Pesticides

Pesticides: Advances in Chemical and Botanical Pesticides

Editor:

Valeria Severino, Ph.D.

Delve Publishing

www.delvepublishing.com

Pesticides: Advances in Chemical and Botanical Pesticides

Editor: Valeria Severino, Ph.D.

708 3rd Avenue, 6th Floor
New York, NY 10017
United States of America
www.delvepublishing.com

ISBN: 978-1-68095-759-4

Library of Congress Control Number: 2016951515

About the Editor

Valeria Severino, Ph.D.

Valeria Severino obtained her PhD in Biological Processes and Biomolecules from the Second University of Naples in 2009. Her scientific interests concern the structural and functional characterization of proteins and peptides by using biochemical approaches. Fields of expertise comprises food chemistry, cellular biology, natural products and related applications. She published more than 40 papers on international peer-reviewed journals.

Contents

List of Contributors

Neela Guha
International Agency for Research on Cancer, 150 cours Albert Thomas, 69008 Lyon, France

Kathryn Z. Guyton
International Agency for Research on Cancer, 150 cours Albert Thomas, 69008 Lyon, France

Dana Loomis
International Agency for Research on Cancer, 150 cours Albert Thomas, 69008 Lyon, France

Dinesh Kumar Barupal
International Agency for Research on Cancer, 150 cours Albert Thomas, 69008 Lyon, France

David R. George
Stockbridge Technology Centre, North Yorkshire YO8 3TZ, UK
Department of Applied Sciences, Northumbria University, Newcastle Upon Tyne NE1 8ST, UK

Jennifer A. Banfield-Zanin
Stockbridge Technology Centre, North Yorkshire YO8 3TZ, UK

Rosemary Collier
Warwick Crop Centre, Warwick CV35 9EF, UK

Jerry Cross
East Malling Research, Kent ME19 6BJ, UK

A. Nicholas E. Birch
James Hutton Institute, Dundee DD2 5DA, UK

Roma Gwynn
Rationale, Duns TD11 3QA, UK

Tim O'Neill
ADAS, Cambridge CB23 4NN, UK

Fariba Taghavian
Department of Biology, Damghan Branch, Islamic Azad University, Semnan, Iran

Gholamhassan Vaezi
Department of Biology, Damghan Branch, Islamic Azad University, Semnan, Iran

Mohammad Abdollahi
Faculty of Pharmacy and Pharmaceutical Sciences Research Center, Tehran University of Medical Science, Tehran, Iran

Ali Akbar Malekirad
Department of Biology, Payame Noor University, Tehran, Iran

Xusheng Huang
Department of Neurology, Chinese PLA General Hospital, 28 Fuxing Road, Beijing 100853, P. R. China.

Chao Zhang
School of Management and Economics, Beijing Institute of Technology, 5 South Zhongguancun Street, Beijing 100081, P. R. China.

Ruifa Hu
School of Management and Economics, Beijing Institute of Technology, 5 South Zhongguancun Street, Beijing 100081, P. R. China.

Yifan Li
Department of Neurology, the First Affiliated Hospital of PLA General Hospital, No. 51 Fucheng Road, Beijing 100048, P. R. China.

YanhongYin
Department of Neurology, Chinese PLA General Hospital, 28 Fuxing Road, Beijing 100853, P. R. China.

ZhaohuiChen
Department of Neurology, Chinese PLA General Hospital, 28 Fuxing Road, Beijing 100853, P. R. China.

JinyangCai
School of Management and Economics, Beijing Institute of Technology, 5 South Zhongguancun Street, Beijing 100081, P. R. China.

Fang Cui
Department of Neurology, Chinese PLA General Hospital, 28 Fuxing Road, Beijing 100853, P. R. China.

Ruifa Hu
School of Management and Economics, Beijing Institute of Technology, Beijing, China

Xusheng Huang
Department of Neurology, Chinese PLA General Hospital, Beijing, China

Jikun Huang
Center for Chinese Agricultural Policy, Institute of Geographic Sciences and Natural Resources Research, Chinese Academy of Sciences, Beijing, China

Yifan Li
Department of Neurology, Chinese PLA General Hospital, Beijing, China

Chao Zhang
School of Management and Economics, Beijing Institute of Technology, Beijing, China

Yanhong Yin
Department of Neurology, Chinese PLA General Hospital, Beijing, China

Zhaohui Chen
Department of Neurology, Chinese PLA General Hospital, Beijing, China

Yanhong Jin
Department of Agricultural, Food and Resource Economics, Rutgers University, New Brunswick, New Jersey, United States of America

Jinyang Cai
School of Management and Economics, Beijing Institute of Technology, Beijing, China

Fang Cui
Department of Neurology, Chinese PLA General Hospital, Beijing, China

Lúcia Maria de Miranda Adad
Cytogenetics Laboratory, Federal Institute of Education, Science and Technology of Piauí, Teresina, PI, Brazil
Graduate Program in Genetics and Applied Toxicology, Lutheran University of Brazil, Canoas, RS, Brazil

Heloísa Helena Rodrigues de Andrade
Stomatology Laboratory, Hospital das Clinicas de Porto Alegre, Federal University of Rio Grande do Sul, Porto Alegre, RS, Brazil

Kátia Kvitko
Graduate Program in Genetics and Molecular Biology, Federal University of Rio Grande do Sul, Porto Alegre, RS, Brazil

Mauricio Lehmann
Graduate Program in Genetics and Applied Toxicology, Lutheran University of Brazil, Canoas, RS, Brazil
Graduate Program in Cellular and Molecular Biology Applied to Health, Lutheran University of Brazil, Canoas, RS, Brazil

Ana Amélia de Carvalho Melo Cavalcante
UNINOVAFAPI University Center, Medical School, Teresina, PI, Brazil

Rafael Rodrigues Dihl
Graduate Program in Genetics and Applied Toxicology, Lutheran University of Brazil, Canoas, RS, Brazil
Graduate Program in Cellular and Molecular Biology Applied to Health, Lutheran University of Brazil, Canoas, RS, Brazil

Bettina Seeger
Institute for Food Toxicology and Analytical Chemistry, University of Veterinary Medicine Hannover, Foundation, Bischofsholer Damm 15, 30173, Hannover, Germany

Frank Klawonn
Biostatistics Group, Helmholtz Centre for Infection Research, Inhoffenstr. 7, 38124, Braunschweig, Germany
Department of Computer Science, Ostfalia University of Applied Sciences, Salzdahlumerstr. 46/48, 38302, Wolfenbüttel, Germany

Boris Nguema Bekale
Department of Computer Science, Ostfalia University of Applied Sciences, Salzdahlumerstr. 46/48, 38302, Wolfenbüttel, Germany

Pablo Steinberg
Institute for Food Toxicology and Analytical Chemistry, University of Veterinary Medicine Hannover, Foundation, Bischofsholer Damm 15, 30173, Hannover, Germany

João Bosco de Salles
State University of West Zone, Rio de Janeiro, RJ, Brazil

Renato Matos Lopes
Mobile Communication Laboratory, Oswaldo Cruz Foundation, Instituto Oswaldo Cruz Avenue Brazil 4365, Manguinhos, 21045-900 Rio de Janeiro, RJ, Brazil

Cristiane M. C. de Salles
Section of Biochemistry, Department of Chemistry, Federal Rural University of Rio de Janeiro, Seropédica, RJ, Brazil

Vicente P. F. Cassano
Department of Biology, Loyola Marymount University, Los Angeles, CA, USA

Manildo Marcião de Oliveira
Ecotoxicology and Environmental Microbiology Laboratory (LEMAM), Federal Institute of Education, Science and Technology Fluminense, Cabo Frio, RJ, Brazil

Vera L. F. Cunha Bastos
Department of Biochemistry, Institute of Biology Roberto Alcantara Gomes, State University of Rio de Janeiro, Rio de Janeiro, RJ, Brazil

Jayme Cunha Bastos
Department of Biochemistry, Institute of Biology Roberto Alcantara Gomes, State University of Rio de Janeiro, Rio de Janeiro, RJ, Brazil

Nelson Mpumi
School of Materials, Energy, Water and Environmental Sciences, Nelson Mandela African Institution of Science and Technology, Arusha, Tanzania

Kelvin Mtei
School of Materials, Energy, Water and Environmental Sciences, Nelson Mandela African Institution of Science and Technology, Arusha, Tanzania

Revocatus Machunda
School of Materials, Energy, Water and Environmental Sciences, Nelson Mandela African Institution of Science and Technology, Arusha, Tanzania

Patrick A. Ndakidemi
School of Materials, Energy, Water and Environmental Sciences, Nelson Mandela African Institution of Science and Technology, Arusha, Tanzania

Manas Sarkar
Research and Development Division, Godrej Consumers Product Ltd., Mumbai 400079, India

Rajendra Kshirsagar
Research and Development Division, Godrej Consumers Product Ltd., Mumbai 400079, India

Haijing Yi
Plant Science and Technology College, Beijing University of Agriculture, Beijing, 102206, China
Beijing Key Laboratory of New Technology in Agriculture Application, Beijing University of Agriculture, 102206, Beijing, China

Yan Chen
Plant Science and Technology College, Beijing University of Agriculture, Beijing, 102206, China
Beijing Key Laboratory of New Technology in Agriculture Application, Beijing University of Agriculture, 102206, Beijing, China

Jun Liu
Institute of Microbiology, Chinese Academy of Sciences, 100101, Beijing, China

Jie Zhang
Plant Science and Technology College, Beijing University of Agriculture, Beijing, 102206, China
Beijing Key Laboratory of New Technology in Agriculture Application, Beijing University of Agriculture, 102206, Beijing, China

Wei Guo
Plant Science and Technology College, Beijing University of Agriculture, Beijing, 102206, China
Beijing Key Laboratory of New Technology in Agriculture Application, Beijing University of Agriculture, 102206, Beijing, China

Weilie Xiao
State Key Laboratory of Phytochemistry and Plant Resources in West China, Kunming Institute of Botany, Chinese Academy of Sciences, Kunming, 650201, China

Yuncong Yao
Plant Science and Technology College, Beijing University of Agriculture, Beijing, 102206, China
Beijing Key Laboratory of New Technology in Agriculture Application, Beijing University of Agriculture, 102206, Beijing, China

Lina Lu
Institute of Pesticide Science, College of Plant Protection, Northwest A & F University, Yangling 712100, Shaanxi, China

Zhijun Qi
Institute of Pesticide Science, College of Plant Protection, Northwest A & F University, Yangling 712100, Shaanxi, China
Key Laboratory of Botanical Pesticide R & D in Shaanxi Province, Yangling 712100, Shaanxi, China

Qiuli Li
Institute of Pesticide Science, College of Plant Protection, Northwest A & F University, Yangling 712100, Shaanxi, China

Wenjun Wu
Institute of Pesticide Science, College of Plant Protection, Northwest A & F University, Yangling 712100, Shaanxi, China
Key Laboratory of Botanical Pesticide R & D in Shaanxi Province, Yangling 712100, Shaanxi, China

Abhishek Singh Rathore
Faculty of Life Sciences and Biotechnology, South Asian University, New Delhi 110021, India

Rinkoo D. Gupta
Faculty of Life Sciences and Biotechnology, South Asian University, New Delhi 110021, India

Yan Wu
Technical Center of China Tobacco Guangxi Industrial Co., Ltd, Nanning 530001, Guangxi, China

Wen-Juan Zhang
Beijing Key Laboratory of Traditional Chinese Medicine Protection and Utilization, Beijing Normal University, No.19 Xinjiekouwai Street, Beijing 100875, China

Ping-Juan Wang
Technical Center of China Tobacco Guangxi Industrial Co., Ltd, Nanning 530001, Guangxi, China

Kai Yang
Beijing Key Laboratory of Traditional Chinese Medicine Protection and Utilization, Beijing Normal University, No.19 Xinjiekouwai Street, Beijing 100875, China

Dong-Ye Huang
Technical Center of China Tobacco Guangxi Industrial Co., Ltd, Nanning 530001, Guangxi, China

Jian-Yu Wei
Technical Center of China Tobacco Guangxi Industrial Co., Ltd, Nanning 530001, Guangxi, China

Zhao-Fu Tian
Technical Center of China Tobacco Guangxi Industrial Co., Ltd, Nanning 530001, Guangxi, China

Jia-Feng Bai
Technical Center of China Tobacco Guangxi Industrial Co., Ltd, Nanning 530001, Guangxi, China

Shu-Shan Du
Beijing Key Laboratory of Traditional Chinese Medicine Protection and Utilization, Beijing Normal University, No.19 Xinjiekouwai Street, Beijing 100875, China

Linlin Wei
College of Plant Protection, Anhui Agricultural University, Hefei 230036, China

Rimao Hua
College of Resources and Environment, Anhui Agricultural University, Hefei 230036, China

Maoye Li
College of Plant Protection, Anhui Agricultural University, Hefei 230036, China

Yanzhang Huang
College of Plant Protection, Anhui Agricultural University, Hefei 230036, China

Shiguang Li
College of Plant Protection, Anhui Agricultural University, Hefei 230036, China

Yujie He
College of Plant Protection, Anhui Agricultural University, Hefei 230036, China

Zonghai Shen
Anhui Zhonggu State Grain Reserve Warehouse, Hefei 231131, China

Suelen L. Reis
Biotechnology Unit, University of Ribeirao Preto, Ribeirao Preto, São Paulo 14096-900, Brazil

Anieli G. Mantello
Biotechnology Unit, University of Ribeirao Preto, Ribeirao Preto, São Paulo 14096-900, Brazil

Jeferson M. Macedo
Biotechnology Unit, University of Ribeirao Preto, Ribeirao Preto, São Paulo 14096-900, Brazil

Erica A. Gelfuso
Biotechnology Unit, University of Ribeirao Preto, Ribeirao Preto, São Paulo 14096-900, Brazil

Cássio P. da Silva
Biotechnology Unit, University of Ribeirao Preto, Ribeirao Preto, São Paulo 14096-900, Brazil

Ana L. Fachin
Biotechnology Unit, University of Ribeirao Preto, Ribeirao Preto, São Paulo 14096-900, Brazil

Alexandre M. Cardoso
Federal Institute of Education, Science and Technology of São Paulo (IFSP), Barretos, São Paulo 14781-502, Brazil

Rene O. Beleboni
Biotechnology Unit, University of Ribeirao Preto, Ribeirao Preto, São Paulo 14096-900, Brazil

Songqing Wu
College of Forestry, Fujian Agriculture and Forestry University, Fuzhou, 350002, People's Republic of China
Fujian-Taiwan Joint Center for Ecological Control of Crop Pests, Fujian Agriculture and Forestry University, Fuzhou, 350002, People's Republic of China
Key Laboratory of Biopesticide and Chemical Biology, Ministry of Education, Fujian Agriculture and Forestry University, Fuzhou, 350002, People's Republic of China

Xiaoli Zhu
Fujian-Taiwan Joint Center for Ecological Control of Crop Pests, Fujian Agriculture and Forestry University, Fuzhou, 350002, People's Republic of China
Key Laboratory of Biopesticide and Chemical Biology, Ministry of Education, Fujian Agriculture and Forestry University, Fuzhou, 350002, People's Republic of China

Zhaoxia Liu
Fujian-Taiwan Joint Center for Ecological Control of Crop Pests, Fujian Agriculture and Forestry University, Fuzhou, 350002, People's Republic of China
Key Laboratory of Biopesticide and Chemical Biology, Ministry of Education, Fujian Agriculture and Forestry University, Fuzhou, 350002, People's Republic of China

Ensi Shao
Fujian-Taiwan Joint Center for Ecological Control of Crop Pests, Fujian Agriculture and Forestry University, Fuzhou, 350002, People's Republic of China
Key Laboratory of Biopesticide and Chemical Biology, Ministry of Education, Fujian Agriculture and Forestry University, Fuzhou, 350002, People's Republic of China

CarballarLejarazú Rebeca
Department of Molecular Biology and Biochemistry, University of California Irvine, Irvine, CA, 92697, United States of America

Yajie Guo
College of Forestry, Fujian Agriculture and Forestry University, Fuzhou, 350002, People's Republic of China

Yueting Xiong
College of Forestry, Fujian Agriculture and Forestry University, Fuzhou, 350002, People's Republic of China
Fujian-Taiwan Joint Center for Ecological Control of Crop Pests, Fujian Agriculture and Forestry University, Fuzhou, 350002, People's Republic of China

Yani Mou
College of Forestry, Fujian Agriculture and Forestry University, Fuzhou, 350002, People's Republic of China
Fujian-Taiwan Joint Center for Ecological Control of Crop Pests, Fujian Agriculture and Forestry University, Fuzhou, 350002, People's Republic of China

Runxue Xu
Fujian-Taiwan Joint Center for Ecological Control of Crop Pests, Fujian Agriculture and Forestry University, Fuzhou, 350002, People's Republic of China

Xia Hu
College of Forestry, Fujian Agriculture and Forestry University, Fuzhou, 350002, People's Republic of China

Guanghong Liang
College of Forestry, Fujian Agriculture and Forestry University, Fuzhou, 350002, People's Republic of China

Shuangquan Zou
College of Forestry, Fujian Agriculture and Forestry University, Fuzhou, 350002, People's Republic of China

Xiong Guan
Fujian-Taiwan Joint Center for Ecological Control of Crop Pests, Fujian Agriculture and Forestry University, Fuzhou, 350002, People's Republic of China
Key Laboratory of Biopesticide and Chemical Biology, Ministry of Education, Fujian Agriculture and Forestry University, Fuzhou, 350002, People's Republic of China

Feiping Zhang
College of Forestry, Fujian Agriculture and Forestry University, Fuzhou, 350002, People's Republic of China

Guangjie Han
Department of Biological Pesticides, Jiangsu Lixiahe Institute of Agricultural Sciences, Yangzhou, 225007, PR China

Jian Xu
Department of Biological Pesticides, Jiangsu Lixiahe Institute of Agricultural Sciences, Yangzhou, 225007, PR China

Qin Liu
Department of Biological Pesticides, Jiangsu Lixiahe Institute of Agricultural Sciences, Yangzhou, 225007, PR China

Chuanming Li
Department of Biological Pesticides, Jiangsu Lixiahe Institute of Agricultural Sciences, Yangzhou, 225007, PR China

Hongxing Xu
State Key Laboratory Breeding Base for Zhejiang Sustainable Pest and Disease Control, Institute of Plant Protection and Microbiology, Zhejiang Academy of Agricultural Science, Hangzhou, 310021, PR China

Zhongxian Lu
State Key Laboratory Breeding Base for Zhejiang Sustainable Pest and Disease Control, Institute of Plant Protection and Microbiology, Zhejiang Academy of Agricultural Science, Hangzhou, 310021, PR China

Jodel S. Alves
Genetics Laboratory Toxicology, Lutheran University of Brazil, ULBRA, Av. Ragamuffin, 8001, Building 22, Room 22, 4th floor, 92425-900 Canoas, RS, Brazil

Fernanda R. Da Silva
University Centre La Salle, Av. Victor Barreto, 2288, 92010-000 Canoas, RS, Brazil

Gabrieli F. Da Silva
Genetics Laboratory Toxicology, Lutheran University of Brazil, ULBRA, Av. Ragamuffin, 8001, Building 22, Room 22, 4th floor, 92425-900 Canoas, RS, Brazil

Mirian Salvador
Institute of Biotechnology, University of Caxias do Sul, Rua Francisco Vargas, 1130, 95070-560 Caxias do Sul, RS, Brazil

Kátia Kvitko
Graduate Program in Genetics and Molecular Biology, Department of Genetics, Federal University of Rio Grande do Sul, Av. Bento Gonçalves, 9500, 91501-970 Porto Alegre, RS, Brazil

Paula Rohr
Laboratory of Cell Biology and Molecular, Universidade do Extremo Sul Catarinense, Av. Universitaria, 1105, 88806-000 Criciuma, SC, Brazil

Carla E. I. Dos Santos
Institute of Physics, University of São Paulo, Matão Street, Lane R, 187, University City, 05508-900 São Paulo, SP, Brazil

Johnny F. Dias
Institute of Physics, Federal University of Rio Grande do Sul, Av. Bento Gonçalves, 9500, 91501-970 Porto Alegre, RS, Brazil

João A. P. Henriques
Institute of Biotechnology, University of Caxias do Sul, Rua Francisco Vargas, 1130, 95070-560 Caxias do Sul, RS, Brazil
Biotechnology Center, Federal University of Rio Grande do Sul, Av. Bento Gonçalves, 9500, 91501-970 Porto Alegre, RS, Brazil

Juliana Da Silva
Genetics Laboratory Toxicology, Lutheran University of Brazil, ULBRA, Av. Ragamuffin, 8001, Building 22, Room 22, 4th floor, 92425-900 Canoas, RS, Brazil

Preface

In general, a pesticide is a chemical or biological agent that deters, incapacitates, kills, or otherwise discourages pests. Target pests can include insects, plant pathogens, weeds, mollusks, birds, mammals, fish, nematodes (roundworms), and microbes that destroy property, cause nuisance, or spread disease, or are disease vectors. Although pesticides have benefits, some also have drawbacks, such as potential toxicity to humans and other species. Pesticides belong to the family of agrochemicals and consist in molecules able to act against pests, insects, fungi and various diseases present in agricultural and crop industry. Pesticides can be chemicals, biological agents, also known as bio-pesticides and the relatively recent plant-derived or botanical pesticides. Pesticides of chemical origin are highly effective against the target organism. However, the chemical pesticides have toxic effects on several non-target species and the growing environmental concerns raises the issues of safety and evaluation of toxicity. Even if commonly used in several fields, pesticides and especially the chemical ones, have toxic effects and can be highly dangerous on non-target species. Thus, several concerns have been raised on the safety and toxicity of pesticides.

In this book some advances in chemical and botanical pesticides are discussed. In particular, in a first part of the book new chemical pesticides and their persistence and toxic effects on human population are presented. In the second part botanical pesticides from different plant sources along with their principal challenges are discussed. Botanical pesticides are agricultural pest management agents which are based on plant extracts. In modern times these have been used as alternatives to synthetic chemicals in organic pest management. The practice of using plant materials against field and storage pests however has a long history in many indigenous and traditional farming communities across the world. Finally, interesting genetic and biochemical approaches used to study genes involved in resistance to pesticides and biomarkers in pesticide assays are also presented.

Editor
Valeria Severino, Ph.D.

INTRODUCTION

A pesticide is any substance or mixture of substances used to destroy, suppress or alter the life cycle of any pest. A pesticide can be a naturally derived or synthetically produced substance. A pesticide can also be an organism, for example, the bacterium Bacillus thuringiensis which is used to control a number of insect pests, or even a genetically modified crop.

Some plants contain components that are toxic to insects. When extracted from the plants and applied on infested crops, these components are called botanical pesticides or botanicals. The use of plant extracts to control pests is not new. Rotenone (Derris sp.), nicotine (tobacco), and pyrethrins (Chrysanthemum sp.) have been used widely both in small-scale subsistence farming as well as in commercial agriculture. Most botanical pesticides are contact, respiratory, or stomach poisons. Therefore, they are not very selective, but target a broad range of insects. This means that even beneficial organisms can be affected. Yet the toxicity of botanical pesticides is usually not very high and their negative effects on beneficial organisms can be significantly reduced by selective application. Furthermore, botanical pesticides are generally highly bio-degradable, so that they become inactive within hours or a few days. This reduces again the negative impact on beneficial organisms and they are relatively environmentally safe.

Pesticides are also sometimes broken down into chemical classes and modes of action. For example, fumigants are pesticides applied as gases to "sterilize" soil, and systemics work their way through a plant's tissue after being taken up at the root. Major chemical classes include: carbamates, organochlorines, organophosphates (mostly developed 70 or more years ago for chemical warfare) and triazines. Newer classes include pyrethroids and neonicotinoids, synthesized to mimic nature's pest protection.

A pesticide is any substance or mixture of substances used to destroy, suppress or alter the life cycle of any pest. A pesticide can be a naturally derived or synthetically produced substance. A pesticide can also be an organism, for example, the bacterium Bacillus thuringiensis which is used to control a number of insect pests, or even a genetically modified crop.

Chemical or biological substance designed to kill or retard the growth of pests that damage or interfere with the growth of crops, shrubs, tress, timber and other vegetation desired by humans. Practically all chemical pesticides, however, are poisons and pose long-term danger to the environment and humans through their persistence in nature and body tissue. Most of the pesticides are non-specific, and may kill life forms that are harmless or useful.

A pesticide is defined as a substance or mixture of substances intended for destroying, repelling or mitigating any pest. A pest is any living thing that exists where it is not wanted. Ideally, a pesticide should control a pest but not be detrimental to humans or other non-target organisms in the environment.

Preparation and Use of Botanical Pesticides

The preparation and use of botanicals requires some know-how, but not much material and infrastructures. It is a common practice under many traditional agricultural systems.

Neem

Neem, derived from the neem tree (Azadiracta indica) of arid tropical regions, contains several insecticidal compounds. The main active ingredient is azadiractin, which both deters and kills many species of caterpillars, thrips and whitefly. Both seeds and leaves can be used to prepare the neem solution. Neem seeds contain a higher amount of neem oil, but leaves are available all year. A neem solution looses its effectiveness within about 8 hours after preparation, and when exposed to direct sunlight. It is most effective to apply neem in the evening, directly after preparation, under humid conditions or when the plants and insects are damp. High neem concentrations can cause burning of plant leaves. Also, natural enemies can be affected by neem applications.

Pyrethrum

Pyrethrum is a daisy-like Chrysanthemum. In the tropics, pyrethrum is grown in mountain areas because it needs cool temperatures to develop its flowers. Pyrethrins are insecticidal chemicals extracted from the dried pyrethrum flower. The flower heads are processed into a powder to make a dust. This dust can be used directly or infused into water to make a spray. Pyrethrins cause immediate paralysis to most insects. Low doses do not kill but have a ′knock down effect. Stronger doses kill. Pyrethrins are not poisonous for humans and warm-blooded animals. However, human allergic reactions are common. It can cause rash, and breathing the dust can cause headaches and sickness.

Pyrethrins break down very quickly in sunlight so they should be stored in darkness. Both highly alkaline and highly acid conditions speed up degradation so pyrethrins should not be mixed with lime or soap solutions. Liquid formulations are stable in storage but powders may lose up to 20 percent of their effectiveness in one year.

Chacals Baobab

Farmers in Eastern Senegal are involved in cotton production. Some years ago, some of them turned to organic farming. To fight cotton pests, they use a natural insecticide based on chacals baobab (Adenium obesum, French: baobab du chacal). It has been used basically against Cotton ballworm (Heliotis sp.), but also against the spiny ballworm (Earias sp.) and Sudan ballworm (Diparopsis watersi). Chacals baobab is one of the most toxic plants in Africa.

Types of Pesticides

Some families or groups of chemical products which are considered pesticides under current are:

- *Bactericides*: These destroy, suppress or prevent the spread of bacteria. Examples are swimming pool chemicals containing chlorine, and products used to control black spot (bacterial blight) on garden plants or in orchards. Household disinfectants and some industrial disinfectants are excluded and not considered pesticides.
- *Baits*: These may be ‘ready to use’ products or products which need to be mixed with a food to control a pest. This category includes baits prepared for the control of large animals, such as foxes, wild dogs and rabbits, and baits for insects (such as cockroaches and ants) and molluscs (snail and slug pellets).
- *Fungicides*: These control, destroy, make harmless or regulate the effect of a fungus. Examples include chemicals used to treat Grey mould on grape vines and fruit trees, or Downy Mildew on cucumbers.
- *Genetically Modified Organisms (GMOs)*: Agricultural crops can be genetically modified to make them more resistant to pests and diseases, or tolerant to certain herbicides. For example, a gene from the bacterium Bacillus thuringiensis can be incorporated into cotton to provide protection against the larval stages of the cotton bollworm and native bollworm.
- *Herbicides*: These destroy, suppress or prevent the spread of a weed or other unwanted vegetation, for example, the

herbicide glyphosate is used to control a range of weeds in home gardens, bushland and agricultural situations.

- *Insecticides:* These destroy, suppress, stupefy, inhibit the feeding of, or prevent infestations or attacks by, an insect. Insecticides are used to control a wide variety of insect pests, including thrips, aphids, moths, fruit flies and locusts. Pesticides include products used on animals to control external parasites if they require dilution or mixing with water. Products applied directly to animals without dilution, injections or other medicines administered internally to treat animals are veterinary medicines.
- *Lures*: These are chemicals that attract a pest to a pesticide for the purpose of its destruction. Solely food-based lures, for example cheese in a mousetrap, are excluded and are not considered pesticides.
- *Rodenticides*: These are chemicals used specifically for controlling rodents such as mice and rats.
- *Repellents*: These repel rather than destroy a pest. Included in this category are personal insect repellents used to repel biting insects.

A number of living organisms that can control pests have also been registered as pesticides. Rabbit Haemorrhagic Disease, for example, has been used to control rabbit numbers; and bacteria that act as biological insecticides have been used to control various insect larvae, such as moths and mosquitoes.

Pesticides Work

Pesticides control pest organisms by physically, chemically or biologically interfering with their metabolism or normal behavior.

Most pesticides are lethal to target pests when applied at the rate specified on the pesticide label.

Some pesticides are non-lethal to the target pest. These include repellents or attractants (such as personal insect repellents), sterilising agents (which interfere with the reproductive ability of a pest), some defoliants (that cause leaf drop without killing the plant) and some pesticide products that boost the action of another pesticide without being particularly toxic themselves (such as piperonyl butoxide which may form part of pyrethrum-based insecticides).

Pesticides Reach the Target Pests

The route that brings a pesticide in contact with the target pest depends on the nature of the pesticide, how it is applied and the type of environment in which the pesticide is placed. Common application methods include spraying, fumigating and baiting.

Many pesticides are 'contact' pesticides. This means to be effective they must be absorbed through the external body surface or the exposed plant tissue; for example, tetramethrin used in household fly sprays, and bipyridillium herbicides such as paraquat.

Other pesticides are systemic in action. Systemic pesticides can be moved (translocated) from the site of application to another site within the plant or animal where they become effective; for example, insecticides that are absorbed by foliage and translocated throughout the plant where they kill chewing or sucking insects; or nematicides that are applied to the leaves of plants and are transferred to the roots of the plant to kill worms or caterpillars that are attacking the plant's roots.

Similarly, blood anticoagulant rodenticides in baits take effect once they have been transferred from the digestive system to the bloodstream of rats or mice.

Insecticides Enter the Insect Body

Insecticides enter the insect body by three common ways; by contact, as stomach poisons or as fumigants. Many pesticides may enter the body by more than one of these possible routes.

Contact Insecticides

Insecticides in this class kill pests by contacting and entering their bodies either directly through the insect integument (skin) into the blood or by entering the respiratory system through the spiracles (air valves along the side of an insect's body which control gas exchange). These materials may be applied directly to the insect body or as a residue on plant or animal surfaces, habitations or other places frequented by insects. In cases where residues are used, pests usually contact the pesticide through their tarsi (feet). Generally speaking, coarse sprays or dusts are a more effective means of applying contact insecticides than are mists or fogs. Most of the synthetic organic compounds act as contact insecticide, although many also confer stomach and fumigant activity.

Stomach Poisons

Stomach poisons must be swallowed in order to cause death. They may be formulated as liquids, dusts, pastes, granules or baits. In the case of liquids and dusts, the pesticide is usually applied to some substance on which the animal will chew or walk through. Pastes, bait and granules may be formulated with a feeding attractant which is consumed by the pest. Inorganic and botanical insecticides in general are predominantly stomach poisons. Some synthetic organic insecticides may also act in this capacity.

Fumigant Pesticides

Gaseous poisons used to kill pests are called fumigants. Their applications are usually limited to materials, structures or organisms that can be or are enclosed in a tight enclosure. There are many fumigants, some distinctly odorless, while others are used in conjunction with odorless fumigants as a warning agent because of their odor.

Many can be used safely about food products and most are very stable at stipulated temperatures. Some combine with commodities to produce corrosion or undesirable gases. When properly used, a fumigant is non-flammable and unlike any other forms of pesticides, kills all the developmental stages of an animal. In the case of insects, this includes the egg, larvae, pupae, and adults. When a fumigant reaches the appropriate concentration, it will kill pests quicker than any other pesticide.

Miscellaneous Pesticides

This group includes pesticides which may enter the body through any one or more of the previously mentioned means, but have some specialized mode of action or means of application.

Systemic Insecticides

Systemic insecticides act mainly as stomach poisons. However these chemicals are typically applied to one area of a plant or animal and are translocated to another area. In the case of plants, systemics are typically applied to the roots or stems of a plant and move through the vascular system to the leaves. Examples are Systox, Thimet and Disyston.

Desiccant Dusts

When applied to an insect, these materials absorb or abrade off the outer waxy layer of the insect's cuticle, causing the loss of body fluids and death by dehydration. The common desiccating dusts used by pest control operators are silica aerogel and boric acid.

Chemosterilants

There are chemicals which cause sexual sterility in either the male or female insect or animal. Chemosterilants have not yet been used on a commercial basis.

Pheromones

These are chemicals which may be of potential use in preventing insect mating when used to saturate the habitat.

Antimetabolites

These are chemicals which interfere with the normal metabolic or physiological processes of animals.

Pesticide Groups

Pesticide classification or grouping may be based on any of several criteria. One of the most common means of classifying a pesticide is on the basis of similarities in chemical structure. Based on this mode of classification, there are 3 classes of pesticides commonly used in the structural pest control industry, the inorganic, botanical, and synthetic organic insecticides.

Inorganic Pesticides

Inorganic pesticides are typically derived from minerals or chemical compounds that occur as deposits in nature. Most of these compounds are quite stable and tend to accumulate in the environment. Some act as stomach poisons (borates and boric acid). Others are considered sorptive dusts (silica aerogel, diatomaceous earth) that absorb the waxy layer from the cuticle of pests. Many of the inorganic pesticides are relatively expensive and are only moderately effective in controlling insects and other pests. Common inorganic pesticides are silica aerogel, boric acid, borates, diatomaceous earth, cryolite, copper, and sulfur.

Botanicals

The botanical pesticides are extracted from various parts (stems, seeds, roots, flower heads) of different plant species. Botanical insecticides usually have a short residual activity and do not accumulate in the environment or in fatty tissues of warm blooded animals. Many botanical pesticides act as stomach poisons, although pyrethrins act mainly as a contact poison. Common examples of botanical pesticides are pyrethrins, sabidilla, rotenone, nicotine, ryania, neem, and limonene.

Synthetic Organic Insecticides

Synthetic organic insecticides do not naturally occur in the environment, but are synthesized by man. Since all these compounds have carbon and hydrogen atoms as the basis of their molecule (as do living plants and animals), they are referred to as organic compounds. The four basic types of synthetic organic insecticides are the chlorinated hydrocarbons, organophosphates, carbamates, and pyrethroids.

Chlorinated Hydrocarbons

This large group of insecticides varies considerably in toxicity to mammals. Most are only moderately toxic, however, a few are very toxic to mammals. The use of these materials has been severely criticized for their effect on the environment. Most chlorinated hydrocarbons are very stable and do not readily decompose in the environment. Most of these insecticides accumulate in the environment and in fatty tissues of birds and mammals. The use of most chlorinated hydrocarbons has been prohibited in the U.S.. Examples of the chlorinated hydrocarbons are DDT, BHC, dieldrin, chlordane, aldrin, endrin, heptachlor and methoxychlor.

Organophosphates

The organophosphates are an extremely large and diverse group of insecticides. Their toxicity to mammals range from extremely toxic to some of the least toxic pesticides known. Most organophosphoates are not persistent and will break down to non-toxic materials in one to 30 days, depending on the compound. Organophosphates do not accumulate in fatty materials and do not accumulate in food chains. These compounds act mainly as contact insecticides although they may also act as stomach poisons and fumigants. Common organophosphates are malathion, chlorpyrifos (Dursban), diazinon, dichlorvos (Vapona), acephate (Orthene), and propetamphos (Safrotin).

Carbamates

The carbamate compounds are also a large group of insecticides. As a rule, these compounds are slightly more persistent in the environment than the organophorphrous compounds, but do not accumulate in the environment or fatty tissues of mammals. Most carbamates are only moderately toxic to mammals. They mainly act as contact insecticides with some stomach activity. Common carbamate insecticides are carbaryl (Sevin) and propoxur (Baygon).

Insect Growth Regulators

Insect growth regulators are chemicals that affect the ability of insects to growth and mature normally. They are based on and often mimic the growth hormones that occur naturally within the insect's body. Because mammals do not molt like insects do, most insect growth regulators are not very toxic to man and domestic animals. Common insect growth regulators are methoprene (Precor), hydroprene (Gentrol, Gencor), fenoxycarb (Torus), and hexaflumuron (Sentricon)

Microbial Pesticides

Microbial pesticides are formulated disease organisms of pests, many of which are grown in large quantities in manufacturing plants. Some of the microorganisms available for pest control are bacteria, fungi, and nematodes. Some of the bacterial pesticides available are Bacillus thuringiensis variety kurstaki (Thuricide, Javelin) for control of caterpillars and Bacillus thuringiensis variety israelensis (Teknar, Vectobac) for control of mosquitoes. Some of the fungi available for pest control are Metarhyzium (Biopath) for control of cockroaches. Some of the nematodes available for pest control are Steinernema feltiae (Vector) for flea control.

Botanical Products

Many plants have insecticidal properties. Extracts of these plants can be sprayed on the crop to either kill or repel insects.

A well-known example is the Neem tree (Azadirachta indica). All parts of this tree, but especially the grains, contain active ingredients, which can be used as insecticide. Neem is found to be active against a large number of pests, including the African bollworm (Helicoverpa armigera), Diamond-back moth (Plutella xylostella), thrips and aphids.

Another plant, well known for its insecticidal properties is Tobacco (Nicotiana tabacum). Extracts of tobacco leaves are effective against aphids and thrips.

Many other plants can also be used to prepare extracts with pesticidal properties. For example garlic, chili pepper (Capsicum) and onion.

An important advantage of botanical pesticides is that they are cheap and affordable for small-scale farmers.

Compared to synthetic pesticides, botanical products are generally safer to use, but care should be taken, as some plants contain very toxic substances. For example, the nicotine in tobacco is a highly toxic chemical.

Another advantage of botanical products is that they are not very persistent. Most of them will break down quickly under influence of high temperature or sunshine. Therefore they don't have a long lasting contaminating effect on the environment.

A disadvantage of botanical products is that they are generally not specific. Many plant extracts will also kill or repel beneficial insects. Just like synthetic pesticides, the botanical products should thus be avoided and only be used as a last resort. However, if a farmer decides, after careful consideration, that active control of a pest is required, botanical extracts are usually a better choice than chemical pesticides.

CHAPTER 1

PRIORITIZING CHEMICALS FOR RISK ASSESSMENT USING CHEMOINFORMATICS: EXAMPLES FROM THE IARC MONOGRAPHS ON PESTICIDES

Neela Guha, Kathryn Z. Guyton, Dana Loomis, and Dinesh Kumar Barupal

INTRODUCTION

The Monographs Programme of the International Agency for Research on Cancer (IARC) has been instrumental in identifying "environmental" factors that can increase the risk of human cancer. Since its inception in 1971, the Monographs programme has evaluated nearly 1000 agents (as of 2016) and classified them with respect to the strength of scientific evidence that they cause cancer in humans.

The IARC Monographs Programme convenes international Working Groups to identify and classify environmental cancer hazards. The evaluations are based on systematic reviews of epidemiological evidence and cancer bioassay data in experimental animals, with supporting evidence concerning the carcinogenic mechanisms that may act in humans. The sources and extent of human exposure, as well as existing regulations, are also reviewed (IARC preamble 2015). Agents are selected for evaluation by the IARC Monographs Programme through a process that traditionally has relied, to a large extent, on expert recommendations. A public call for nominations of agents is posted on the IARC website and additional nominations are solicited from participating states, the scientific community (including IARC staff), and the general public. An advisory group is then assembled every 5 years to review the nominated agents (Straif et al. 2014) and assign them low, medium or high priority for eventual evaluation. These priority levels reflect the committee's ranking based on the availability of new data, evidence of carcinogenicity, extent of human exposure, and public health importance (IARC preamble 2015). This method has proved useful for identifying agents of public health importance as priorities for the evaluation of their carcinogenic potential. Following this advice, the Programme selects, groups and orders the priority agents into a series of Monographs, based on a systematic and objective review of the available evidence. Considerations include the priority level assigned by the IARC advisory group, whether and when a compound was last reviewed by IARC and the potential for the classification to change, usage data (including regulations on use) and the extent of human exposure worldwide, the compound's classification by other agencies and the volume and complexity of informative data that can be reasonably considered during the course of an IARC Monograph meeting. Another important consideration is public health concern, including possible impacts in low- and middle-income countries.

The most recent IARC Advisory Group recommended that the Monographs programme evaluate pesticides. Nominations included particular compounds, chemical classes, related occupations, as well as systematic consideration of cancer-relevant information across a wide range of related exposures. Some pesticides have been identified as potential human carcinogens by authoritative bodies, including IARC, but many are either lacking evaluations or the evaluations may be outdated. Specifically compounds of the organophosphate, organochlorine, triazine, carbamate, dinitroaniline, and pyrethroid classes were accorded priority in 2014 (Straif et al. 2014).

Pesticides include thousands of unique chemical structures distributed across broad chemical and functional classes. Many are chemically or functionally related, but the extent to which they have been studied and the amount of information available from public databases (e.g. PubMed, the Tox21 Program, the PubChem bioactivity assay database) differs markedly across compounds. Given the large amount of data and the structural diversity between compounds, manual review may be prone to incomplete coverage, bias and low efficiency.

Citation: Neela Guha, Kathryn Z. Guyton, Dana Loomis, and Dinesh Kumar Barupal. Prioritizing Chemicals for Risk Assessment Using Chemoinformatics: Examples from the IARC Monographs on Pesticides, Environ Health Perspect; DOI: 10.1289/EHP186.

Automation of literature mining, integration of electronically available databases and advanced data visualization could be employed as a complimentary approach to systematically incorporate chemical similarity as well as to identify the extent of available information. To address the challenge of appropriately grouping agents and ordering recommended priorities for hazard assessment, we present a systematic and objective approach using chemoinformatics that has been used to select pesticides for evaluation in recent IARC Monographs (Guyton et al. 2015; Loomis et al. 2015).

METHODS

Overview

A bioinformatics approach was undertaken to systematically assemble and visualize the extent of available information according to chemical similarity across pesticide active compounds. The ranking obtained from the bioinformatics approach was later compiled manually (see Tables 1 and 2) with other important factors considered in selecting priorities (please see the considerations listed in the Introduction above) as presented in Table 1, particularly the assigned priority and the availability of new data to update a previous IARC evaluation.

Table 1. High ranking organophosphate, organochlorine and chlorophenoxy pesticides identified using a chemoinformatics approach.

Name	Rank	PubMed cancer hits	PubMed human cancer hits	IARC Advisory Group Priority	Other classifications	Usage notes	Prior IARC classification (year)	2015 IARC classification
Organophosphates								
Parathion	1	42	6	-	US EPA Group C (1991)[a]	Restricted[b]	3 (1987)	2B
Malathion	2	40	12	High	US EPA Suggestive (2000)[a]	High[c]	3 (1987)	2A
Chlorpyrifos	3	38	14	Medium	US EPA Group E (1993)[a]	High[d,e]	3 (1987)	
Dichlorvos	4	35	12	-	US EPA Suggestive (2000)[a]	Some current uses[f]	2B (1991)	
Diazinon	5	30	16	High	US EPA Not likely (1997)[a]	High[c]	2A (2015)	2A
Glyphosate	7	21	9	Medium	US EPA Group C (1985), Group E (1991)[a]	High[d,e]	None	2A
Tetrachlorvinphos	13	6	1	-	US EPA Likely (2002)[a]	Currently used	3 (1987)	2B
Organochlorines								
DDT	1	494	190	Medium	POP[g], RoC-RA[h]	Restricted[b]	2B (1991)	2A
Lindane	2	189	51	High	POP[g], RoC-RA[h]; US EPA Suggestive (2001)[a]	Restricted[b]	2B[i] (1987)	1
Dieldrin	3	151	57	-	POP[g]	Restricted[b]	3 (1987)	
Aldrin	7	56	25	-	POP[g]	Restricted[b]	3 (1987)	
Chlorophenoxy								
2,4-Dichlorophenoxy acetic acid	1	145	84	-	US EPA Group D (2004)[a]	Currently used	None	2B

[a]For a description of US EPA cancer classifications, see http://www.epa.gov/pesticides/health/cancerfs.htm#terms.

[b]Severely banned or restricted for health or environmental reasons (Rotterdam Convention, Annex III)

[c]Among the most commonly used OP insecticides in all US market sectors (2001 to 2007), malathion and chlorpyrifos are listed 1 or 2; diazinon is listed 3 (2001, 2003) or 8 (2005, 2007) (http://www.epa.gov/pesticides/pestsales/).

[d]Chlorpyrifos and glyphosate are among the most commonly used conventional pesticide active ingredients in the US (http://www.epa.gov/pesticides/pestsales/).

[e]Glyphosate was the most commonly used conventional pesticide in the agricultural market sector from 2001 to 2007; in this same period chlorpyrifos ranked 13 (2003), 14 (2007), or 15 (2001, 2005) (http://www.epa.gov/pesticides/pestsales/).

[f]Many domestic and other uses of dichlorvos in the US have been discontinued (http://www.gpo.gov/fdsys/pkg/FR-1995-04-19/pdf/95-9166.pdf)

[g]POP, Listed as a persistent organic pollutant under the Stockholm Convention

[h]RoC-RA, Listed as "reasonably expected to be a human carcinogen" in the US Report on Carcinogens

A first step was to compile a list of all pesticide compounds, which was then organized into chemical similarity network maps. To visualize the availability of data on all pesticides, information by topic area that is considered for an IARC Monograph evaluation (cancer epidemiology, cancer bioassays in animals, mechanistic studies) was then overlaid onto the network maps. Chemical network maps were generated by integrating lists of pesticide compounds with their chemical structure and subsequently mining public databases of the published literature. This process is documented in Figure 1 and detailed in the sections that follow.

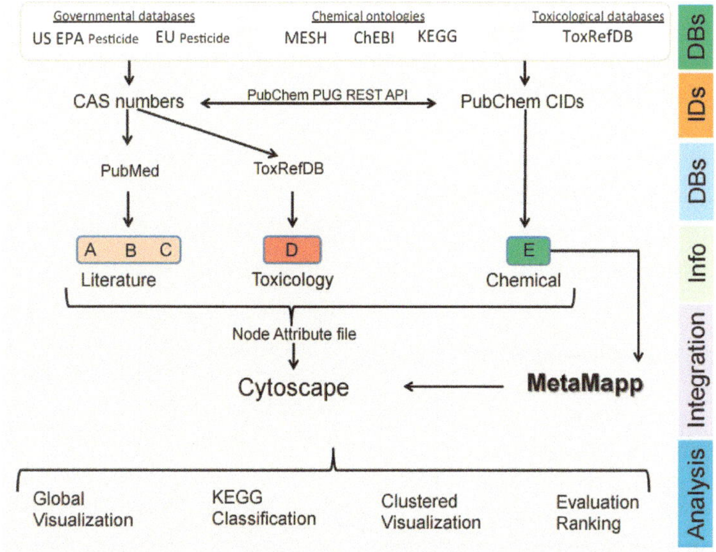

Figure 1. Overall scheme of the chemoinformatics approach for data retrieval and visualization for the prioritization of pesticides for the evaluation of their carcinogenic potential. A. PubMed Cancer All; B. PubMed Cancer Epidemiology; C. PubMed animal cancer bioassays; D. ToxRefDB carcinogenicity; E. Chemical Similarity Scores.

Creation of a Master List of Pesticides

We compiled a master list of pesticide compounds - including chemical name, chemical class and Chemical Abstract Service (CAS) number- from the Kyoto Encyclopedia of Genes and Genomes (KEGG), MESH Pesticides, EU pesticides, ChEBI pesticides, USEPA Pesticides, and the United States Environmental Protection Agency (USEPA) Toxicity Reference (ToxRef) databases. Chemical structures for each pesticide compound were obtained by linking CAS numbers to PubChem Compound Identifiers (CID) (https://pubchem.ncbi.nlm.nih.gov/search/help_search.html).

Data Retrieval from NCBI Databases

To assess the scope of the published literature for each pesticide, we searched the titles and abstracts of publications catalogued in PubMed on cancer epidemiology and animal bioassays using the CID, CAS number and various search terms. The MeSH term "neoplasm" was used for these searches, since the keyword "cancer" frequently retrieves false positive hits. The papers manually retrieved in our study were also retrieved by the "neoplasms[MeSH]" query, indicating that it covers relevant papers. Searches of ToxRefDB and of PubChem bioassays were also conducted. NCBI Eutils and PubChem PUG REST web services were used to systematically query the databases to obtain results of literature and bioassays searches (see Figure 1). Automation for retrieval of data from APIs (Application Programming Interface) was achieved in NodeJS software using JavaScript programming language. To rank pesticides using the chemoinformatics approach, pesticides were sorted by chemical class, the number of publications on cancer and that pesticide overall, the number of cancer epidemiology publications, and the information in ToxRefDB (present or absent).

Chemical Similarity Network Visualization

Network graphs for the chemicals were created using MetaMapp software (Barupal et al. 2012) and visualized in Cytoscape software version 3.1 (U.S. National Institute of General Medical Sciences [http://www.cytoscape.org/release_notes_3_1_1.html]). Individual pesticides are represented as nodes on the chemical similarity maps. Two nodes were linked in the chemical network graph if their Tanimoto similarity score (a coefficient of similarity between two molecules, a measure commonly used in chemoinformatics) was above 0.60, indicating more than 60% chemical similarity. The length of the line connecting the nodes had no meaning itself; it was drawn in reference to the nodes it connected. The node positions within the network maps were controlled by the organic layout algorithm in Cytoscape software which considered a node's degree (the number of connections to a node) and its clustering coefficient (the ratio of the number of actual connections to the total number of possible connections among the node and its neighbors).

A global network of all the pesticides and two focused network graphs of the organophosphorus (OP) and organochlorine (OC) pesticide classes were created. Beyond the KEGG classification, we broadened the pesticide categories for visualization by including those pesticides with at least 1 phosphorous atom or 2 chlorine atoms to the OP and OC pesticide classes, respectively. These network graphs along with the data table used to generate the graph are provided online http://pesticide.barupal.org/.

Automated Text Mining Versus Directed Literature Searches

For the top ranking chemicals identified through the chemoinformatics approach, we compared the results from the automated searches to directed PubMed searches. The comparison focused on the cancer epidemiology, as most such studies are found in the published literature. It also considered any published animal cancer bioassays and studies of key mechanistic evidence (Smith et al. 2015) relating to carcinogenicity of the compound. The manual literature searches and screening were performed using The Health Assessment Workspace Collaborative (hawcproject.org) (Shapiro A 2015).

RESULTS

Creation of a Master Pesticide List for Literature Mining

A master list of nearly 6000 pesticide compounds was created from governmental databases, ontologies and databases providing toxicological data on chemicals: KEGG (n = 916 pesticides), MESH Pesticides (n = 451 pesticides), ChEBI (n = 2448 pesticides), USEPA Pesticides (n = 5774 pesticides), EU Pesticides (n = 1318 pesticides) and ToxRefDB (n=474). Entries that were imported from these databases were excluded from the final list if: 1) the structures represented additives such as ethanol 2) they did not have a Chemical Abstracts Service Registry Number (CASRN) or a PubChem Compound Identifier (CID) 3) they were not present in at least 3 of the aforementioned databases and 4) they were compounds that have applications in multiple industries, such as phenol, nicotine, acrolein, and bisphenol A. All the compounds from KEGG, ToxRef, and MESH Pesticide databases were included in the analysis but a number of entries were excluded from the USEPA (n=5024), ChEBI (n=2033), and EU Pesticides databases (n=643). KEGG provided chemical classification information and ToxRefDB provided toxicological data, especially for cancer bioassay data for around 400 selected pesticides. The final list contained 980 pesticide structures (see http://pesticide.barupal.org/ and http://pesticide.barupal.org/dataTable.html).

Selection of Pesticides for Evaluation in IARC Monograph Volumes 112, 113 and 117

The preceding approach was a starting point for selecting pesticides for evaluation in IARC Monographs volumes 112, 113 and 117 (described below). Using this approach, many of the top ranked pesticides belonged to the organophosphate (OP) and organochlorine (OC) classes; therefore these pesticide classes were accorded priority. To rank pesticides using the chemoinformatics approach, pesticides were sorted by chemical class, the overall number of publications on cancer and that pesticide, the number of cancer epidemiology publications, and the information in ToxRefDB (present or absent) (see Table 1). The chemical network maps of OPs (Figure 2a) and OCs (Figure 2b) were informative as to the chemical similarity across potential candidates and for identifying related compounds that might be evaluated as a mechanistic class.

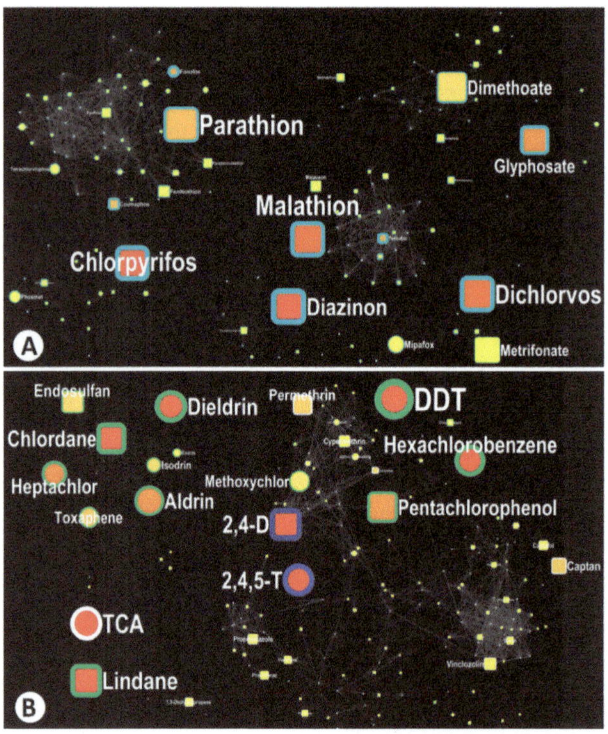

Figure 2. Focused visualization of the chemical similarity network maps for A) organophosphorus and other pesticides with at least 1 phosphorous atom and B) organochlorines and other pesticides with more than 2 chlorine atoms.

In addition to the ranking of OPs and OCs obtained using the chemoinformatics approach (Table 1), several other criteria were considered in order to select a subset of pesticides for evaluation in Monographs 112 and 113 (please see the considerations listed in the Introduction above) as presented in Table 1, particularly the assigned priority and the availability of new data to update a previous IARC evaluation. The volume and complexity of informative data is an important determinant of the number and diversity of agents that can be evaluated in a Monograph meeting. The chemoinformatics approach was therefore useful for visualizing the volume of literature by topic area (cancer epidemiology, animal cancer bioassays, supporting mechanistic evidence). To further refine the list of agents for evaluation, additional directed PubMed searches for epidemiologic and mechanistic data were conducted using standard search strings developed for the IARC Monographs.

Organophosphate Pesticides

Based on the preceding criteria, parathion, malathion, diazinon, glyphosate and tetrachlorvinphos emerged as promising candidates for new evaluation in IARC Monograph Volume 112 (Figure 2B). Parathion and malathion, the top-ranked OPs identified from the chemoinformatics approach, were previously evaluated by the IARC Monographs in 1987 and were then assigned to Group 3 (not classifiable). However, they were later (in 1991 and 2000) classified by the US EPA as potential carcinogens on the basis of positive bioassay data. The availability of newly published epidemiologic studies, particularly for malathion, also supported their selection for re-evaluation in Volume 112. Diazinon, like malathion, was assigned high priority for evaluation by an international advisory group to the IARC Monograph Progamme. It ranked fifth by the chemoinformatics approach, had the most cancer epidemiologic studies among the OPs, and had not been previously evaluated by the IARC Monographs Programme.

On the other hand, several candidate agents did not appear to have animal bioassay or other evidence to support a different (e.g., dichlorvos, Group 2B; chlorpyrifos, Group 3) or new (e.g., dimethoate) IARC evaluation. This included several compounds classified by the US EPA in "Group E- Evidence non-carcinogenicity" (terbufos, fonofos, chlorpyrifos and phorate), indicating that animal bioassays had been conducted but none had positive findings [http://npic.orst.edu/chemicals_evaluated.pdf]. An additional factor, noted in Advisory Group recommendations, is that while new epidemiological evidence has emerged it remains incomplete for these agents. Ongoing analyses (e.g., as being conducted by the Agricultural Health Study or AGRICOH), would be important to await before any new or updated evaluation. Thus, these candidates were accorded a lower ranking overall for near-term evaluation.

In contrast, for glyphosate, a recent meta-analysis identified relevant epidemiologic findings(Schinasi and Leon 2014). Additionally, an earlier 1985 classification by the US EPA in Group C indicated the possible availability of pertinent bioassay data (http://www.epa.gov/iris/subst/0057.htm). Glyphosate, ranked seventh by the chemoinformatics approach, was assigned medium priority for near-term evaluation by the Advisory Group. The high production volume of glyphosate leads all OPs and all herbicides, and exposures are widespread, which was another factor in its inclusion among agents in Volume 112. While having some structural similarity to other OPs, glyphosate is toxicologically dissimilar and lacks cholinesterase-inhibiting activity.

Another compelling candidate that emerged from the chemoinformatics approach was tetrachlorvinphos (ranked 13). Tetrachlorvinphos is in current use although overall production volume is low. It was previously evaluated by the IARC Monographs in 1987 and was then assigned to Group 3 (not classifiable). However, tetrachlorvinphos was later (in 2002) classified by the US EPA as a likely human carcinogen based on positive cancer bioassays. Additionally, because tetrachlorvinphos is a direct-acting oxon, in vitro tests for bioactivity might be more informative than for other compounds (e.g. malathion) that require metabolic activation to their oxon forms. These mechanistic considerations together with the positive cancer bioassay findings were the basis for its inclusion in Volume 112. The selection also took into account the overall volume of literature for the other four compounds, with the relatively small size overall of the tetrachlorvinphos literature making it feasible to include.

For the five selected compounds, IARC queried governments and requested public release of government reports on animal cancer bioassay and other relevant data (e.g., genotoxicity) that had been developed by the industry. Direct literature searches identified recently reported epidemiological data, including case–control and cohort studies in the US, Canada, Europe and Sweden. Directed literature searches also identified studies examining relevant carcinogenic mechanisms, including genotoxicity, for both the parent compounds (e.g., malathion, diazinon) and their oxon metabolites. Recent high-throughput data also provided new insights into the extent of biological activity.

In all, these considerations supported the selection of compounds accorded high (malathion, diazinon) or medium (glyphosate) priority for evaluation by the Advisory Group, as well as two others (parathion and tetrachlorvinphos) that were not specifically highlighted in the broad recommendation to evaluate pesticides.

Organochlorine Pesticides and 2,4-D

Among the OC pesticides, DDT, lindane, aldrin, and dieldrin were identified as promising candidates for new evaluation according to the criteria described above (Figure 2B). DDT was particularly notable as the pesticide with the largest number of human cancer studies and the largest overall number of PubMed articles retrieved (see table online athttp://pesticide.barupal.org/dataTable.html). DDT was previously evaluated by IARC in 1991 and had a large number of new studies, lindane had only been evaluated as part of the broader class of hexachlorocyclohexanes and also had new data, while aldrin and dieldrin were last evaluated in 1987 but had relatively few human studies (Table 1). All four are listed as persistent organic pollutants under the Stockholm Convention. DDT and lindane were assigned medium and high priority, respectively, for evaluation by the IARC advisory group, and both had previously been listed as "reasonably anticipated to be a human carcinogen" in the US NTP Report on Carcinogens(USNTP 2014). No additional classifications were identified for aldrin or dieldrin.

A notable literature database was also available for the herbicides 2,4-D and 2,4,5-T (Figure 2B), which were classified in Group 2B by IARC in 1987 as part of the class of chlorophenoxy herbicides (IARC 1987). However, since 2,4,5-T is frequently contaminated with dioxin, which is already classified in IARC Group 1(IARC 2012), it was not considered further.

As was done for the OP pesticides, directed literature searches identified relevant epidemiology, cancer bioassay and mechanistic studies for the OC pesticides. Additionally, we requested public release of information on cancer bioassays conducted with these compounds that were not available in the public domain.

After considering feasibility, including the unusually large volume of data retrieved for DDT, in addition to the preceding scientific issues, lower ranked aldrin and dieldrin were set aside from the list of potential candidates for later evaluation in IARC Monograph Volume 117 (October 2016), while DDT and lindane were selected for evaluation in Volume 113. 2,4-D was also selected after considering its widespread use and the volume of published literature available (Table 1).

Directed Literature Searches for Pesticides Prioritized for Evaluation

For the pesticides prioritized for evaluation by IARC in Monograph Volumes 112 and 113 based on the chemoinformatics approach, manual searching and screening of the epidemiological literature was performed using HAWCproject.org. Such manual validation is supported because the size of the published literature—particularly concerning epidemiology or carcinogenic mechanisms—does not always predict the need for a new or updated evaluation. On the other extreme, even a single new well-conducted cancer bioassay could justify further evaluation. Accordingly, the findings of manual screening were compared to the results of the automated searches to determine the relevance of retrieved articles for any resulting evaluation (Table 2).

Table 2. Cancer epidemiology literature retrieved through automated mining (chemoinformatics) and manual PubMed searches for pesticides evaluated in IARC Monographs 112 and 113.

	Chemoinformatics	Manual searches		
Pesticide	Retrieved	Retrieved	Included	Excluded
Malathion	12	80	28	52
Parathion	6	12	9	3
Diazinon	16	39	22	17
Tetrachlorvinphos	1	4	4	0
Glyphosate	9	50	19	31
DDT	190	224	116	107
Lindane	51	46	22	24
2,4-D	84	76	62	11

Generally, the chemoinformatics approach retrieved fewer cancer epidemiology papers than identified through manual searches. For example, for DDT, 190 cancer epidemiology papers were identified through automated searches in PubMed. In comparison, 224 were identified through targeted manual searches, of which 116 were included after manual review as relevant to an evaluation. Automated searches were not performed concerning the literature on cancer mechanisms as methods to comprehensively identify the broad range of cancer-relevant mechanistic data for potential carcinogens have only recently been advanced (Smith et al. 2015). Nonetheless, targeted searches developed according to the principles outlined by Smith et al (2015) identified a substantial volume of articles on each selected compound. For instance, targeted manual searches and screening of the mechanistic literature for the evaluation of organophosphate pesticides identified relevant publications on malathion (n=370), parathion (n=578), diazinon (n=215), tetrachlorvinphos (n=40) and glyphosate (n=204), respectively. Yet more articles were included for the subsequent evaluation of DDT (n=953), lindane (n=545) and 2,4-D (n=420). Overall, this exercise demonstrated that the chemoinformatics approach provided an efficient and accurate indication of not only the size of the relevant literature, but also identified studies that would be relevant for any resulting evaluation.

DISCUSSION

We illustrate a novel method for the selection of agents for hazard identification that has been applied in the IARC Monographs Programme. Although the data used to construct the chemical network graphs are publicly available, they had not been previously organized in a unified manner that would allow for the simultaneous analysis of the volume of literature on a particular chemical or group of related chemicals. Beyond the KEGG classification, we broadened the pesticide categories for visualization by including those pesticides with at least 1 phosphorous atom or 2 chlorine atoms to the OP and OC pesticide classes, respectively. Doing so enabled us to map pesticides by chemical similarity and include pesticides that may have been missed by pesticide databases or that may have been discarded due to an error in assignment of chemical class. Accordingly, using a chemoinformatics approach, we were able to integrate information on chemical structure similarity for 980 compounds with the results of systematic, automated text mining of cancer-relevant published information in public databases. The use of web technologies streamlined the integration of information retrieved from different databases/sources and improved efficiency through creating network maps to visualize key chemicals as well as those less studied but chemically related, that may act through a similar mechanism. By enhancing visualization of large-scale public data, our chemoinformatics approach can complement other technologies that employ biomedical text mining strategies to support cancer risk assessment and research (Korhonen et al. 2012).

Using this chemoinformatics approach, pesticides in the organophosphate and organochlorine classes were accorded priority for evaluation in IARC Monograph Volumes 112 and 113: malathion, parathion, diazinon, glyphosate, tetrachlorvinphos (Guyton et al. 2015) and DDT, lindane, and 2,4-D (Loomis et al. 2015). In the resulting evaluations, all of these pesticides were assigned a new or higher IARC classification, reflecting the adequacy of the identified evidence to support these cancer hazard evaluations. In particular, three pesticides previously assigned to Group 3 (not classifiable) were classified in Group 2B (parathion, tetrachlorvinphos) or Group 2A (malathion). Likewise, two others were re-classified from Group 2B to Group 2A (DDT) or Group 1 (lindane). 2,4-D was newly classified in Group 2B; previously IARC had classified the entire class of chlorophenoxy herbicides as Group 2B. Finally, two pesticides, diazinon and glyphosate, both assigned Group 2A, had not been previously classified by IARC. For several of these compounds (including malathion, diazinon, glyphosate, DDT, lindane, and 2,4-D) strong mechanistic evidence supported the resulting evaluations. In all, these results affirm the utility of the prioritization method for identifying compounds that have evidence warranting new or updated IARC Monograph evaluations.

In addition to the pesticides selected for evaluation in IARC Monograph Volumes 112 and 113, the chemoinformatics approach highlighted several other compounds or compound classes (see http://pesticide.barupal.org/). Several of these have been previously evaluated by the IARC Monographs Programme (http://monographs.iarc.fr/ENG/Classification/latest_classif.php) in Group 2B or higher including trichloroactetic acid (Group 2B, 2014), inorganic arsenic compounds (Group 1, 2012), hexachlorobenzene (Group 2B, 2001) and polychlorphenols (Group 2B, 1999). Others previously assigned to Group 3 and of high use, including atrazine (Group 3, 1999), could be immediately prioritized for re-evaluation together with related compounds (i.e., simizine). Indeed, atrazine was accorded medium priority by the expert advisory group (Straif et al. 2014) based on extensive use and exposures, as well as suspicion of carcinogenicity from newly published information; furthermore, the chemoinformatics method also indicated an extensive literature base. Similarly, diverse compounds currently assigned to IARC Group 3 also emerged (e.g., captan, 1987; methyl bromide, 1999; piperonyl butoxide, 1987), evidently based on information published since the last IARC evaluation. These include some of the most used conventional pesticide active ingredients (e.g., the fumigant methyl bromide, ranked as the 8th (in 2007, 2005 and 2003) or 7th (in 2001) most commonly used conventional pesticide active ingredient by the US EPA (http://www.epa.gov/sites/production/files/2015- 10/documents/market_estimates2007.pdf). As noted above for the agents selected for evaluation in Volume 112, chlorpyrifos and other compounds of the organophosphate class may merit reevaluation following completion and publication of important epidemiological evaluations. Interestingly, some compounds that emerged as having relevant studies for cancer hazard evaluation have not been previously evaluated or specifically nominated for IARC evaluation (e.g., paraquat).

In general, the chemoinformatics approach retrieved fewer cancer epidemiology papers than identified through the directed literature searches. There are several possible explanations for this discrepancy. Some articles may have been missed because automated searches were of the article title and abstract whereas epidemiology papers sometimes report on multiple pesticides, and specific compounds may not be listed in the title or abstract. The automated searches relied on MeSH annotation using "neoplasms[mesh]" as a more precise search term instead of the keyword "cancer". This keyword could potentially retrieve irrelevant papers (e.g. that do not describe a laboratory or epidemiological finding on cancer) that MeSH terms would filter. Nonetheless, there may be some delay in assigning publications MeSH annotations and thus more recent but still relevant papers may not be retrieved. The directed searches scanned the full text of publications, enabling identification of publications not retrieved by automated searches using only MeSH terms. A potential limitation of the automated text mining approach is that the search is more specific but less sensitive, sometimes necessitating a manual validation of the literature base to ensure that all relevant publications are captured. However, this limitation affects primarily the later stages of literature retrieval, rather than the initial planning phase. For future efforts using the chemoinformatics approach, we could increase sensitivity of the automated search by broadening the search terms to include key terms as identified from the most informative studies found in a manual search.

Demand for the evaluation of potential chemical hazards is currently increasing, while the resources for testing these chemicals are decreasing (Benigni et al. 2013). The use of long term cancer bioassays in animals, which have previously played a fundamental role in the hazard assessment of chemicals, is declining for ethical and practical reasons (e.g. concern for animal welfare, expense, time). Therefore alternative, more cost-effective strategies for predicting the toxicological properties of chemicals, such as (Q)SAR (Quantitative) Structure–Activity Relationships), are being proposed and supported by regulatory initiatives such as REACH (http://ec.europa.eu/environment/chemicals/reach/reach_en.htm) (Benigni et al. 2013;van et al. 2009). These approaches capitalize on the wealth of data already captured in publicly available databases.

By employing technological advances in bioinformatics and computational toxicology, we demonstrate that the use of chemoinformatics is a powerful and complementary approach for prioritizing chemicals for risk assessment. This approach could be further extended to support prediction of emerging risks and informed substitution of hazardous chemicals by "safer" alternatives (Jacobs et al. 2016) wherein bioinformatics approaches could compare compounds that are not yet tested, but structurally similar, to agents already classified for their carcinogenic potential by the IARC Monographs Programme. Epidemiologists and other researchers assessing associations between numerous chemicals and outcomes may also be able to employ this strategy to identify agents for further investigation when designing large-scale studies of human health. Since national health agencies use the information from the IARC Monographs as scientific support for their actions to prevent or reduce exposure to potential carcinogens, efficiently prioritizing agents for risk assessment and predicting emerging hazards are important steps towards protecting public health.

CONCLUSION

Using a novel chemoinformatics approach, we integrated information on chemical structure similarity with the results of systematic, automated text mining of cancer-relevant information in public databases to select chemical agents for hazard identification. We demonstrate this as an efficient method for grouping of chemicals within class in selecting agents for hazard evaluation in the IARC Monographs. This systematic approach, accounting for chemical similarity and overlaying multiple data sources, can be used to systematize, inform and increase efficiency in selecting and prioritizing agents for hazard identification, risk assessment and regulation or further investigation. Further, by overlaying new chemicals on to a network map of agents already classified by the IARC Monographs, emerging risks and potential cancer hazards (e.g occupational carcinogens, drugs, environmental pollutants, nutritional compounds) might be identified. This innovation could be extended to an array of outcomes and agents and may prove particularly useful to national regulatory agencies for prioritizing agents for risk assessment and regulation.

ACKNOWLEDGMENTS

The IARC Monographs Programme gratefully acknowledges support from the US National Cancer Institute (U01CA033193), the National Institute of Environmental Health Sciences (LAC/IMO/2015/01) and the European Commission (VS/2015/0156).

REFERENCES

1. Barupal DK, Haldiya PK, Wohlgemuth G, Kind T, Kothari SL, Pinkerton KE, et al. 2012. MetaMapp: mapping and visualizing metabolomic data by integrating information from biochemical pathways and chemical and mass spectral similarity. BMC Bioinformatics 13: 99.
2. Benigni R, Bossa C, Battistelli CL, Tcheremenskaia O. 2013. IARC classes 1 and 2 carcinogens are successfully identified by an alternative strategy that detects DNA-reactivity and cell transformation ability of chemicals. Mutat Res 758: 56-61.
3. Guyton KZ, Loomis D, Grosse Y, El GF, Benbrahim-Tallaa L, Guha N, et al. 2015. Carcinogenicity of tetrachlorvinphos, parathion, malathion, diazinon, and glyphosate. Lancet Oncol 16: 490-491.
4. IARC. 1987. Overall evaluations of carcinogenicity: an updating of IARC Monographs volumes 1 to 42. IARC Monogr Eval Carcinog Risks Hum Suppl 7: 1-440.
5. IARC. 2012. Chemical agents and related occupations. IARC Monogr Eval Carcinog Risks Hum 100: 9-562.
6. IARC preamble. 2015.
7. Jacobs MM, Malloy TF, Tickner JA, Edwards S. 2016. Alternatives Assessment Frameworks: Research Needs for the Informed Substitution of Hazardous Chemicals. Environ Health Perspect 124: 265-280.
8. Korhonen A, Seaghdha DO, Silins I, Sun L, Hogberg J, Stenius U. 2012. Text mining for literature review and knowledge discovery in cancer risk assessment and research. PLoS One 7: e33427.
9. Loomis D, Guyton K, Grosse Y, El GF, Bouvard V, Benbrahim-Tallaa L, et al. 2015. Carcinogenicity of lindane, DDT, and 2,4-dichlorophenoxyacetic acid. Lancet Oncol 16: 891- 892.
10. Schinasi L, Leon ME. 2014. Non-Hodgkin lymphoma and occupational exposure to agricultural pesticide chemical groups and active ingredients: a systematic review and meta-analysis. Int J Environ Res Public Health 11: 4449-4527.
11. Shapiro A. 2015. Health Assessment Workspace Collaborative (HAWC). 2013. [accessed 1 August 2015].
12. Smith MT, Guyton KZ, Gibbons CF, Fritz JM, Portier CJ, Rusyn I, et al. 2015. Key Characteristics of Carcinogens as a Basis for Organizing Data on Mechanisms of Carcinogenesis. Environ Health Perspect.
13. Straif K, Loomis D, Guyton K, Grosse Y, Lauby-Secretan B, El Ghissassi F, et al. 2014. Future priorities for the IARC Monographs. The Lancet Oncology 15: 683-684.
14. USNTP. 2014. Report on Carcinogens, Thirteenth Edition. Available: http://ntp.niehs.nih.gov/pubhealth/roc/roc13/.
15. van LK, Schultz TW, Henry T, Diderich B, Veith GD. 2009. Using chemical categories to fill data gaps in hazard assessment. SAR QSAR Environ Res 20: 207-220.

CHAPTER 2

IDENTIFICATION OF NOVEL PESTICIDES FOR USE AGAINST GLASSHOUSE INVERTEBRATE PESTS IN TOMATOES AND PEPPERS

David R. George, Jennifer A. Banfield-Zanin, Rosemary Collier, Jerry Cross, A. Nicholas E. Birch, Roma Gwynn, and Tim O'Neill

Keywords: Pesticide, thrips, whitefly, two-spotted spider mite, aphid, tomato, pepper, SCEPTRE project

INTRODUCTION

In the UK, some 818 hectares are devoted to the growing of protected vegetables. Tomato is one of the most important crops in terms of both value and area cropped, and is classified as a major crop by Defra (the Department for Environment, Food and Rural Affairs). Over 200 hectares of glasshouse area are used to produce British tomatoes and the total tonnage of fresh tomatoes consumed in the UK is increasing. The value per planted hectare of tomato is estimated at £474,000 [1], with the retail value of British tomato production around £175 million. Bell peppers, by contrast, are considered a minor crop, and take up some 85 hectares of UK glasshouse area. Despite this, they are a high value crop, with an estimated worth over £200,000 per planted hectare [1].

Glasshouse production has many benefits, not least that it allows production of crops such as tomato and pepper under temperate climates in countries, including the UK. In addition, glasshouse systems may afford some level of protection from pests, limiting access to the crop through creation of a physical barrier to inward pest movement that may be fortified through screening with appropriate mesh sizes [2]. Even with screening in place, however, pest penetration is often significant [3] and supplementary control measures are typically required [4]. Indeed, the artificial system created, with controlled and buffered temperature and humidity and without wind and rain, exacerbates pest problems where they do occur by providing relatively optimal conditions for pest population growth [3,5].

Where pest outbreaks occur on protected edible crops in Europe, conventional chemical pesticides are still widely utilised in their control, the cost of which exceeds (US)$1 billion for two-spotted spider mite (*Tetranychus urticae*) alone, albeit across multiple sectors [6]. Nevertheless, reliance on biological control of pests through release of pest natural enemies is commonplace in glasshouse production systems worldwide [7], having been developed since the 1960s to provide in excess of 100 commercially-available biocontrol species today [5]. However, despite these advances in biocontrol of glasshouse pests, aphid, thrips, spider mite and whitefly populations in particular remain an ever-present threat, especially where resistance to conventional chemical pesticides has been documented (e.g., [8,9]). Though biological control may be effective against these pests, results may be variable and dependent upon factors such as release timings, grower experience and incidence of hyper-parasitism/predation, which can lead to biocontrol failure [10]. Where biocontrol does break down, growers often turn to conventional chemical pesticides to curtail pest populations, further disrupting biological or integrated pest management (IPM) programmes (D. George, personal observation). Consequently, the predicted and complete phasing-out of conventional chemical pesticides from the glasshouse sector by 2010 has not been realised as some predicted [5], at least in the majority of cases. In contrast, there remains a need to identify new products to target pest outbreaks under glass. Biopesticides are likely to be particularly important here; as future use of numerous conventional chemical pesticides is threatened in the UK and Europe [11], alternative products are likely to become ever more utilised in crop production. The Compound Annual Growth Rate of biopesticides (16%) already far exceeds that of conventional chemical pesticides, albeit from a much smaller base, with the global market for biopesticides (including

Citation: David R. George, Jennifer A. Banfield-Zanin, Rosemary Collier, Jerry Cross, A. Nicholas E. Birch, Roma Gwynn, and Tim O'Neill. Identification of novel pesticides for use against glasshouse invertebrate pests in tomatoes and peppers, Insects 2015, 6(2), 464-477. DOI:10.3390/insects6020464.

fungicides, herbicides and nematicides) expected to reach (US)$4.4 billion by 2019, from (US)$1.8 billion in 2013 [12].

Benefits of biopesticides (semiochemicals, botanicals and microbials) and biorationals (e.g., physical pest disruptors such as fatty acids), are reported as the potential for reduced occurrence of pest resistance [13,14], higher amenability to combined use with biological control [15,16], and lowered environmental persistence with reduced product residues [13,17,18]. Such products also align better with modern consumer/retailer demands, with many, though not all, meeting organic production standards [19]. Biopesticide and biorational (henceforth grouped as "biopesticide") products are not without issue, however, typically achieving lower and slower overall kill rates than conventional chemical pesticides [20], being more sensitive to environmental degradation [14], and often having a requirement for direct pest contact that necessitates high spray volumes and more informed application procedures [19]. Thorough testing of both "new" conventional chemical and biopesticide products against industry standard regimes is therefore critical to identify those products with the greatest potential to contribute to future management of glasshouse pests. This is especially true in the UK where the approved number of biopesticides is relatively small in comparison to countries such as the Netherlands [21].

With the above in mind, the aim of the current project was to assess the efficacy of a number of new experimental conventional chemical and biopesticide products against selected glasshouse pests of tomato and bell pepper in order to provide data to support new product approvals for the protected edibles sector. Products and target pests were selected through industry consultation, and all products tested against industry standard pesticides and negative controls.

EXPERIMENTAL SECTION

From the outset of the project, a number of key priority target pests were identified following discussion with the "Protected Crops Panel" of the Horticultural Development Company, individual growers and crop consultants. Two glasshouse crops were selected for study, tomato (*Solanum lycopersicum*, var. Dometica) and bell pepper (*Capsicum annuum*, var. Ferrari), with experiments conducted on five key pests of these crops: western flower thrips (*Frankliniella occidentalis*(Pergande)), glasshouse whitefly (*Trialuerodes vaporariorum* (Westwood)), two-spotted spider mite (*Tetranychus urticae*(Koch)) and the glasshouse-potato and peach-potato aphids (*Aulacorthum solani* (Kaltenbach) and *Myzus persicae* (Sulzer), respectively).

All trials utilised plants grown in rockwool blocks under glasshouse conditions at Stockbridge Technology Centre, North Yorkshire, UK (Figure 1). Plants were transferred to rockwool slabs (three plants per slab) when sufficiently mature and thereafter maintained according to standard commercial practises, with the exception of pest control measures. Treatments were arranged in a randomised block design, utilising two rockwool slabs as an experimental plot. Plots were always separated by at least one rockwool slab to serve as a "guard" and limit pest movement between plots. Guard blocks were also positioned at the ends of any row of plants to limit edge effects, with rows adjacent to glasshouse walls being made up entirely of guard plants for the same reason. Spaces between rows consisted either of concrete pathways or fleece partitions (Figure 1), erected to prevent spray drift and limit pest movement between plots. Plants were always transferred to glasshouses at commercially applicable densities, with six replicates of each treatment, including an industry standard pesticide, plus a water-only control, always included.

Figure 1. Tomato crop showing rockwool blocks, slabs and path/fleece divides.

Once transferred to glasshouses, plants were infested with pests that had been reared in either an insect growth room or bio-houses (small glasshouses) present at the trial site. Pests were transferred to crop plants either by releasing adult insects into plots (*T. vaporariorum* only), or by inoculation with infested plant material (all other pests). Once infested, pests were given a short period to settle before pre-treatment counts of pest numbers (at date "D0") were made according to guidelines provided by the European Plant Protection Organisation, which were consulted at all stages of trial design. Counts of *T. vaporariorum* were made from top, middle and lower leaves of each plant in a plot, recording adults and larvae (all stages) separately; counts of *F. occidentalis* were made from two excised flower heads per plant (Figure 2), for each plant in a plot, also counting adults and nymphs separately. Counts of *T. urticae* (adults and nymphs) were made from each plant in a plot, from marked leaves on which mites had been previously released (Figure 3), and counts of aphids (all stages combined) were made from whole plants on each plant in a plot, taking care to examine both adaxial and abaxial leaf surfaces (Figure 4). Where pre-treatment counts were deemed too low to commence treatment, the infestation process was repeated.

Figure 2. Pepper flower head prior to excision with *Frankliniella occidentalis* visible.

Figure 3. Tomato leaf showing signs of *Tetranychus urticae* feeding damage and rubber bands used to mark leaves on which mites were released.

Figure 4. Pepper leaf with *Aulacorthum solani* on abaxial surface (NB: aphids were recorded from all plant parts).

Once sufficient pest levels had been reached, plots were treated with pesticide products at concentrations recommended by manufacturers, deployed in water such that all plots in a given experiment received the same overall spray volume (750 L/Ha/mch). Each treatment was applied by qualified personnel using a single-lance Oxford Precision Sprayer, fitted with a hollow-cone nozzle, at approximately 3 bar pressure. Due to the nature of the biopesticides being tested, care was taken to ensure complete plant coverage and thorough mixing of products (in water) prior to and during treatment delivery, with pest counts made six days after each application (6DAA). The timing and number of applications for some pesticides varied within trials to conform to existing or anticipated regulatory requirements and product labelling. This was particularly so for industry standard pesticides, which were often applied on a reduced number of occasions to conform to existing constraints on their use (see Table 1), as informed by product information held in the LIAISON pesticide database <<liaison.fera.defra.gov.uk/>>.

Table 1. Target pests, crops and treatments for glasshouse pest trials conducted between 2012 and 2014. Experimental treatments (*i.e.*, products) are coded for confidentiality, though a distinction is made between experimental conventional chemical pesticides (codes starting with "C") and biopesticides: B = botanical; M = microbial; P = physically-acting. Control treatments were always water-only and industry standards are denoted first by trade name, then by the primary active ingredient.

YEAR	**2012**		**2013**	**2014**	
CROP	**Tomato**	**Tomato**	**Pepper**	**Pepper**	
PEST	*T. vaporariorum*	*T. urticae*	*M. persicae*	*F. occidentalis*	*A. solani*
PRODUCTS TESTED	Control	Control	Control	Control	Control
	Chess WG [3] (pymetrozine)	Borneo [1] (etoxazole)	Pyrethrum 5EC (pyrethrin)	Calypso [2] (thiacloprid)	Chess WG [3] (pymetrozine)
	B-130	P-091	B-130	C-200	B-062
	B-062	B-062	B-062	B-062	B-130
	B-001	M-051	M-051	B-130	M-051
	C-054	C-131	-	M-209	P-208
	C-106	M-092	-	-	-
APPLICATION *	Wk(3)**	Wk(2)	Wk(3)	Wk(4)	Wk(4)
REPLICATION	$n = 6$	$n = 6$	$n = 6$	$n = 6$	$n = 6$
SUB-UNIT	6 plants (18 leaves)/plot	6 plants (6 leaves)/plot	6 (whole) plants/plot	12 flowers (6 plants)/plot	6 (whole) plants/plot

* Wk = Weekly application, with number of applications in parenthesis. NB: Standards (and other treatments where labels existed) were applied according to commercial practice in the UK, where the maximum number of applications specified and used is provided in superscript. ** Except P-001 which was applied 4 times at 4–5 day intervals.

Statistical analysis was run in SPSS (v22, IBM, New York, U.S.; 2013) considering sampling dates independently and with treatment (*i.e.*, product) as the main and only factor. Data were checked for normality and homoscedasticity and transformed where possible to fit the assumptions of ANOVA and *post hoc* Tukey's tests. Where data could not be made to conform to the requirements of parametric testing, analysis was run using a Kruskal Wallis test for main effects, with Bonferroni-Dunn tests used to conservatively explore pairwise differences. Further details of tests/transformations used are provided within Table 2, Table 3, Table 4, Table 5 and Table 6, where for ease of reference all data are displayed as means with standard errors.

RESULTS

Results of trials are presented below by crop and target species.

Tomato: Trialuerodes Vaporariorum

Trialuerodes vaporariorum populations established well, with both adults and larvae evenly infesting treatments at the start of the trial. Numbers increased in the control treatment up to six days after the second application (6DDA2), with reduced, but still high, numbers in the control six days after the third treatment application (6DDA3). For adults, a treatment effect was only observed 6DAA2, at which time all experimental pesticides, and the industry standard, had similarly reduced adult counts as compared to the control. The same pattern was observed for *T. vaporariorum* larvae 6DAA2, also being repeated at the final assessment 6DAA3.

Table 2. Mean (±SE) number of *Trialuerodes vaporariorum* per plot (6 plants) under varying treatments pre- and post-treatment application. $n = 6$ for all means. p values display the significance of the overall treatment effect on any given sampling date. Means followed by a different letter within a row are significantly different at $p < 0.05$.

Adults	Control	Chess WG	B-130	B-062	B-001	C-054	C-106	$F_{(6,35)}/\chi^2_{(6)}$; p value
D0	17.8 ± 3.5	17.3 ± 4.0	18.2 ± 4.3	12.0 ± 2.3	13.2 ± 2.9	13.3 ± 2.0	13.8 ± 2.6	(F)0.653; 0.687
6DAA1	16.2 ± 3.5	16.8 ± 3.8	11.5 ± 2.7	12.3 ± 1.8	19.7 ± 7.3	13.7 ± 1.3	10.8 ± 2.8	(χ^2)2.507; 0.868
6DAA2	64.0 ± 9.7 ^{a}	21.2 ± 2.9 ^{b}	18.3 ± 6.6 ^{b}	26.2 ± 9.6 ^{b}	22.3 ± 4.8 ^{b}	18.0 ± 6.3 ^{b}	9.2 ± 2.8 ^{b}	(F)7.107; <0.001
6DAA3	30.7 ± 13.9	17.5 ± 4.1	24.5 ± 5.2	18.7 ± 6.9	14.3 ± 4.1	14.7 ± 5.1	15.0 ± 5.9	(F)0.563; 0.756 *

* Data square root transformed prior to analysis.

Larvae	Control	Chess WG	B-130	B-062	B-001	C-054	C-106	$F_{(6,35)}$; p value
D0	71.2 ± 15.8	35.7 ± 12.9	93.5 ± 37.1	27.8 ± 10.7	44.3 ± 20.0	54.0 ± 14.1	36.8 ± 9.2	(F)1.475; 0.215 *
6DAA1	108.2 ± 25.2	50.7 ± 21.0	31.3 ± 8.3	28.7 ± 4.5	65.7 ± 22.8	72.2 ± 24.0	51.0 ± 20.2	(F)1.509; 0.204 *
6DAA2	201.8 ± 28.7 ^{a}	35.0 ± 3.0 ^{b}	47.7 ± 10.2 ^{b}	54.3 ± 7.4 ^{b}	44.5 ± 11.5 ^{b}	59.0 ± 11.7 ^{b}	38.3 ± 11.2 ^{b}	(F)13.863; <0.001 *
6DAA3	165.2 ± 24.8 ^{a}	22.7 ± 8.5 ^{b}	36.0 ± 8.6 ^{b}	67.7 ± 14.9 ^{b}	39.7 ± 9.5 ^{b}	66.0 ± 18.0 ^{b}	56.2 ± 14.4 ^{b}	(F)9.733; <0.001

* Data square root transformed prior to analysis.

Tomato: Tetranychus Urticae

Adults and nymphs of *T. urticae* evenly infested treatments at the start of the trial, with experimental populations establishing well thereafter. For adults, a treatment effect was only observed 6DAA1, at which time the experimental pesticide C-131 had reduced numbers relative to the P-091 treatment. No other differences between treatments existed, though C-131 came close to reducing numbers as compared to the control 6DAA1 ($p = 0.092$), with no adults recovered from treated plants. Mean control counts were actually higher at this time than in the P-091 treatment, but notably more variable. For nymphs, the industry standard exerted a suppressive effect, this being statistically significant in comparison to the control 6DAA2. Notably, lowered counts against the control were also seen at this time for other experimental products, though control counts were highly variable and differences were not statistically significant.

Pepper: Myzus persicae

For *M. persicae*, an even infestation of treatments was achieved at the start of the trial, with numbers in the control treatment generally holding at a stable level thereafter. Treatment effects were observed 6DAA1 and 6DAA3, at which points the biopesticide B-130 had reduced numbers of *M. persicae* relative to the control, also outperforming other treatments, including the industry standard. No other differences between treatments existed, including for the industry standard as compared to the control. Consistently, albeit non-significantly, reduced aphid numbers were also observed in B-062 treated plots relative to the control.

Table 3. Mean (±SE) number of *Tetranychus urticae* per plot (6 plants) under varying treatments pre- and post-treatment application. $n = 6$ for all means. *p* values display the significance of the overall treatment effect on any given sampling date. Means followed by a different letter within a row are significantly different at $p < 0.05$.

Adults	Control	Borneo	P-091	B-062	M-051	C-131	M-092	$F_{(6,35)}/\chi^2_{(6)}$; *p* value
D0	9.8 ± 1.9	4.5 ± 1.2	16.0 ± 7.5	9.0 ± 3.1	10.2 ± 5.0	11.3 ± 4.0	10.7 ± 2.2	(F)0.853; 0.538 *
6DAA1	4.0 ± 1.9 ^{ab}	2.0 ± 1.4 ^{ab}	3.2 ± 0.9 ^{a}	1.3 ± 0.3 ^{ab}	2.0 ± 0.9 ^{ab}	0.0 ± 0.0 ^{b}	1.2 ± 0.5 ^{ab}	(χ^2)12.969; 0.044
6DAA2	18.0 ± 14.0	0.5 ± 0.5	0.8 ± 0.5	2.7 ± 1.1	0.3 ± 0.2	1.2 ± 0.5	2.8 ± 1.6	(χ^2)9.218; 0.162
D0	56.2 ± 22.2	64.7 ± 15.4	82.3 ± 22.9	76.2 ± 15.8	84.8 ± 32.5	47.0 ± 7.8	62.0 ± 11.2	(F)0.502; 0.802
6DAA1	53.7 ± 20.3	16.5 ± 6.9	18.7 ± 4.8	17.7 ± 10.7	19.8 ± 6.7	6.8 ± 2.3	15.0 ± 6.2	(F)2.238; 0.062
6DAA2	42.0 ± 22.7 ^{a}	1.3 ± 0.5 ^{b}	4.7 ± 2.1 ^{ab}	5.2 ± 1.7 ^{ab}	6.5 ± 2.0 ^{ab}	3.2 ± 1.9 ^{ab}	5.3 ± 2.6 ^{ab}	(χ^2)13.609; 0.034

* Data square root transformed prior to analysis.

Table 4. Mean (±SE) number of *Myzus persicae* per plot (6 plants) under varying treatments pre- and post-treatment application. $n = 6$ for all means. Means are displayed with ± SEs. *p* values display the significance of the overall treatment effect on any given sampling date. Means followed by a different letter within a row are significantly different at $p < 0.05$.

Motiles	Control	Pyrethrum 5EC	B-130	B-062	M-051	$F_{(4,25)}$; *p* value
D0	105.8 ± 30.9	94.5 ± 44.7	35.8 ± 15.0	35.0 ± 7.9	61.2 ± 16.7	1.156; 0.354 *
6DAA1	158.0 ± 41.8 ^{a}	48.5 ± 13.7 ^{ab}	14.4 ± 5.4 ^{b}	52.7 ± 14.0 ^{ab}	65.5 ± 25.9 ^{ab}	3.441; 0.023 *
6DAA2	118.6 ± 28.6	97.8 ± 46.8	11.2 ± 4.4	68.5 ± 23.7	54.8 ± 19.2	2.441; 0.073 *
6DAA3	92.6 ± 13.8 ^{a}	119.0 ± 39.4 ^{a}	14.0 ± 3.8 ^{b}	71.5 ± 13.5 ^{ab}	116.8 ± 38.2 ^{a}	5.390; 0.003 *

* Data square root transformed prior to analysis.

Pepper: Frankliniella Occidentalis

Populations of *F. occidentalis* adults and nymphs established well, evenly infesting treatments following release. For adults, a treatment effect was observed 6DAA4, at which time the biopesticide B-062 had reduced pest numbers relative to the control. No other differences between treatments existed, including for the industry standard against the control. For nymphs, the industry standard did exert a significant effect, though with highest pest numbers observed under this treatment as compared to the control (6DAA3). Statistically significant reductions in nymph counts, compared to the control, were only observed for C-200, being evident 6DAA1, 6DAA2 and 6DAA4.

Pepper: Aulacorthum solani

Even establishment of *A. solani* populations was recorded across treatments prior to application of products. Numbers in control treatments initially increased between D0 and 6DAA1, though at this time the trial was infested by parasitoids. Thereafter aphid numbers declined rapidly, with counts in the control treatment being so low at the end of the trial that data collected 6DAA4 were not considered suitable for analysis. Six days after application 1 (6DAA1), when *A. solani* populations were growing in the control, reduced numbers of aphids, in comparison to the control, were observed under the industry standard treatment, where counts were also lower than in the P-208 treatment. This pattern continued 6DAA2, with numbers reduced in the B-062 treatment, relative to the P-208 treatment at this time. Reductions under the B-062 treatment were not significant in comparison to the control 6DAA2, though were more than an order of magnitude lower. By 6DAA3, differences between treatments could not be identified by (relatively conservative) *post-hoc* testing, though an overall treatment effect persisted.

Table 5. Mean (±SE) number of *Frankliniella occidentalis* per plot (6 plants) under varying treatments pre- and post-treatment application. $n = 6$ for all means. *p* values display the significance of the overall treatment effect on any given sampling date. Means followed by a different letter within a row are significantly different at $p < 0.05$.

Adults	Control	Calypso	C-200	B-062	B-130	M-209	$F_{(5,30)}$; *p* value
D0	18.8 ± 3.2	12.8 ± 3.2	10.2 ± 2.5	12.5 ± 2.9	6.5 ± 1.4	13.5 ± 2.6	1.935; 0.118
6DAA1	24.8 ± 3.9	16.8 ± 3.4	17.0 ± 2.1	19.0 ± 3.8	23.2 ± 5.3	15.6 ± 5.3	0.776; 0.575 *
6DAA2	39.0 ± 4.9	32.3 ± 8.9	34.8 ± 5.5	26.8 ± 7.7	38.6 ± 8.7	17.7 ± 7.2	1.195; 0.335

6DAA3	41.5 ± 8.6	49.0 ± 17.8	30.0 ± 5.8	34.7 ± 8.4	44.5 ± 13.6	45.7 ± 6.3	0.521; 0.758 *
6DAA4	90.4 ± 9.1 ab	116.5 ± 12.1 a	87.4 ± 11.4 ab	57.2 ± 6.1 b	110.8 ± 13.2 a	74.5 ± 12.0 ab	4.160; 0.005

* Data square root transformed prior to analysis.

Nymphs	Control	Calypso	C-200	B-062	B-130	M-209	$F_{(6,35)}$; *p* value
D0	11.4 ± 4.7	14.2 ± 4.9	31.6 ± 6.3	24.2 ± 3.9	23.5 ± 7.4	11.3 ± 2.5	1.881; 0.127
6DAA1	28.2 ± 3.8 a	14.7 ± 2.9 ab	3.0 ± 1.7 b	11.0 ± 5.4 ab	10.8 ± 3.6 ab	12.4 ± 4.8 ab	3.246; 0.018 *
6DAA2	35.0 ± 4.6 ab	34.3 ± 10.9 ab	6.4 ± 1.9 ab	16.8 ± 4.5 ab	14.2 ± 5.5 ab	17.1 ± 5.8 ab	3.842; 0.008 *
6DAA3	10.5 ± 2.0 ab	34.7 ± 7.6 ab	7.4 ± 1.8 ab	22.1 ± 5.1 ab	15.2 ± 4.1 ab	15.8 ± 4.9 ab	4.196; 0.005
6DAA4	30.6 ± 5.5 ab	44.8 ± 8.1 ab	9.4 ± 1.0 ab	35.5 ± 8.7 ab	23.7 ± 4.0 ab	27.5 ± 2.9 ab	9.254; <0.001 **

* Data square root transformed prior to analysis; ** Data log transformed prior to analysis.

Table 6. Mean (±SE) number of *Aulacorthum solani* per plot (6 plants) under varying treatments pre- and post-treatment application. $n = 6$ for all means. *p* values display the significance of the overall treatment effect on any given sampling date. Means followed by a different letter within a row are significantly different at $p < 0.05$.

Motiles	Control	Chess WG	B-062	B-130	M-051	P-208	$F_{(5,30)}/\chi^2_{(5)}$; *p* value
D0	15.8 ± 2.7	20.8 ± 4.9	23.2 ± 6.4	15.2 ± 7.0	13.2 ± 3.8	14.3 ± 5.3	(F)0.829; 0.539 *
6DAA1	36.2 ± 9.8 a	1.5 ± 0.6 b	7.8 ± 4.7 ab	19.8 ± 5.6 ab	18.3 ± 4.4 ab	35.2 ± 8.6 a	(χ^2)18.080; 0.003
6DAA2	7.8 ± 2.2 ac	0.0 ± 0.0 b	0.6 ± 0.5 bc	1.7 ± 0.9 ab	8.2 ± 3.4 ab	20.7 ± 5.9 a	(χ^2)22.420; <0.001
6DAA3	3.0 ± 0.9	0.0 ± 0.0	0.8 ± 0.5	1.8 ± 1.3	6.2 ± 3.9	5.5 ± 2.0	(χ^2)11.329; 0.045

* Data square root transformed prior to analysis.

DISCUSSION

The aim of this study was to deliver applied research on high priority pests of glasshouse tomato and pepper crops in order to confirm relative efficacy of standard conventional chemical pesticides and support approval of "new" conventional and/or biopesticide products. In all experiments, at least one of the pesticide products tested provided control of the target pest on at least one post-treatment sampling date, relative to the control treatment. In some cases, the control provided by biopesticides was greater than that achieved through use of the industry standard, where the latter performed poorly as compared to the control in multiple trials. Thus, in addition to providing data in support of approval for effective pesticides in glasshouse tomato and pepper, this work also suggests that current use of some "standard" pesticides may need to be reviewed, particularly if these results are repeatable, or supported by industry trends for reduced or failing product efficacy.

Of those biopesticides that significantly suppressed pests on at least one sampling date in comparison to the control, B-130 demonstrated notable potential, being statistically effective against *T. vaporariorum* (adults and larvae) and *M. persicae*, and resulting in notable reductions in *A. solani* (to a maximum of 78%). Other botanical products also performed well when compared to controls, with B-062 reducing *A. solani* on peppers by more than an order of magnitude and statistically lowering adult and larval *T. vaporariorum* on tomato. It was ineffective against *M. persicae* and *F. occidentalis*, though did cause statistically non-significant reductions in average *T. urticae* counts on tomato of up to 88%. B-001 was also effective, reducing adult and larval *T. vaporariorum* counts on tomato in the only experiment that included this product.

Of the other pesticides tested in the current study, only experimental conventional chemical pesticides exerted any statistically significant effect on pest numbers, with microbials and physically acting products thus performing relatively poorly. M-092 showed some potential against both stages of *T. urticae*, though after adjusting *post-hoc p*-values for multiple comparisons pest reductions observed after treatment with M-092 were not statistically significant. Being reliant on living organisms, some degree of variability in the efficacy of microbials is perhaps to be expected. Such products often require relatively specific thresholds for temperature, moisture and UV exposure post-application, with sensitivity to these variables noted as a constraint to their effective use [22]. Nevertheless, many microbials may be highly selective, with application challenges arguably outweighed by their consequent compatibility with biocontrol [16]. Physically-acting, and in this case fatty-acid-based, pesticides should be less constrained by such "biological" requirements, though may be influenced by other "non-biological" factors. It is plausible in the current study, for example, that efficacy of these products was negatively affected by relatively high water hardness at the experimental site, which may have resulted in the "active ingredient" precipitating out of solution (D. George, personal observation).

The efficacy of industry standards in the current study varied greatly, with some being entirely ineffective against target pests. This was particularly so for Calypso when used against *F. occidentalis* in peppers and Pyrethrum 5EC when deployed against *M. persicae* in peppers. The latter of these pesticides is itself a plant-based formulation, though one widely adopted and with a long history of use. This prolonged use may explain the lack of efficacy of Pyrethrum 5EC in the current study, where it is possible that prior exposure of the *M. persicae* used to pyrethrins/pyrethroids had led to evolution of resistance in the experimental pest population to this group of actives. Pesticide resistance is widely reported in *M. persicae*, being first documented in 1955 (to organophosphates) [23]. *Myzus persicae* has since developed resistance to numerous actives, including pyrethroids, and is currently considered one of the world's most strongly resistant pest species [8]. *Frankliniella occidentalis* are also noted as readily developing resistance [9], though conversely Calypso is an entirely synthetic product with thiacloprid (a neonicotinoid) as its active ingredient. Nevertheless, in a study investigating the contact toxicity of 36 pesticides to *F. occidentalis*, Shan *et al.* [24] reported thiacloprid to be the least effective tested. Using the glass-vial method, the LC_{50} of thiacloprid to larval *F. occidentalis* exceeded 40,000 mg/L in this work, with that of the most effective product (Phoxim) being 0.003 mg/L for comparison. This apparent lack of toxicity of thiacloprid to *F. occidentalis* is supported by the results of the current study. For species such as *M. persicae* and *F. occidentalis* that are clearly sometimes difficult to control using existing industry standards, studies investigating alternative control options are critical. The current study supports that B-130 could be of use for *M. persicae* and that C-200 holds promise to target *F. occidentalis* nymphs. C-106 was also effective in the current study (against *T. vaporariorum*), with C-131 causing large, albeit statistically non-significant, reductions in average *T. urticae* counts.

Even where current industry standards performed well, there is still a need to identify biopestiocides and alternative conventional chemical pesticides to safeguard against product failures and/or withdrawals of existing pesticides, and better align current and future glasshouse production with EU legislation. According to a recent UK report by farm business consultants Andersons, numerous existing conventional chemical pesticides are being, or are expected to be lost as a result of EC 1107/2009 [11], with increased pressure being placed on growers to adopt integrated pest management approaches via the Sustainable Use Directive (Directive 2009/128/EC). One advantage of many biopesticides is that they often perform better than conventional chemical pesticides in an integrated approach; displaying short residual activities that lend well to combined use with subsequent biological control releases [16]. Complex chemistries and potentially multi-facetted modes of action (e.g., certain botanicals) may also provide better long-term resistance management where such products are used [13,14], though immunity of biopesticides to pest resistance should not be assumed [14]. Microbial/physically-acting products are expected to be similarly robust in this regard, though it deserves note that resistance to Bt has already been reported in multiple pest species [25].

Though use of biopesticides requires a more considered approach to pest management than treatment with conventional pesticides, requiring extra care in product handling, application and treatment timings [19], the results presented support that these products, and especially botanical biopesticides, hold promise in glasshouse production. Guidance to optimise the uptake of biopesticides in the UK, and ensure that potential limitations are alleviated by broader benefits, is being well supported through levy board technology transfer.

CONCLUSIONS

The work presented supports that existing industry pesticides may be variable in efficacy, supporting the need to investigate potential alternative products. The current project has identified a number of experimental conventional chemical and botanical products that show significant potential to contribute to tomato and bell pepper production under glasshouse conditions, the latter being particularly attractive in the current legislative climate where existing conventional chemistries are under threat and integrated pest management is prioritised. Work is on-going to ensure that the results of this study translate into increased pesticide product availability for the glasshouse vegetable industry in the UK.

ACKNOWLEDGMENTS

This work formed part of SCEPTRE, a larger multi-partner project to identify potential pesticides, fungicides and herbicides for use in multiple crop production sectors <<http://sceptre.hdc.org.uk/>>. SCEPTRE was sponsored by Defra with financial and in-kind contributions from the Agriculture and Horticulture Development Board-HDC, BASF, Bayer CropScience Ltd., Belchim Crop Protection Ltd., Berry Gardens Growers Ltd., BerryWorld Ltd., Dow Agrosciences Ltd., DuPont (UK) Ltd., The Fresh Produce Consortium, H&H Duncalfe, International Produce Ltd., JE Piccaver & Co., Koppert UK Ltd., Marks and Spencer plc, Stewarts of Tayside Ltd., Syngenta Crop Protection Ltd., and Total World Fresh Ltd. The researchers were ADAS UK Ltd., Allium & Brassica Agronomy Ltd., East Malling Research, Rationale Biopesticide Strategists, James Hutton Institute, STC Research Foundation, and the University of Warwick. The project consortium was chaired by David Piccaver of JE Piccaver & Co. and Harriet Duncalfe of H&H Duncalfe on behalf of HDC; the project was co-ordinated by Tim O'Neill of ADAS on behalf of HDC.

AUTHOR CONTRIBUTIONS

David George and Jennifer Banfield-Zanin conducted experiments at Stockbridge Technology Centre, analysed data and prepared the first draft of the manuscript, on which all other authors provided subsequent comment and input. Rosemary Collier led on pest work *per se* in SCEPTRE, with Jerry Cross and Nick Birch leading pest trials at their respective institutions. Roma Gwynn provided input across the whole SCEPTRE project and Tim O'Neill served as Project Leader for SCEPTRE as a whole. All authors contributed to experimental design, product/target selection and discussion on the direction of the work during the project.

REFERENCES

1. Department for Environment, Food and Rural Affairs. Basic Horticultural Statistics Report 2014. Available online: https://www.gov.uk/government/statistics/basic-horticultural-statistics-2014 (accessed on 25 April 2015).
2. Bell, M.L.; Baker, J.R. Comparison of greenhouse screening materials for excluding whitefly (Homoptera: Aleyrodidae) and thrips (Thysanoptera: Thripidae). *J. Econ. Entomol.* **2000**, *93*, 800–804.
3. Berlinger, M.J.; Jarvis, W.R.; Jewett, T.J.; Lebiush-Mordechi, S. Managing the greenhouse, crop and crop

environment. In *Developments in Plant Pathology: Integrated Pest and Disease Management in Greenhouse Crops*; Albajes, R., Lodovica Gullino, M., van Lenteren, J.C., Elad, Y., Eds.; Kluwer Academic Publishers: Dordrecht, The Netherlands, 2002; Volume 14, pp. 97–123.

4. Berlinger, M.J.; Dalian, R.; Mordechi, S. Integrated pest management of organically grown tomato in Israeli greenhouses. *Appl. Agric. Res.* **1988**, *3*, 233–238.
5. Van Lenteren, J.C. A greenhouse without pesticides: Fact or fantasy? *Crop Prot.* **2000**, *19*, 375–384.
6. Attia, S.; Grissa, K.L.; Lognay, G.; Bitume, E.; Hance, T.; Mailleux, A.C. A review of the major biological approaches to control the worldwide pest *Tetranychus urticae* (Acari: Tetranychidae) with special reference to natural pesticides. *J. Pest Sci.* **2013**, *86*, 361–386.
7. Caspell, N.; Drakes, D.; O'Neill, T. *Food Standards Agency Pesticide Residue Minimisation Crop Guide: Tomatoes*; Bradshaw, N., Ogilvy, S., Eds.; Food Standards Agency: UK, 2006. Available online: http://www.food.gov.uk/sites/default/files/multimedia/pdfs/cropguidetomatodec06.pdf (accessed on 24 April 2015).
8. Bass, C.; Puinean, A.M.; Zimmer, C.T.; Denholm, I.; Field, L.M.; Foster, S.P.; Gutbrod, O.; Nauen, R.; Slater, R.; Williamson, M.S. The evolution of insecticide resistant in the peach potato aphid, *Myzus persicae. Insect Biochem. Molec.***2014**, *51*, 41–51.
9. Gao, Y.; Lei, Z.; Reitz, S.R. Western flower thrips resistance to insecticides: Detection, mechanisms and management strategies. *Pest Manag. Sci.* **2012**, *68*, 1111–1121.
10. Schooler, S.S.; de Baro, P.; Ives, A.R. The potential for hyperparasitism to compromise biological control: Why don't hyperparasitoids drive their primary parasitoid hosts extinct? *Biol. Control* **2011**, *58*, 167–173.
11. Anon. *Healthy Harvest: The Impact of Losing Plant Protection Products on UK Food Production*; NFU: Stoneleigh, UK, 2014.
12. MarketsandMarkets. *AGI 2716: Biopesticides Market by Active Ingredient (Microbials & Biorationals), by Types (Bioinsecticides, Biofungicides, Bionematicides & Bioherbicides), by Application, by Formulation, by Crop Type & by Geography—Global trends & forecasts to 2019*; MarketsandMarkets: Dallas, TX, USA, 2014.
13. Isman, M.B. Plant essential oils for pest and disease management. *Crop Protect.* **2000**, *19*, 603–608.
14. George, D.R.; Finn, R.; Graham, K.; Sparagano, O.A.E. Present and future potential of plant-derived products to control arthropods of veterinary and medical significance. *Parasit. Vectors* **2014**, *7*.
15. El-Wakeil, N.; Gaafar, N.; Sallam, A.; Volkmar, C. Side effects of insecticides on natural enemies and possibility of their integration in plant protection strategies. Available online: http://www.intechopen.com/books/insecticides-development-of-safer-and-more-effective-technologies/side-effects-of-insecticides-on-natural-enemies-and-possibility-of-their-integration-in-plant-protec (accessed on 25 April 2015).
16. George, D.R.; Gwynn, R.; O'Neill, T. Compatibility of biopesticides with pest natural enemies. In Proceedings of the SCEPTRE Conference, Peterborough, UK, 24 February 2015.
17. Isman, M.B. Botanical insecticides, deterrents, and repellents in modern agriculture and an increasingly regulated world. *Annu. Rev. Entomol.* **2006**, *51*, 45–66.
18. Czaja, K.; Góralczyk, K.; Strucinski, P.; Hernik, A.; Korcz, W.; Minorczyk, M.; Łyczewska, M.; Ludwicki, J.K. Biopesticides—Towards increased consumer safety in the European Union. *Pest Manag. Sci.* **2014**, *71*, 3–6.
19. O'Neill, T.; Gwynn, R. Getting the best from biopesticides. In *HDC Factsheet 18/14*; Horticultural Development Company (HDC): Kenilworth, UK, 2014.
20. Chandler, D.; Bailey, A.S.; Tatchell, G.M.; Davidson, G.; Greaves, J.; Grant, W.P. The development, regulation and use of biopesticides for integrated pest management. *Phil. Trans. R. Soc. B* **2011**, *366*, 1987–1998.
21. Dik, A. *Biopesticides Use in The Netherlands: What is Working Well in Protected Edible Crops*; HDC Biopesticides Workshop—Protected Edibles: Kenilworth, UK, 2015.
22. Rai, D.; Updyay, V.; Mehra, P.; Rana, M.; Pandey, A.K. Potential of entomopathogenic fungi as biopesticides. *Ind. J. Sci. Res. Technol.* **2014**, *2*, 7–13.
23. Anthon, E.W. Evidence for green peach aphid resistance to organophosphorous insecticides. *J. Econ. Entomol.* **1955**, *48*, 56–57.
24. Shan, C.; Ma, S.; Wang, M.; Gao, G. Evaluation of insecticides against the western flower thrips, *Frankliniella occidentalis* (Thysanoptera: Thripidae), in the laboratory. *Fla. Entomol.* **2012**, *95*, 454–460.
25. Tabashnik, B.E.; Brévault, T.; Carrìere, Y. Insect resistance to *Bt* crops: Lessons from the first billion acres. *Nat. Biotechnol.* **2013**, *31*, 510–521.

CHAPTER 3

COMPARATIVE TOXICOLOGICAL STUDY BETWEEN EXPOSED AND NON-EXPOSED FARMERS TO ORGANOPHOSPHORUS PESTICIDES

Fariba Taghavian, Gholamhassan Vaezi, Mohammad Abdollahi, and Ali Akbar Malekirad

Keywords: Genotoxicity, Organophosphorus Pesticides, Acetylcholinesterase, Interleukins, Farmers

INTRODUCTION

Nowaday, with the growth of farming and widespread use of chemicals and organophosphorus pesticides, major problems may occur in the nervous, endocrine, reproductive and immune systems of people in addition to numerous chronic diseases such as cancer (1). In this case, the human detoxification system plays a critical function in reducing the harmful effects of pesticides. Toxin levels that increase above the detoxification system's capacity cause worsening health conditions. The intake of antioxidants such as vitamins C and E, as well as organic crops may play an important role in reducing the harmful effects of pesticides through increasing blood antioxidant capacity (2, 3). Also, the evolution of the pesticide industry and the indiscriminate usage of pesticides cause adverse effects on human health, food, and environment. Assessing oxidative stress parameters in plasma is an important means to detect oxidative stress induced by organophosphorus pesticides (3). Organophosphorus pesticides may inhibit acetylcholinesterase (AChE) activity as one of the markers of oxidative stress in workers exposed to these substances (4). In addition to AChE inhibition, 8-hydroxy- 2-deoxy guanosine (8-OHdG) level is an oxidative DNA damage indicator to determine the amount of oxidative stress caused by organophosphorus pesticides (5, 6). Cholinergic syndrome is the most important neurotoxic disorder that results from vulnerability to these pesticides (7). Organophosphorus pesticides, by affecting liver cells, increase the release of glucose through glycogenolysis and gluconeogenesis processes and can be seen as a diabetes risk factor (8, 9). Hence, the genetic risk of pesticide exposure is dangerous for humans. Pesticides cause DNA damage in workers involved in producing these compounds. However, these damages are much less in those who observe safety precautions (10). Persistent organic pollutants (POPs) are chemicals that resist biodegradation and are toxic, and are mostly discharged into the environment by industrial activities. Various animal studies have shown the effects of POPs in different inflammatory markers (11). C-reactive protein (CRP), a plasma protein marker of inflammation in humans, belongs to the pentraxin group of molecules was discovered in 1930. Its gene is located on chromosome 1. CRP is a member of the defense molecules known as acute phase proteins (12). This molecule has been used over a number of years as an inflammatory marker in infectious and inflammatory diseases (13).

The effects of acute exposure to a variety of pesticides on human health are well known. However, the effects of long term exposure to low-dose pesticides on various aspects of human health, including genotoxicity and inflammatory markers, are not clearly defined. Despite the fact that some studies show the effects of pesticides on induced immunity, there is little evidence of severe immune disorders in humans (14). It seems that although the exposure to pesticides can alter immune function, the emergence of immune disorders is related to the dose and duration of exposure to pesticides (15). The current study aims to investigate the effects of chronic exposure to low doses of organophosphate pesticides on genotoxicity and inflammatory markers in farmers. Farmers are exposed to different types of pesticide during the spraying season. Since pesticides can endanger the physical and mental health of people, they impose tremendous expense on individuals and society. Thus, considering the extensive cognitive and physical effects of pesticides on health and the enormous economic

Citation: Taghavian F, Vaezi G, Abdollahi M, Malekirad A. A. Comparative Toxicological Study between Exposed and Non-Exposed Farmers to Organophosphorus Pesticides. Cell Journal (Yakhteh). 2016; 18(1):89-96.

costs imposed on society for treating these disorders, the present work has intended to compare DNA damage, AChE activity and the inflammatory factors in farmers exposed to organophosphorus pesticides to a control group of unexposed individuals living in a village in the vicinity of Arak, Iran.

MATERIALS AND METHODS

This cross-sectional survey enrolled 134 people. The study group consisted of 67 farmers who resided in a village near Arak in the Central province of Iran who were exposed to organophosphorus pesticides for at least 5 years.

The control group consisted of 67 subjects who were not farmers and lived in a nearby village. Control group participants had no exposure to pesticides and were matched with the study group in terms of age, gender, and education. All subjects had expressed their written informed consent before they entered the experiment. The study was conducted in complete accordance with the National Code of Ethics and approved by the Ethics Committee of the Islamic Azad University of Damghan Branch.

Interviews with the farmers determined that they used no proper protective equipment such as aprons, boots, filtered masks, or goggles. All subjects in this study underwent clinical examinations to determine the presence of any symptoms of chronic conditions such as high blood pressure, heart problem, cancer, thyroid disorder, asthma, diabetes and anemia. The data concerning job history, smoking, drug consumption, vitamins and antioxidant supplements were obtained through a questionnaire. After clinical examinations, farmers with chronic diseases, alcohol and antioxidant use, or those under pharmacological therapy were excluded.

A total of 5 ml of venous blood was taken from each person in both the subject and control groups. After serum separation by centrifugation, oxidative DNA damage was studied by measuring 8-OHdG (ng/ml) as a bio-marker of oxidative stress and the product of the DNA repair enzyme with an enzyme-linked immunosorbent assay (ELISA) kit (Cayman Chemical Co., USA). For the measurement of cholinesterase activity in plasma, a total of 3 ml from the incubation solution of a potassium phosphate buffer (75 mmol/L, pH=7.9) and 5,5'-dithiobis-(2-nitrobenzoic acid) (DTNB) reagent (0.25 mmol/L) were added to 10 μl of the serum samples. The activity of the enzyme was induced by the addition of 10 μl of acetylthiocholine iodide (3 mmol/L). Absorption changes were measured with a spectrophotometer at 412 nm. We used the extinction coefficient of 5-thio-2-nitrobenzoic acid (TNB) to measure AChE activity (16).

To assess the level of CRP (ng/ml), we used the highly sensitive ELISA [highly sensitive-CRP, (hs-CRP)] and a kit by Bender Med Systems Inc., Austria. The concentrations (pg/ml) of interleukin-6 (IL6) and IL10 in serum samples were evaluated by ELISA and the kits from the above mentioned company. ELISA test procedures were carried out according to the protocol proposed by each kit.

To assess the impact of gender, the levels of 8-OHdG, serum acetylcholine esterase activity and inflammatory factors in male farmers exposed to pesticides were compared with male farmers with no pesticide exposure. The same analysis was performed for females.

Statistical Analysis

SPSS version 18 (IBM Inc., USA) was used to analyze the data. For this purpose, statistical analysis (independent samples t test) and descriptive statistics (mean ± SD) were performed. The P values lower than 0.05 were considered statistically significant. Pearson's correlation coefficient was used to determine relationships between variables.

RESULTS

The total study population consisted of 134 people, 67 subjects and 67 controls, of whom 52.2% were male and 47.8% were female. The subjects were between 16 to 86 years of age. The majority (98.2%) had less than a diploma.

According to Table 1, there was a significantly greater increase in the level of 8-OH2dG in the subject group (30.41 ± 15.81 ng/ml) compared to the control group (26.04 ± 7.22 ng/ml, P<0.05). Serum AChE activity significantly decreased in the subject group (5098.82 ± 558.81 U/l) compared to the control group (5397.26 ± 574.05 U/l, P<0.05. (There was no significant difference in IL6 levels between the subject (1.80 ± 0.61 pg/ml) and control (1.62 ± 0.49 pg/ml) groups (P>0.05). The level of IL10 in the subjects was 1.98 ± 0.62 pg/ml and 1.38 ± 0.77 pg/ml for the control group. This was a significant increase in the subjects compared with the controls (P<0.05). CRP level significantly increased in the subject group (2.41 ± 0.69 ng/ml) compared to the control group (2.17 ± 0.60 ng/ml, P<0.05, Table 1).

Table 1. Comparison of the levels of 8-hydroxy-2-deoxy guanosine (8-OH2dG), serum acetylcholinesterase activity (AChE) and inflammatory markers in farmers exposed to organophosphorus pesticides to unexposed controls

	Subjects (n=67)	Controls (n=67)	P value
8-OH2dG (ng/ml)	30.41 ± 15.81	26.04 ± 7.22	0.04
AChE (U/l)	5098.82 ± 558.81	5397.26 ± 574.05	0.003
IL6 (pg/ml)	1.80 ± 0.61	1.62 ± 0.49	0.06
IL10 (pg/ml)	1.98 ± 0.62	1.38 ± 0.77	0.001
CRP (ng/ml)	2.41 ± 0.69	2.17 ± 0.60	0.03

According to Table 2, in male farmers exposed to pesticide, the level of 8-OH2dG was 34.61 ± 17.78 ng/ml compared to 24.52 ± 5.14 ng/ml in control group males, which was a significant increase (P<0.05). Serum AChE activity significantly decreased in male subjects (5137.17 ± 560.72 U/l) and compared to the control male (5441.31 ± 536.07 U/l, P<0.05.(IL6 levels in male subjects was 1.60 ± 0.50 pg/ml and in male controls it was 1.62 ± 0.49 pg/ml which showed no significant difference (P>0.05). The level of IL10 in male subjects was 1.91 ± 0.52 pg/ml which significantly increased compared to unexposed control males (1.19 ± 0.76 pg/ml, P<0.05). CRP level in exposed male farmers was 1.41 ± 0.41 ng/ml and 1.48 ± 0.39 ng/ml in males without exposure, which did not significantly differ (P>0.05, Table 2).

Table 2. A comparison of the levels of 8-hydroxy-2-deoxy guanosine (8-OH2dG), serum acetylcholinesterase (AchE) activity and inflammatory factors in male farmers exposed to organophorphorus pesticides with unexposed male farmers

	Exposed male farmers (n=35)	Unexposed male farmers (n=32)	P value
8-OH2dG (ng/ml)	34.61 ± 17.78	24.52 ± 5.14	0.003
AChE (U/l)	5137.17 ± 560.72	5441.31 ± 536.07	0.02
IL6 (pg/ml)	1.60 ± 0.50	1.62 ± 0.49	0.87
IL10 (pg/ml)	1.91 ± 0.52	1.19 ± 0.76	0.001
CRP (ng/ml)	1.41 ± 0.41	1.48 ± 0.39	0.42

According to Table 3 the level of 8-OH2dG was 25.81 ± 11.99 ng/ml in subject females compared to 27.71 ± 8.74 ng/ml in control females, which showed no significant difference (P>0.05). Serum AChE activity in subject females was 5056.89 ± 562.60 U/l and 5349.09 ± 617.91 U/l in female unexposed control farmers, which showed no significant difference (P>0/05). There was a significant increase in IL6 levels in subject females (2.02 ± 0.65 pg/ml) compared to control females (1.63 ± 0.50 pg/ml, P<0.05). IL10 levels also significantly increased in female subjects (2.05 ± 0.71 pg/ml) compared to female controls (1.58 ± 0.74 pg/ml, P<0.05). CRP levels in exposed female farmers was 1.31 ± 0.42 ng/ml and 1.38 ± 0.39 ng/ml in female controls which showed no significant difference (P>0.05, Table 3).

Table 3. A comparison of the levels of 8-hydroxy-2-deoxy guanosine (8-OH2dG), serum acetylcholinesterase (AchE) activity and inflammatory factors in female farmers exposed to organophosphorus pesticides with unexposed female farmers

	Exposed female farmers (n=32)	Unexposed female farmers (n=32)	P value
8-OH2dG (ng/ml)	25.81 ± 11.99	27.71 ± 8.74	0.47
AChE (U/l)	5056.89 ± 562.60	5349.09 ± 617.91	0.052
IL6 (pg/ml)	2.02 ± 0.65	1.63 ± 0.50	0.009
IL10 (pg/ml)	2.05 ± 0.71	1.58 ± 0.74	0.01
CRP (ng/ml)	1.31 ± 0.42	1.38 ± 0.39	0.48

According to Pearson correlation analysis, no significant correlation existed between the levels of 8-OH2dG and serum AChE activity to inflammatory factors (Table 4, P>0.05), nor were there any significant correlations observed between age and the levels of 8-OH2dG, AChE activity, IL6, IL10 and CRP (P>0.05, Table 5).

Table 4. The correlations between 8-hydroxy-2-desoxyguanosine (8-OH2dG) levels and serum acetylcholinesterase (AChE) activity with inflammatory factors

	IL6 (pg/ml)	IL10 (pg/ml)	CRP (ng/ml)
8-OH2dG (ng/ml)	-0.20	-0.15	-0.15
AChE (U/l)	0.09	0.79	0.42

Table 5. The correlations between age to 8-hydroxy-2-deoxy guanosine (8-OH2dG) levels, acetylcholinesterase (AchE) activity and inflammatory factors

	8-OH2dG (ng/ml)	AChE (U/l)	IL6 (pg/ml)	IL10 (pg/ml)	CRP (ng/ml)
Age (Y)	0.006	-0.05	-0.08	-0.08	0.14

DISCUSSION

According to the results of this study, we observed a significant decrease in AChE activity in farmers exposed to organophosphorus pesticides compared to the control group. On the other hand, the extent of DNA damage, IL10 and CRP significantly increased compared to the controls. These results indicated that the farmers' exposure to organophosphorus pesticides could produce free radicals, genomes and tissue damage. Farmers, when interviewed, stated that they did not use proper protective equipment during farming.

There was a substantial decrease in AChE activity in the blood serum of farmers exposed to organophosphorus pesticides compared with the control group, which reflected the high degree of pesticides absorbed. Earlier studies described the inhibition of AChE activity in farmworkers who sprayed pesticides (17). Farmers exposed to organophosphorus pesticides had a 30% reduction in blood AChE levels occurred compared with the control group (6). Manu et al. (18) reported a low activity of serum cholinesterase to be an indispensable component in the high rate of deaths due to acute organophosphorus poisoning.

Shadnia et al. (19) reported no substantial differences in AChE levels between farmers and controls. Cholinesterase levels in their study were consistent with those reported by Rastogi et al. (17) and Atherton et al. (6), however have contradicted results reported by Shadnia et al. (19). This contradiction may be attributed to the differences in pesticide dosage or duration of exposure. The main mechanism of action of organophosphorus compounds is to inhibit carboxylic ester hydrolase, particularly AChE. Organophosphorus compounds inhibit AChE using phosphorylation of the serine hydroxyl group located at the active site of the enzyme (20).

These compounds inhibit the actions of both cholinesterases, chain enzymes in red blood cell (RBC) and serum AChE-resulting in the cholinergic features of organophosphorus toxicity. RBC AChE activity, which is reduced at a lower rate than the serum AChE activity, is a measure of chronic exposure to organophosphates. Therefore, in organophosphorus poisoning, the activity of serum ChE reduces before the activity of RBCs (RBC-ChE) (21).

In the present study, we observed no significant relationship between age and serum AChE activity in the farmers. In the study done by Zheng et al. (22) age had no statistical significance in cholinesterase activity. Fairbrother et al. (23) reported that age significantly affected ChE activity. The present study supported results reported by Zheng et al. (22), on the other hand, it was inconsistent with the results reported by Fairbrother et al. (23). Young people have stated they are less willing to use health and safety equipment such as masks, boots, apron, or perform hand washing, and showering. Increased use of these measures will lead to less absorption of organophosphates. The absorption of organophosphates is associated with AChE activity which may justify the obtained results.

In this study, serum AChE activity in the exposed subject group significantly decreased compared to control, unexposed males. Serum AChE activity in exposed and unexposed female farmers did not significantly differ.

Sidell and Kaminskis (24) reported higher plasma cholinesterase activities in adult men during their first six decades of life compared to women; however, there was no difference between the two sexes after the age of 60. In a study by Fairbrother et al. (23) the plasma cholinesterase activity was affected by sex steroid concentrations and phases of the menstrual cycle. Thus, a significant positive correlation was found between serum cholinesterase activity and progesterone level. The present study agreed with the results by Zheng (22); on the other hand, the current study results were inconsistent with the results obtained by Fairbrother et al. (23). Physiological and genetic factors, lifestyle, nutrition and diet could make a difference in the levels of 8-OH2dG, acetylcholine esterase serum and inflammatory factors in both sexes.

Serum levels of 8-OH2dG in individuals exposed to organophosphorus pesticides significantly increased compared with the control group. Increased 8-OH2dG, as a biomarker of DNA damage, have indicated an increased risk of mutagenesis in these subjects (25, 26). Nuclear and mitochondrial DNA is biologically one of the primary targets of oxidative attack for free radicals. Among the purine and pyrimidine bases, guanine oxidation is more prone to oxidation in such a manner that as the result of a hydroxyl radical attack on the eighth position of guanine molecules results in production of 8-OHdG (27).

The levels of 8-OHdG and other modified bases which are separated from DNA and excreted from cells show a dynamic equilibrium between the rate of oxidative DNA damage and its repair. Thus, the level of oxidized bases changes not only as the result of oxidative damage repair but also as the result of fluctuations in the repair rate (28). Numerous studies have considered 8-OHdG to be a bio-marker of cellular oxidative stress and DNA repair product (29). The mutagenic property of 8-OHdG is known (30). Atherton et al. (6) have reported increased DNA damage in farmers exposed to organophosphorus compounds compared with the control group. In another study, exposure to organophosphorus compounds in young children was linked to increased levels of urinary 8-OHdG (31). In study by Muniz et al. (32) urinary 8-OHdG levels in farmers exposed to organophosphorus pesticides were higher than the control group. Organophosphorus pesticides could cause oxidative DNA damage in farmers, therefore the increased serum levels of 8-OH2dG in the present study were consistent with other surveys. CRP level in people exposed to organophosphorus pesticides was higher compared to the control group.

These results have strengthened the hypothesis that organophosphorus pesticides can affect the immune system and disrupt IL. In surveys conducted by Patil et al. (33), a substantial increase in serum CRP of vineyard spraying workers has been noted compared to controls. According to Tsai et al. (34) CRP levels strongly correlated with the level of severity of organophosphorus poisoning, such that the CRP level was important as an acute phase reactant in severe organophosphorus poisoning.

The outcomes of the present work were consistent with results obtained by Patil et al. (33) and Tsai et al. (34). The CRP molecule is an acute phase serum protein. In case of inflammation or infection this molecule is increasingly made by the liver. Increased levels of CRP are related to infection, age, body volume index, smoking, drug and dietary habits (35).

IL6 levels in both groups did not significantly differ. Kumar et al. (11) have stated that no significant relationship exists between the levels of POPs and IL6 which is consistent with the findings of this study.

Although exposure to pesticides can alter immune function, the emergence of immune disorders are dependent on the dose and duration of exposure to pesticides (15). The studies on the effects of pesticides on human immune system have many limitations, such as a shortage of data on exposure levels, heterogeneity of the methods employed, difficulty in providing a prognostic significance to the slight changes often observed and the interpretation of reported findings (36).

In this study, the exposure to organophosphorus pesticides led to a significantly increased IL10 serum level in farmers. Alluwaimi and Hussein (37) reported that the immunotoxicity caused by organophosphorus pesticides such as diazinon led to increased levels of IL10 in mice. The outcomes of this study were consistent with those reported by Alluwaimi and Hussein which indicated the effects of organophosphorus pesticides on the immune system. In studies by Neishabouri et al. (38) the dose dependent induction of diazinon affected the actions of both humoral and cellular immunities. In another study, malathion was used as an organophosphorus pesticide to induce increased activity of liver enzymes in plasma and pro-inflammatory cytokines. Thus, malathion has been shown to induce liver inflammation and hepatotoxicity (39). Organophosphorus compounds have toxic effects on immune systems and immune functions of vertebrates, invertebrates and fish. Immunotoxicity may be directed via inhibition of serine hydrolases or esterases in components of the immune system through oxidative damage to immune organs or by modulation of signal transduction pathways that control immune function. Indirect effects involve modulating of the nervous system or chronic effects of altered metabolism/nutrition on the immune system. Immunotoxicities are varied and include pathology of immune organs, and decreased humoral and/or cell mediated immunity. Modified non-specific resistance, reduced host resistance, hypersensitivity and autoimmunity are among the symptoms of immunotoxicity (40).

According to Pearson correlation analysis, no significant relation existed between 8-OHdG and AChE activity to inflammatory factors.

In a study performed by Kumar et al. (11), there was no significant relationship observed between POPs and inflammatory factors such as IL6 and CRP. Karanlık compared experimental data from two groups of patients who either died or were discharged with organophosphate poisoning. There was no statistically significant differences between the levels of IL6 in these patients (41). The findings in present study were consistent with those by Karanlık et al. (41) and Kumar et al. (11). Despite the large number of experimental data which have shown that induced immune safety is affected by pesticides, there is little or no evidence of severe immune disorders in humans caused by pesticides (14). Although exposure to pesticides can alter immune function, immune disorders are related to dose and duration of exposure to pesticides (15) and the low activity of cholinesterase is an essential factor for mortality rates by organophosphate poisoning (41).

CONCLUSION

Exposure to organophosphorus pesticides can cause oxidative DNA damage, inhibit the activity of AChE, and increase the levels of IL10 and CRP which place people's health in danger. Further studies can help identify the types and mechanisms of contaminants in workplaces and the environment, which will help promote individual and social healthcare.

ACKNOWLEDGMENTS

The authors wish to thank the Research and the Education Departments at Damghan Azad University for their financial support for this study as a part of the Ph.D. thesis of the first author. The authors declare no conflict of interest.

REFERENCES

1. Mostafalou S, Abdollahi M. Pesticides and human chronic diseases: evidences, mechanisms, and perspectives. Toxicol Appl Pharmacol. 2013; 268(2):157–177.
2. Xavier R Jr, Rekha K, Bairy K. Health perspective of pesticide exposure and dietary management. Malays J Nutr. 2004; 10(1):39–51.
3. Soltaninejad K, Abdollahi M. Current opinion on the science of organophosphate pesticides and toxic stress: a systematic review. Med Sci Monit. 2009; 15(3):RA75–90.
4. Ranjbar A, Pasalar P, Abdollahi M. Induction of oxidative stress and acetylcholinesterase inhibition in organophosphorous pesticide manufacturing workers. Hum Exp Toxicol. 2002; 21(4):179–182.
5. Karami Mohajeri S, Abdollahi M. Toxic influence of organophosphate, carbamate, and organochlorine pesticides on cellular metabolism of lipids, proteins, and carbohydrates: a systematic review. Hum Exp Toxicol. 2011; 30(9):1119–1140.
6. Atherton KM, Williams FM, Egea González FJ, Glass R, Rushton S, Blain PG, et al. DNA damage in horticultural farmers: a pilot study showing an association with organophosphate pesticide exposure. Biomarkers. 2009; 14(7):443–451.
7. Jokanović M, Kosanović M. Neurotoxic effects in patients poisoned with organophosphorus pesticides. Environ Toxicol Pharmacol. 2010; 29(3):195–201.
8. Teimouri F, Amirkabirian N, Esmaily H, Mohammadirad A, Aliahmadi A, Abdollahi M. Alteration of hepatic cells glucose metabolism as a non-cholinergic detoxication mechanism in counteracting diazinon-induced oxidative stress. Hum Exp Toxicol.2006; 25(12):697–703.
9. Rahimi R, Nikfar S, Larijani B, Abdollahi M. A review on the role of antioxidants in the management of diabetes and its complications. Biomed Pharmacother.2005; 59(7):365–373.
10. Undeğer U, Başaran N. Assessment of DNA damage in workers occupationally exposed to pesticide mixtures by the alkaline comet assay. Arch Toxicol.2002; 76(7):430–436.
11. Kumar J, Lind PM, Salihovic S, van Bavel B, Ingelsson E, Lind L. Persistent organic pollutants and inflammatory markers in a cross-sectional study of elderly Swedish people: the PIVUS cohort. Environ Health Perspect. 2014; 122(9):977–983.
12. Anderson GP. COPD, asthma and C-reactive protein. Eur Respir J. 2006; 27(5):874–876.
13. Pinto-Plata VM, Müllerova H, Toso JF, Feudjo-Tepie M, Soriano JB, Vessey RS, et al. C-reactive protein in patients with COPD, control smokers and non-smokers. Thorax.2006; 61(1):23–28.
14. Vial T, Nicolas B, Descotes J. Clinical immunotoxicity of pesticides. J Toxicol Environ Health. 1996; 48(3):215–229.
15. Banerjee BD, Koner BC, Ray A. Immunotoxicity of pesticides: perspectives and trends. Indian J Exp Biol. 1996; 34(8):723–733.
16. Ellman GL, Courtney KD, Anders V Jr, Fether-Stonerm RM. A new and rapid colorimetric determination of acetylcholinesterase activity. Biochem Pharmacol.1961; 7:88–95.
17. Rastogi SK, Satyanarayan PV, Ravishankar D, Tripathi S. A study on oxidative stress and antioxidant status

of agricultural workers exposed to organophosphorus insecticides during spraying. Indian J Occup Environ Med. 2009; 13(3):131–134.

18. Manu MS, Prashant V, Akila P, Suma MN, Basavanagowdappa H. A retrospective analysis of serial measurement of serum cholinesterase in acute poisoning with organophosphate compounds. Toxicol Int. 2012; 19(3):255–259.
19. Shadnia S, Azizi E, Hosseini R, Khoei S, Fouladdel S, Pajoumand A, et al. Evaluation of oxidative stress and genotoxicity in organophosphorus insecticide formulators. Hum Exp Toxicol. 2005; 24(9):439–445.
20. Yurumez Y, Durukan P, Yavuz Y, Ikizceli I, Avsarogullari L, Ozkan S, et al. Acute organophosphate poisoning in university hospital emergency room patients. Intern Med.2007; 46(13):965–969.
21. Jaga K, Dharmani C. Sources of exposure to and public health implications of organophosphate pesticides. Rev Panam Salud Publica. 2003; 14(3):171–185.
22. Zheng J, Zhou ZJ, Zheng G, Dai XF, Gu XA. Relationship between the polymorphism of carboxylic esterases and genetic susceptibility to organophosphates pesticides exposure. Zhonghua Lao Dong Wei Sheng Zhi Ye Bing Za Zhi.2005; 23(2):83–86.
23. Fairbrother A, Wagner SL, Welch S, Smith BB. Influence of menstrual cycle on serum cholinesterase. Environ Res. 1989; 49(2):181–189.
24. Sidell FR, Kaminskis A. Influence of age, sex, and oral contraceptives on human blood cholinesterase activity. Clin Chem. 1975; 21(10):1393–1395.
25. Bayrami M, Hashemi T, Malekirad AA, Ashayeri H, Faraji F, Abdollahi M. Electroencephalogram, cognitive state, psychological disorders, clinical symptom, and oxidative stress in horticulture farmers exposed to organophosphate pesticides. Toxicol Ind Health. 2012; 28(1):90–96.
26. Malekirad AA, Faghih M, Mirabdollahi M, Kiani M, Fathi A, Abdollahi M. Neurocognitive, mental health, and glucose disorders in farmers exposed to organophosphorus pesticides. Arh Hig Rada Toksikol. 2013; 64(1):1–8.
27. Schneider JE, Price S, Maidt L, Gutteridge JM, Floyd RA. Methylene blue plus light mediates 8-hydroxy 2'-deoxyguanosine formation in DNA preferentially over strand breakage. Nucleic Acids Res. 1990; 18(3):631–635.
28. Halliwell B. Why and how should we measure oxidative DNA damage in nutritional studies? How far have we come. 2000; 72(5):1082–1087.
29. Cooke MS, Evans MD, Herbert KE, Lunec J. Urinary 8-oxo-2'-deoxyguanosine--source, significance and supplements. Free Radic Res. 2000; 32(5):381–397.
30. Kasai H. Analysis of a form of oxidative DNA damage, 8-hydroxy-2'-deoxyguanosine, as a marker of cellular oxidative stress during carcinogenesis. Mutat Res. 1997; 387(3):147–163.
31. Ding G, Han S, Wang P, Gao Y, Shi R, Wang G, et al. Increased levels of 8-hydroxy-2'-deoxyguanosine are attributable to organophosphate pesticide exposure among young children. Environ Pollut. 2012; 167:110–114.
32. Muniz JF, McCauley L, Scherer J, Lasarev M, Koshy M, Kow YW, et al. Biomarkers of oxidative stress and DNA damage in agricultural workers: a pilot study. Toxicol Appl Pharmacol. 2008; 227(1):97–107.
33. Patil JA, Patil AJ, Sontakke AV, Govindwar SP. Occupational pesticides exposure of sprayers of grape gardens in western Maharashtra (India): effects on liver and kidney function. J Basic Clin Physiol Pharmacol. 2009; 20(4):335–355.
34. Tsai JR, Sheu CC, Cheng MH, Hung JY, Wang CS, Chong IW, et al. Organophosphate poisoning: 10 years of experience in southern Taiwan. Kaohsiung J Med Sci. 2007; 23(3):112–119.
35. Chenillot O, Henny J, Steinmetz J, Herbeth B, Wagner C, Siest G. High sensitivity C-reactive protein: biological variations and reference limits. Clin Chem Lab Med.2000; 38(10):1003–1011.
36. Corsini E, Liesivuori J, Vergieva T, Van Loveren H, Colosio C. Effects of pesticide exposure on the human immune system. Hum Exp Toxicol. 2008; 27(9):671–680.
37. Alluwaimi AM, Hussein Y. Diazinon immunotoxicity in mice: modulation of cytokines level and their gene expression. Toxicology. 2007; 236(1-2):123–131.
38. Neishabouri EZ, Hassan ZM, Azizi E, Ostad SN. Evaluation of immunotoxicity induced by diazinon in C57bl/6 mice. Toxicology. 2004; 196(3):173–179.
39. Lasram MM, Lamine AJ, Dhouib IB, Bouzid K, Annabi A, Belhadjhmida N, et al. Antioxidant and anti-inflammatory effects of N-acetylcysteine against malathion-induced liver damages and immunotoxicity in rats. Life Sci. 2014; 107(1-2):50–58.
40. Galloway T, Handy R. Immunotoxicity of organophosphorous pesticides.Ecotoxicology. 2003; 12(1-4):345–363.
41. Karanlık MG, Sebe A, Oğuzhan Ay M, Açıkalın A, Atlı M, Çitilcioğlu S, et al. Relationship between mortality and acute inflammatory markers, erythrocyte cholinesterase, serum cholinesterase levels in the acute organic phosphorus intoxication.Cukurova Med J. 2013; 38(2):233–240.

CHAPTER 4

ASSOCIATION BETWEEN OCCUPATIONAL EXPOSURES TO PESTICIDES WITH HETEROGENEOUS CHEMICAL STRUCTURES AND FARMER HEALTH

Xusheng Huang, Chao Zhang, Ruifa Hu, Yifan Li, YanhongYin, ZhaohuiChen, JinyangCai, and Fang Cui

INTRODUCTION

Although pesticides are important for agricultural production throughout the world[1], the widespread application of pesticides has induced well-documented adverse health effects in humans[2,3,4,5,6,7]. Most epidemiological studies have analyzed the health effects of exposure to a single pesticide or a specific group of pesticides with similar chemical structure, measuring exposure level by using the sum of all pesticides, or classifying the study subjects into exposed (or high-exposed) and non-exposed (or low-exposed) groups[8,9,10,11,12,13]. In addition, the previous studies have often analyzed the associations of pesticide exposure with human health using a single parameter or a group of parameters for a specific health type[14,15,16,17].

In reality, the examined health parameters may be probably affected by a complex mixture of pesticides. For example, farmers in China not only use a wide variety but also a great amount of pesticides to protect their crops, which greatly differs from farmers in other regions[18,19]. However, few previous studies have associated human health with exposures to pesticides with heterogeneous chemical structures in an integrated framework, which would generate the biased results. Moreover, most previous studies have analyzed the associations of pesticide exposure with the risk ratio (RR), odds ratio (OR) and rate difference (RD). However, it should be noted that some irreversible sub-clinical damages to human health induced by pesticide exposure would also be unobservable. Unfortunately, few studies have focused on this important issue. This study aimed to analyze and compare the sub-clinical health effects of occupational exposures to pesticides with heterogeneous chemical structures among 224 farmers in China in 2012. Farmer health was measured using 52 parameters in terms of the complete blood cell counts (CBC), a blood chemistry panel (BCP), and the conventional nerve conduction studies (CNCS).

METHODS

Study Sample

Data for this study were collected from farmers in three Chinese provinces (Guangdong, Jiangxi and Hebei) in 2012, which has been previously described[11]. In brief, we randomly selected two counties within each province, two villages within each county, and 20–25 farmers within each village. An eligible farmer was defined as the person who was over 18 years old and mainly sprayed pesticides in the household. A total of 246 farmers were initially enrolled, but 22 farmers were excluded due to their failure to provide information of pesticide application or their absence from the health examinations. This study was approved by the Ethics Committee of Chinese PLA General Hospital, and all farmers provided informed consent before the health examinations. The methods in this study were carried out in accordance with the approved guidelines. A baseline face-to-face survey was conducted to collect data regarding the individual characteristics, which included household population, number of labors, history of farm activity and

Citation: Huang, X. et al. Association between occupational exposures to pesticides with heterogeneous chemical structures and farmer health. Sci. Rep. 6, 25190. DOI: 10.1038/srep25190 (2016).

pesticide application, cropping structure, age, sex, smoking status, and alcohol consumption.

Health Examinations and Outcomes

Two health examinations were performed by professional doctors and technicians from Chinese PLA General Hospital in 2012. The first examination was performed before farmers began planting crops in March, and the second examination was performed before the end of the crop harvest. A total of 52 health parameters regarding CBC, BCP and CNCS were examined. Moreover, farmers' height and weight were also obtained at two health examinations.

A blood sample was collected in the morning after a 12-hour fast for each farmer. All blood samples were subsequently centrifuged and refrigerated until they were submitted for laboratory testing in Beijing within 8 hours. Moreover, all blood samples from the first and second examinations were tested at the same laboratory to ensure consistency.

The CNCS were performed using the surface electrodes with standard placements on the non-dominant side of the farmers. The upper and lower limbs were warmed to maintain temperatures of 32–34 °C and 30–33 °C, respectively. The CNCS parameters consisted of the motor conduction velocities, the distal motor latencies, and the compound muscle action potential amplitudes of the median, ulnar, tibial, and common peroneal nerves, as well as the sensory conduction velocities and the sensory action potential amplitudes of the median, ulnar, and sural nerves.

Assessing Pesticide Exposure

Occupational pesticide exposures were measured using the amount (in kilograms) of the pesticides' active ingredients applied by farmers. All farmers were asked to record information of all pesticides they applied. The essential information included the chemical name, the active ingredient percentage(s), the applied amount, and the purchase price of each applied pesticide, as well as the target crop, date, and duration of each pesticide application. To ensure the data accuracy and reliability, we provided several training sessions regarding how to make effective records, and conducted a follow-up visit semimonthly to check if they made accurate records. Note that all farmers were asked to keep the package bags or bottles of pesticide products for each follow-up check to minimize the recall biases. Using these data, we were able to calculate the amount of each active ingredient.

In this study, all pesticides were classified into organophosphates (OPs), organosulfurs (OSs), organonitrogens (ONs), pyrethroids (PYRs) and others according to the chemical structures of the active ingredients. The adopted standard criteria for classification were based on the recommendations from the Institute for the Control of Agrochemicals, Ministry of Agriculture of China[20,21]. To specify the associations of pesticide exposure with health parameters during different durations, pesticides applied within and more than 10 days before the second health examination was defined as the short- and medium-term exposure, respectively.

Statistical Analysis

In this study, multiple linear regression analyses were used to evaluate the associations between pesticide exposures and the health parameters. Both the short- and medium-term exposures for each group of pesticides were integrated in the multivariate model separately. The multivariate model was also adjusted for age, sex, body mass index, smoking status, alcohol consumption, and two province dummy variables. We included the baseline parameter from the first examination in the final multivariate model to control for systematic heterogeneity. Thus, the final model was described as:

$$HI_i^2 = \beta_0 + \beta_1 \cdot ShortExp_i + \beta_2 \cdot MediumExp_i + \beta_3 \cdot Characteristics_i + \beta_4 \cdot Region_i + \beta_5 \cdot HI_i^1 + \varepsilon_i \quad (1)$$

In equation (1), HI_i^2 and HI_i^1 indicate the health parameter of the second and first health examination of the i-th farmer, respectively. $ShortExp_i$ and Medium Exp_i indicate farmers' exposure in the short and medium term, respectively. $Characteristics_i$ consists of age, sex, body mass index, smoking status and alcohol consumption. $Region_i$ is a vector of dummies for provinces, including Guangdong and Jiangxi. ε_i indicates the random error term. β_0–β_5 indicate the coefficients to be estimated. Thus, equation (1) is similar to the first-order auto-regression model described in the econometric literature[22].

All tests were two-tailed, and $p<0.05$ was considered statistically significant. All statistical analyses were performed using STATA version 13.1 (StataCorp LP, College Station, Texas).

RESULTS

Demographic Characteristics and Exposure Levels

Table 1 summarizes the demographic characteristics, and levels of pesticide exposures. The final sample consisted of 224 farmers with a mean age of 51.77 years and a mean body mass index of 23.53 kg/m^2. Among these farmers, 26.8% were female, 46.9% had a smoking habit, and 42.4% consumed alcohol.

Table 1: Summary of demographic characteristics and exposure levels (n=224).

Variable	Mean±SD or *n* (%)
Age (year)	51.77±10.05
Body mass index (kg/m^2)	23.53±3.64
Female (yes=1, no=0)	60 (26.8)
Smoking habit (yes=1, no=0)	105 (46.9)
Alcohol consumption (yes=1, no=0)	95 (42.4)
Total pesticide exposure (kg)	4.54±5.52
Short-term exposures (kg)	
OPs	0.09±0.27
OSs	0.08±0.20
ONs	0.04±0.14

PYRs	0.01±0.03
Others	0.06±0.32
Medium-term exposures (kg)	
OPs	2.11±3.36
OSs	0.51±1.21
ONs	1.15±2.13
PYRs	0.07±0.19
Others	0.41±0.85

SD: standard deviation.

As described in Table 1, the sample farmers applied an average of approximately 4.54 kg of pesticides between two health examinations. In the short term, the average amount of OPs and OSs was 0.09 kg and 0.08 kg, respectively, higher than the amounts of ONs, PYRs and other pesticides. In the medium term, the average amount of OPs was approximately 2.11 kg, which was the largest amount among these five groups of pesticides. The average amount of ONs was 1.15 kg, which was second to OPs but larger than that of OSs (0.51 kg) and other pesticides (0.41 kg). Among ONs, the most commonly used pesticides were carbamate and neonicotinoid insecticides. In addition, PYRs remained the least applied group of pesticides with an average amount of 0.07 kg.

Associations of Pesticide Exposures with the Complete Blood Cell Counts

Table 2 reports the associations between pesticide exposures and CBC parameters. None of the CBC parameters was associated with OPs exposure in either short or medium terms. The decreased platelet distribution width (PDW) was associated with the short-term OSs and PYRs exposures. Following the medium-term exposure, the increased PDW was positively associated with PYRs, but insignificantly associated with OSs. In terms of the white blood cells, each kilogram increase in the medium-term OSs exposure was associated with an increase in lymphocyte counts (Lym) of 0.07×10^9/L (95% CI: 0.00, 0.14). The short-term ONs exposure was also associated with the increased Lym. However, the short-term exposure to other pesticides was associated with a decrease in Lym of 0.32×10^9/L (95% CI: −0.62, −0.02) or 5.37% (95% CI: −9.34, −1.40). Moreover, the positive association between the short-term exposure to other pesticides and neutrophil counts (Neu) was also seen. As for the parameters of the red blood cells, we found that the short-term ONs exposure was associated with the increased hematocrit (Hct), and the medium-term ONs exposure was then associated with the increased hemoglobin (Hb).

Table 2: Association between pesticide exposure and complete blood cell counts (n=214).

Group	Parameter (Unit)	Short term		Medium term	
		β_1 (95% CI)	*p*-value	β_2 (95% CI)	*p*-value
OSs	PDW (fL)	−0.38 (−0.68, −0.08)*	0.01	0.01 (−0.04, 0.06)	0.76
	Lym (10^9/L)	0.03 (−0.41, 0.46)	0.90	0.07 (0.00, 0.14)*	<0.05
ONs	Lym (10^9/L)	0.58 (0.00, 1.15)*	<0.05	−0.01 (−0.05, 0.03)	0.55
	Hb (g/L)	7.90 (−1.96, 17.75)	0.12	0.66 (0.02, 1.30)*	0.04
	Hct (%)	8.11 (4.51, 11.70)**	0.00	0.20 (−0.02, 0.41)	0.06
PYRs	PDW (fL)	−2.91 (−5.43, −0.39)*	0.02	0.32 (0.02, 0.63)*	0.04
Others	Neu (10^9/L)[b]	0.83 (0.19, 1.47)*	0.01	−0.08 (−0.35, 0.20)	0.59
	Neup (%)	5.71 (1.50, 9.92)**	<0.01	−1.16 (−2.95, 0.63)	0.20
	Lym (10^9/L)	−0.32 (−0.62, −0.02)*	0.03	0.12 (−0.01, 0.25)	0.06
	Lymp (%)	−5.37 (−9.34, −1.40)**	<0.01	1.23 (−0.45, 2.92)	0.15

The models were adjusted for age, sex, body mass index, smoking habit, alcohol consumption, and two province dummy variables for Guangdong and Jiangxi. The number of observations for Neu and Neup is 213 due to data non-availability of indices. CI: confidence interval.

Associations of Pesticide Exposures with the Blood Chemistry Panel

As shown in Table 3, there were considerable associations between occupational pesticide exposures and nine BCP parameters. The short-term exposures to both OPs and OSs were associated with the increased level of alanine aminotransferase (ALT) and aspartate aminotransferase (AST). Moreover, each kilogram increase in the short-term OSs exposure was also associated with 0.98 U/L (95% CI: −1.84, −0.13) decrease in cholinesterase (ChE), while the medium-term OSs exposure was positively associated with an increase in total protein (TP) of 0.39 g/L (95% CI: 0.01, 0.76). In addition, the medium-term ONs exposure was also found to be positively associated with TP. In terms of parameters of renal function, only a positive association between the medium-term OSs exposure and blood urea nitrogen (BUN) was seen. Moreover, we also found that the medium-term OSs exposure was also associated with a decrease in vitamin B_{12} (VB_{12}) of 26.89 ng/L (95% CI: −51.73, −2.06). The magnitudes of serum sodium (Na) and glucose (GLU) were only seen to be positively associated with the medium-term ONs exposure. PYRs was only associated with the increased C-reactive protein (CRP) following the medium-term exposure.

Table 3: Association between pesticide exposure and blood chemistry panel (n=214).

Group	Parameter (Unit)	Short term		Medium term	
		β_1 (95% CI)	*p*-value	β_2 (95% CI)	*p*-value
OPs	ALT (U/L)	11.93 (5.14, 18.72)**	<0.01	0.04 (−0.46, 0.54)	0.88
	AST (U/L)	9.79 (5.07, 14.51)**	<0.01	−0.06 (−0.41, 0.29)	0.74
OSs	ALT (U/L)	13.15 (4.65, 21.65)**	<0.01	0.88 (−0.52, 2.28)	0.22
	AST (U/L)	8.84 (2.89, 14.79)**	<0.01	0.92 (−0.07, 1.90)	0.07
	ChE (10^3 U/L)	−0.98 (−1.84, −0.13)*	0.02	0.06 (−0.08, 0.20)	0.39
	TP (g/L)	−1.80 (−4.09, 0.49)	0.12	0.39 (0.01, 0.76)*	<0.05
	BUN (mmol/L)	−0.48 (−1.48, 0.52)	0.35	0.22 (0.06, 0.38)**	<0.01
	VB_{12} (ng/L)	53.54 (−98.78, 205.80)	0.49	−26.89 (−51.73, −2.06)*	0.03
ONs	TP (g/L)	−0.58 (−3.61, 2.45)	0.71	0.22 (0.02, 0.41)*	0.03
	Na (mmol/L)	−1.51 (−4.09, 1.08)	0.25	0.19 (0.02, 0.36)*	0.03
	GLU (mmol/L)	−0.23 (−1.05, 0.60)	0.59	0.05 (0.00, 0.11)*	<0.05
PYRs	CRP (mg/L)	−27.56 (−59.86, 8.74)	0.14	4.33 (0.18, 8.49)*	0.04

The models were adjusted for age, sex, body mass index, smoking habit, alcohol consumption, and two province dummy variables for Guangdong and Jiangxi. CI: confidence interval.

Associations of Pesticide Exposures with Peripheral Nerve Conduction

Table 4 shows the results for CNCS parameters associated with occupational pesticide exposures. OPs exposure was the most commonly associated with the CNCS parameters. The medium-term OPs exposure was associated with the reductions in the motor (UNMCV) and sensory conduction velocity (UNSCV) of the ulnar nerve of 0.20 m/s (95% CI: −0.39, −0.02) and 0.26 m/s (95% CI: −0.45, −0.06), respectively. Moreover, we also found that the medium-term OPs exposure was also negatively associated with the motor conduction velocity (TNMCV) and the compound muscle action potential amplitudes of the tibial nerve at the proximal (TNPCMAPA) and distal ends (TNDCMAPA). In addition, the decreased motor conduction velocity of the common peroneal nerve (PNMCV), and the decreased compound muscle action potential amplitudes of the median nerve at the distal ends (MNDCMAPA) were also seen following the medium- and short-term OPs exposure, respectively. However, OSs was only associated with a decrease in the sensory nerve action potential amplitude of the median nerve (MNSNAPA) of 2.11 mV (95% CI: −3.75, −0.46) and ulnar nerve (UNSNAPA) of 1.26 mV (95% CI: −2.47, −0.04) follow the short-term exposure. In contrast, the associations of ONs with the CNCS parameters were only seen following the medium-term exposure. For example, a decrease in the sensory conduction velocity (MNSCV) and the action nerve potential amplitude (MNSNAPA) of the median nerve were associated with the medium-term ONs exposure. Moreover, the medium-term ONs exposure was also associated with a decrease in the motor conduction velocity of the tibial nerve of 0.24 m/s (95% CI: −0.47, −0.01), and a decrease in the sensory conduction velocity of the sural nerve of 0.33 m/s (95% CI: −0.64, −0.01). PYRs was only associated with the decreased UNSNAPA following the short-term exposure, while other pesticides were only associated with the decreased UNSCV following the short-term exposure, and the decreased sensory nerve action potential amplitude of the sural nerve (SNSNAPA) following the medium-term exposure.

Table 4: Association between pesticide exposure and nerve conduction (n=218).

Group	Parameter (Unit)	Short term		Medium term	
		β_1 (95% CI)	*p*-value	β_2 (95% CI)	*p*-value
OPs	MNDCMAPA (mV)	−1.77 (−3.47, −0.07)*	0.04	0.06 (−0.07, 0.19)	0.36
	UNMCV (m/s)	0.64 (−1.80, 3.08)	0.61	−0.20 (−0.39, −0.02)*	0.03
	UNSCV (m/s)	1.31 (−1.18, 3.80)	0.30	−0.26 (−0.45, −0.06)**	<0.01
	TNMCV (m/s)	−0.20 (−2.32, 1.92)	0.85	−0.21 (−0.37, −0.04)*	0.01
	TNPCMAPA (mV)	−0.34 (−1.88, 1.21)	0.67	−0.13 (−0.25, −0.01)*	0.03
	TNDCMAPA (mV)	0.07 (−1.92, 2.06)	0.94	−0.18 (−0.33, −0.03)*	0.02
	PNMCV (m/s)	−0.58 (−2.49, 1.33)	0.55	−0.27 (−0.41, −0.12)**	0.00
OSs	MNSNAPA (mV)	−2.11 (−3.75, −0.46)*	0.01	0.19 (−0.08, 0.46)	0.17
	UNSNAPA (mV)	−1.26 (−2.47, −0.04)*	0.04	−0.06 (−0.26, 0.14)	0.56
ONs	MNSCV (m/s)	−1.46 (−5.88, 2.97)	0.52	−0.44 (−0.73, −0.14)**	0.00

	MNSNAPA (mV)	1.13 (−1.02, 3.27)	0.30	−0.21 (−0.36, −0.07)**	<0.01
	TNMCV (m/s)	−3.06 (−6.61, 0.49)	0.09	−0.24 (−0.47, −0.01)*	0.05
	SNSCV (m/s)	−3.05 (−7.83, 1.74)	0.21	−0.33 (−0.64, −0.01)*	0.04
PYRs	UNSNAPA (mV)	−11.54 (−20.49, −2.59)*	0.01	0.20 (−1.02, 1.42)	0.75
Others	UNSCV (m/s)	−2.27 (−4.40, −0.13)*	0.04	0.43 (−0.45, 1.30)	0.34
	SNSNAPA (mV)	0.76 (−1.93, 3.45)	0.58	1.22 (−2.32, −0.12)*	0.03

The models were adjusted for age, sex, body mass index, smoking habit, alcohol consumption, and two province dummy variables for Guangdong and Jiangxi. CI: confidence interval.

DISCUSSION

Our results demonstrated that occupational exposures to pesticides with heterogeneous chemical structures were associated with farmer health in different patterns, and the association between a specific group of pesticides and farmer health also differed between the short- and medium-term exposures. Although we found associations between multiple groups of pesticides and several CBC parameters, it was not enough to provide strong evidence of hematological disorders. However, occupational exposures to OPs, OSs and ONs might be associated with liver damage, renal dysfunction, the loss of vitamins and the increased glucose. In addition, the associations of OPs and ONs with peripheral nerve conduction among farmers were much more common and severe than that of OSs and PYRs.

In this study, OPs accounted for the largest proportion (approximately 48.5%) of the total amount of applied pesticides, which was similar to the findings of previous studies[23,24]. Most OPs applied by farmers were moderately, highly, and even extremely toxic (e.g., dichlorvos, chlorpyrifos, omethoate, and terbufos). Here it appeared that the adverse health effects of OPs exposure were the most severe. Our results demonstrated that the short-term OPs exposure might induce acute liver damage, which was consistent with the previous human and animal studies[25,26]. Moreover, our results also provided evidence that occupational OPs exposure was the most commonly associated with chronic neurotoxicity. The reduced nerve conduction velocities and the compound muscle action potential amplitudes associated with the medium-term OPs exposure suggested a increased risk of demyelination and axonal damage, respectively. In this study, the adverse effects of the medium-term OPs exposure on the ulnar and tibial nerves seemed to be more severe than on the other nerves. Similar findings were also reported in substantial previous epidemiological studies[27,28,29].

As for OSs exposure, the decreased PDW in the short term was not a strong evidence of hematological disorders, since there was no association between the short-term OSs exposure and mean volume of platelets (MPV). Similar situation was also found on the association between the medium-term OSs exposure and Lym. In terms of BCP parameters, OSs might impair farmers' hepatic function due to its associations with ALT, AST and ChE following the short-term exposure, and with TP following the medium-term exposure. Consistently, the liver damages associated with several most commonly used organosulfur pesticides were previously reported in the animal studies[30,31,32]. Here our results also showed that the medium-term OSs exposure might also increase the risk of chronic renal impairment, which was also consistent with the conclusions previously[33,34]. Moreover, OSs might lead to the loss of VB_{12} following the medium-term exposure. In terms of the neurotoxic effects, our results suggested that the short-term OSs exposure might induce acute axonal damage of the median and ulnar nerves. However, the neurotoxicity of OSs exposure was inappreciable compared with that of OPs exposure.

In this study, farmers' exposure to ONs might increase Hct only in the short term and increase Hb only in the medium term. However, none associations between ONs exposure and these two hematological parameters were simultaneously seen in the same duration, which failed to provide convincing evidence of hematological disorders. Similar to OSs exposure, the medium-term ONs exposure might also induce chronic renal impairment indicated by the increased TP. Moreover, the medium-term ONs exposure might increase the level of GLU, which was consistent with the findings in previous rat or fish studies[35,36]. In addition, our results showed that the neurotoxicity of ONs exposure was second only to that of OPs exposure. In particular, the reduced sensory conduction velocity and sensory nerve action potential amplitude of the median nerve indicated the demyelination and axonal damage. Note that these findings were in accordance with the previously documented neurotoxicity of carbamate and neonicotinoid insecticides[37].

Occupational PYRs exposure appeared to be the least toxic to humans. The increased CRP associated with the medium-term PYRs exposure might be a useful biomarker for inflammation. However, similar result was not found in the previous studies. Although the short-term PYRs exposure was also associated with the reduced UNSNAPA, its neurotoxicity was much slighter than that of OPs and ONs exposures. Similarly, a previous study indicated that the PYRs exposure during pregnancy might not induce considerable neurotoxic effect on the infants[38]. However, the estimated least neurotoxicity of PYRs exposure in this study might also be attributed to the low dose of application among farmers since the effect was probably dose-dependent.

There are several strengths and limitations to be considered. We included different groups of pesticides with heterogeneous chemical structures in an integrated framework, which allowed us to analyze the net health effect of each group of pesticides simultaneously. Moreover, subject to the application dose and duration, farmers' exposure to pesticides may not induce the observable clinical symptoms, but the sub-clinical damages often ignored in the previous studies may probably occur following pesticide exposure. Thus, the analyses on the dose-dependent effects of pesticide exposure on farmer health, rather than the associations between pesticide exposure and the RR, OR and RD, could produce more useful findings. However, compared with some previous epidemiological studies, this study had a relatively small sample size, which might not perfectly avoid the biased estimation of the associations between pesticide exposure and health parameters. Moreover, our analyses did not include farmers' exposure to other non-pesticide toxic substances or pesticides applied by their neighbors as the confounders in the estimation models.

In conclusion, occupational exposures to pesticides with heterogeneous chemical structures may induce different adverse health effects. The short- and medium-term OPs exposure may mainly associated with liver damage and peripheral nerve impairment, respectively, while OSs exposure may induce both liver damage and renal dysfunction. The neurotoxicity of ONs exposure is second only to OPs exposure in addition to its potential liver damage and the induced alterations in glucose. In comparison, the estimated results show that PYRs may be the least toxic in terms of the low dose of application.

ACKNOWLEDGEMENTS

The authors thank all the subjects who participated in this study. In addition, this study was funded by grants from the National Natural Sciences Foundation of China (71210004 and 71173014).

REFERENCES

1. Oerke, E. C. Crop losses to pests. J. Agr. Sci. 144, 31–43 (2006).
2. Antle, J. M. & Pingali, P. L. Pesticides, productivity, and farmer health: a Philippine case study. Am. J. Agr. Econ. 76, 418–430 (1994).
3. Bulgaroni, V., Rovedatti, M. G., Sabino, G. & Magnarelli, G.Organophosphate pesticide environmental exposure: analysis of salivary cholinesterase and carboxilesterase activities in preschool children and their mothers. Environ. Monit. Assess. 184, 3307–3314 (2012).
4. Jayasinghe, S. S., Pathirana, K. D. & Buckley, N. A. Effects of acute organophosphorus poisoning on function of peripheral nerves: a cohort study. PLos ONE 7, e49405 (2012).
5. Narayan, S. et al. Household organophosphorus pesticide use and Parkinson's disease. Int. J. Epidemiol. 42, 1476–1485 (2013).
6. Rauh, V. et al. Seven-year neurodevelopment scores and prenatal exposure to chlorpyrifos, a common agricultural pesticide.Environ. Health Perspect. 119, 1196–1201 (2011).
7. Recio-Vega, R., Ocampo-Gómez, G., Borja-Aburto, V. H., Moran-Martínez, J. & Cebrian-Garcia, M. E. Organophosphorus pesticide exposure decreases sperm quality: association between sperm parameters and urinary pesticide levels. J. Appl. Toxicol. 28, 674–680 (2008).
8. Aminov, Z., Haase, R. F., Pavuk, M., Carpenter, D. O. & Anniston Environment Health Research Consortium. Analysis of the effects of exposure to polychlorinated biphenyls and chlorinated pesticides on serum lipid levels in residents of Anniston, Alabama.Environ. Health 12, 108 (2013).
9. Chevrier, J., Eskenazi, B., Holland, N., Bradman, A. & Barr, D. B.Effects of exposure to polychlorinated biphenyls and organochlorine pesticides on thyroid function during pregnancy.Am. J. Epidemiol. 168, 298–310 (2008).
10. Raanan, R. et al. Early-life exposure to organophosphate pesticides and pediatric respiratory symptoms in the CHAMACOS cohort. Environ. Health Perspect. 123, 179–185 (2015).
11. Hu, R. et al. Long- and short-term health effects of pesticide exposure: a cohort study from China. PLos ONE 10, e0128766 (2015).
12. Ndlovu, V., Dalvie, M. A. & Jeebhay, M. F. Asthma associated with pesticide exposure among women in rural Western Cape of South Africa. Am. J. Ind. Med. 57, 1331–1343 (2014).
13. Parrón, T., Requena, M., Hernández, A. F. & Alarcón, R.Environmental exposure to pesticides and cancer risk in multiple human organ systems. Toxicol. Lett. 230, 157–165 (2014).
14. Clary, T. & Ritz, B. Pancreatic cancer mortality and organochlorine pesticide exposure in California, 1989–1996. Am. J. Ind. Med. 43, 306–313 (2003).
15. Cocco, P. et al. Lymphoma risk and occupational exposure to pesticides: results of the Epilymph study. Occup. Environ. Med.70, 91–98 (2013).
16. Eriksson, M., Hardell, L., Carlberg, M. & Åkerman, M. Pesticide exposure as risk factor for non-Hodgkin lymphoma including histopathological subgroup analysis. Int. J. Cancer 123, 1657–1663 (2008).
17. Fenster, L. et al. Association of in utero organochlorine pesticide exposure and fetal growth and length of gestation in an agricultural population. Environ. Health Pespect. 114, 597–602 (2006).
18. Zhang, C. et al. Overuse or underuse? An observation of pesticide use in China. Sci. Total Environ. 538, 1–6 (2015).
19. Alavanja, M. C. R., Hoppin, J. A. & Kamel, F. Health effects of chronic pesticide exposure: cancer and neurotoxicity. Annu. Rev. Public Health 25, 155–197 (2004).
20. Institute for the Control of Agrochemicals, Ministry of Agriculture, China (ICAMA). In A New Pesticide Manual. (China Agriculture Press, 1996).
21. of Agrochemicals, Ministry of Agriculture, China (ICAMA). In A New Pesticide Manual (Continue). (China Agriculture Press, 1998).
22. Angrist, J. D. & Pischke, J. S. In Mostly Harmless Econometrics: An Empiricist's Companion. (Princeton University Press, 2009).
23. Awad, O. M., Ei-Fiki, S. A., Abou-Shanab, R. A. I., Hassanin, N. M. A.& Abd Ei Rahman, R. Influence of exposure to pesticides on liver enzymes and cholinesterase levels in male agricultural workers.Global NEST J. 16, 1006–1015 (2014).
24. Hernandez, A. F. et al. Pesticide exposure and genetic variation in xenobiotic-metabolizing enzymes interact to induce biochemical liver damage. Food Chem. Toxicol.

61, 144–151 (2013).

25. Cecchi, A., Rovedatti, M. G., Sabino, G. & Magnarell, G. G.Environmental exposure to organophosphate pesticides: assessment of endocrine disruption and hepatotoxicity in pregnant women. Ecotox. Environ. Saf. 80, 280–287 (2012).
26. Gomes, J., Dawodu, A., Lloyd, O. D. & Anilal, S. Hepatic injury and disturbed amino acid metabolism in mice following prolonged exposure to organophosphorus pesticides. Hum. Exp. Toxicol. 18, 33–37 (1999).
27. Starks, S. E. et al. Peripheral nervous system function and organophosphate pesticide use among licensed pesticide applicators in the agricultural health study. Environ. Health Perspect. 120, 515–520 (2012).
28. Jayasinghe, S. S., Pathirana, K. D. & Buckley, N. A. Effects of acute organophosphorus poisoning on function of peripheral nerves: a cohort study. PLos ONE 7, e49405 (2012).
29. Karami-Mohajeri, S., Nikfar, S. & Abdollahi, M. A systematic review on the nerve-muscle electrophysiology in human organophosphorus pesticide exposure. Hum. Exp. Toxicol. 33, 92–102 (2014).
30. Siddiqui, A., Ali, B. & Srivastava, S. P. Effect of mancozeb on hepatic glutathione s-transferase in rat. Toxicol Lett. 68, 301–305 (1993).
31. Meneguz, A. & Michalek, H. Effects of zineb and its metabolite, ethylenethiourea, on hepatic microsomal systems in rats and mice. Bull Environ Contam Toxicol. 38, 862–867 (1987).
32. Dalvi, P. S., Wilder-Kofie, T., Mares, B., Dalvi, R. R. & Billups, L. H.Toxicologic implications of the metabolism of thiram, dimethyldithiocarbamate and carbon disulfide mediated by hepatic cytochrome p450 isozymes in rats. Pestic Biochem Physiol.74, 85–90 (2002).
33. Paiva, F. et al. Kidney histological alterations and metallothionein and heat shock protein expression in wistar rats after fungicide thiram exposure. Arq. Bras. Med. Vet. Zootec. 65, 95–102 (2013).
34. Maita, K., Tsuda, S. & Shirasu, Y. Chronic toxicity studies with thiram in wistar rats and beagle dogs. Fundam. Appl. Toxicol. 16, 667–686 (1991).
35. Giri, S. N., Curry, D. L., Hollinger, M. A. & Freywald, M. Effect of paraquat on plasma enzymes, insulin, glucose, and liver glycogen in the rat. Environ. Res. 20, 300–308 (1980).
36. Nascimento, C. R. B., Souza, M. M. & Martinez, C. B. R. Copper and the herbicide atrazine impair the stress response of the freshwater fish prochilodus lineatus. Comp. Biochem. Phys. C. 155, 456–461 (2011).
37. Pathak, M. K., Fareed, M., Bihari, V., Patel, D. K. & Mathur, N. Nerve conduction studies in sprayers occupationally exposed to mixture of pesticides in a mango plantation at Lucknow, North India.Toxicol. Environ. Chem. 93, 188–196 (2011).
38. Xue, Z. et al. Effect of synthetic pyrethroid pesticide exposure during pregnancy on the growth and development of infants.Asia-Pacific J. Public He. 25, 72S–79S (2013).

CHAPTER 5 LONG- AND SHORT-TERM HEALTH EFFECTS OF PESTICIDE EXPOSURE

Ruifa Hu, Xusheng Huang, Jikun Huang, Yifan Li, Chao Zhang, Yanhong Yin, Zhaohui Chen, Yanhong Jin, Jinyang Cai, and Fang Cui

INTRODUCTION

Pesticide use in crop production worldwide increased nearly twentyfold from 1960 to 2000 [1] and further increased from 1.0 billion tons in 2002 to 1.7 billion tons in 2007 [2]. China is the largest producer of pesticides and one of the most intensive pesticide users in crop production in the world [3, 4]. Despite the well-documented deleterious effects of pesticides on biological pest control function, the environment, and food safety [5–7], the health effects of these agents have also attracted substantial attention [8].

Previous studies showed that pesticide exposure often induces acute and chronic neurological toxicity [9–14] and dysfunctional lipid, protein, and carbohydrate metabolism [15]. However, most surveys used subjective or qualitative measurements, such as symptoms, psychological scales, or clinical signs, to evaluate the health effects of pesticide use [16–18]. For example, one study found that farmers who used a greater amount of pesticides were more likely to suffer from headache, nausea, and skin problems [13]. These outcomes had limitations in the following two aspects. First, the reported symptoms may not be necessarily pathological. Second, there may be subclinical neuroelectrophysiological changes when no clinical symptoms or signs were found. Our study relies on objective and quantitative measures of neuroelectrophysiology to evaluate health effects of pesticide exposure. Furthermore, the relationship between pesticide exposure and human health remains unclear because the time and extent of pesticide exposure were not sufficiently estimated in many previous studies [19]. The majority of studies used one-time household survey data to examine the health effects of pesticide exposure [9, 13, 20], and used nonspraying workers [21] or the general population [16] as non-exposed controls. While these studies did provide evidence of the adverse health effects of pesticide exposure, it was difficult to effectively control for the potentially confounding health-related factors of each participant and to accurately define exposure level. The objective of the present study is to use the most detailed neuroelectrophysiological methods to examine the clinical and subclinical health effects of pesticide exposures based on a unique dataset that covers 2 time periods in China.

MATERIALS AND METHODS

Sample Selection

This study was conducted in Guangdong, Jiangxi, and Hebei provinces, which represent high, middle, and low levels of pesticide applications in China, respectively. In each province, 2 counties and 2 villages from each county were randomly selected. In each village, 20–25 households were randomly sampled. In total, 246 households participated in this study. The samples differ slightly in terms of health examinations because a few participants were absent during health investigations.

Citation: Hu R, Huang X, Huang J, Li Y, Zhang C, Yin Y, et al. (2015) Long- and Short-Term Health Effects of Pesticide Exposure. PLoS ONE 10(6): e0128766. DOI:10.1371/journal.pone.0128766.

Exposure Estimation

Pesticide exposure information was obtained from 2 datasets. We first collected baseline data regarding pesticide use history, household and individual characteristics, and pesticide-related poisoning events via face-to-face interviews. Participants were asked to report the frequency of pesticide application, and whether they had suffered from headache, nausea, skin irritation, or digestive discomfort that was severe enough to interfere with their work during the past three years (2009–2011). These data were used to assess the long-term health effects of pesticide exposure. All participants were divided into 2 groups. One group had been exposed to relatively high pesticide levels (more than 50 times of pesticide application in 2009–2011, Group H), and the other group had been exposed to lower pesticide levels (less than 50 times of pesticide application in 2009–2011, Group L). Secondly, participants were asked to record each pesticide application in 2012 on the specially designed calendar we had provided them. Moreover, this recording process was supervised by the village leaders under the guidance of the research team. We checked the record at least once a month to ensure its validity and reliability. We used the frequency of pesticide application in the previous 3 days and 4–10 days to analyze short-term effect of pesticide application.

Health Investigations

Two rounds of health investigation were conducted. The first was in March 2012, before participants had applied pesticides for their crop production. The second one was during the course of crop production but prior to crop harvesting. In total, 66 health indicators were examined during health investigations. For each health indicator, abnormality was defined as the corresponding value falls out of the normal range.

Blood Tests

All blood samples were obtained after participants had fasted for 12 hours. Samples were centrifuged immediately and transported in a refrigerated state to the same laboratory in Beijing within 8 hours. The tests included a complete blood count (CBC), blood chemistry panel (e.g., liver function, cholinesterase, renal function, electrolytes, vitamin B_{12}, folic acid, fasting plasma glucose) and measurement of C-reactive protein.

Clinical Examinations

The clinical examinations consisted of general and neurological examinations. The former comprised height, weight, and blood pressure measurements. The latter consisted of the clinical total neuropathy score (TNSc) and the mini-mental state examination (MMSE). The TNSc was administered to all subjects by two certified neurologists who were trained to ensure the results were comparable. The TNSc is composed of seven items: sensory symptoms, motor symptoms, autonomic symptoms, pin sensibility, vibration sensibility, muscle strength, and deep tendon reflexes [22]. For each item, the score ranged between 0 (normal) and 4 (worst), so that the total score range was 0–21. Participants who received a score of 2 or greater were judged to be abnormal.

Nerve Conduction Studies

The conventional nerve conduction studies were performed on four motor (median, ulnar, tibial, and peroneal) and three sensory (median, ulnar, and sural) nerves on the participant's non-dominant side using surface electrodes with standard placement. The parameters included motor conduction velocity (MCV), distal motor latency (DML), sensory conduction velocity, compound muscle action potential (CMAP), and sensory nerve action potential (SNAP) amplitudes. Participants' limbs were heated: the lower limb was maintained at 30–33°C, and the upper limb was maintained at 32–34°C during the evaluation. The normal reference ranges were developed at the Neuroelectrophysiology Unit, Department of Neurology, Chinese People's Liberation Army General Hospital. The study was approved by the Ethics Committee of Chinese PLA General Hospital and all the participants gave their informed, written consent.

Statistical Methods

To control for the impact of confounding factors, multivariate regression analysis was performed. The model used to evaluate the long-term effects of pesticide use on farmers' health is as follows:

$$Indicator_i = a_0 + a_1 * Pesticide_{it} + a_2 * Characteristic_{it} + a_3 * \mathrm{Re}gion_{it} + e_{it} \quad (1)$$

where $Indicator_i$ indicates an health indictor for farmer *i*. The independent variable of interest, *pesticide*, takes value of 1 if farmer i is in Group H and zero for Group L.*Characteristic* is a vector of the following farmer characteristics: gender (*female = 1; male = 0*), age (*year)*, education (*year*), height (*cm*), weight (*kg*), current smoker (*smoker = 1, non-smoker = 0*), and current drinker (*drinker = 1, non-drinker = 0*). Finally, regional dummies, Guangdong and Jiangxi, were added into the equation. When, Eq (1) is estimated by the OLS model when *indicator* is a continuous variable. Otherwise, it is an ordered probit model (Dprobit) if *indicator* is a dichotomous variable.

The health indicators from the 2 rounds of health investigations comprise panel data, which allowed us to estimate the short-term effects of pesticide use on farmers' health, using the farmer's fixed effect model

$$\Delta Indicator_i = \beta_0 + \beta_1 * \Delta Pesticide_i + \varepsilon_i \quad (2)$$

where $\Delta Indicator_i$ is the change in the ith farmer's health indicator during the two rounds of health investigations and $\Delta Pesticide_i$ represents additional pesticide use over the same period. All analyses were conducted with STATA 11.0 software. Statistical tests were two-sided, and p-values of < 0.05 were considered statistically significant.

RESULTS

Overview of Crop Pesticide Application in China

In general, pesticides are extensively used by farmers in China. The means of the measured participant characteristics are similar to that of the average farmer in China. On average, participants in the present study applied pesticides on 12.8 separate occasions (total amount of 11.5 kg) in 2012 (Table 1). About 14% of the participants took protective measures (e.g., wearing masks and/or gloves) during pesticide application (Table 1).

Table 1. Average Pesticide Applications Per Person in 2012 and Percentage of Participants Reported Poisoning in 2009–2011.

Province	Time or frequencies	Total hours	Total amount of pesticide used (kg)	Percentage of wearing masks and/or gloves	Percentage of farmers reported poisoning in 2009–2011
Average	12.8	43.9	11.5	14	13
Guangdong	15.9	47.3	13.4	17	23
Jiangxi	14.7	56.3	8.9	11	8
Hebei	8.2	30.0	12.0	13	8

Data are from authors' survey.

doi:10.1371/journal.pone.0128766.t001

There were 183 varieties of pesticides applied by the farmers between the two rounds of health investigations. Most of them are pyrethroid, organophosphates and others. The high toxicity and heavy application of these pesticides produced effects on farmers' health. The survey results show that 13% of participants had suffered from at least one acute health problem during pesticide application in the farm fields during 2009–2011 (Table 1). The percentage of participants who suffered from such health problems in Guangdong (23%) is higher than that of Jiangxi and Hebei (8%). This may be explained by the fact that farmers in Guangdong used a greater amount of pesticides than those in the other two provinces (Table 1).

Long-Term Effects

Table 2 shows the results of two rounds of health investigations. CBC analysis in the first round showed that 70.25% of participants had at least one abnormal indicator (Table 2). Blood chemistry tests indicated abnormal renal function in 5.79% of subjects, and aberrant hepatic function, electrolyte levels, vitamin levels, and fasting plasma glucose levels in 10%–15% of subjects. Nerve conduction studies showed high percentages of abnormal DML (45.93%). Similar findings were observed in the TNSc (45.71%) and MMSE (20.82%).

Table 2. Percentage (%) of Abnormalities in Health Assessments in 2012.

Indicators	First round	Second round
Blood examinations		
Complete blood count	70.25	95.40
Blood chemistry tests		
Hepatic function	10.33	30.13
Renal function	5.79	20.92
Electrolytes	9.50	25.94
Vitamins	14.88	8.37
Fasting plasma glucose	13.22	5.44
C-reactive protein	6.61	2.93
Nerve conduction studies		
Conduction velocity	19.51	18.26
Motor nerves	10.57	10.37
Sensory nerves	12.60	12.86
Distal motor latency	45.93	31.54
Amplitude	10.57	9.13
Motor nerves	8.13	5.39
Sensory nerves	3.25	4.56
Neurological examinations		
TNSc	45.71	31.93
MMSE	20.82	10.50

Details of these results are presented in S5 Table. Data are from authors' survey.

doi:10.1371/journal.pone.0128766.t002

The number of abnormal health indicators increased significantly in the second round of investigation. For example, results of CBC analysis showed that nearly all participants (95.40%) examined had at least one abnormal indicator in the second round (Table 2). A large percentage of subjects also demonstrated abnormalities in hepatic function, renal function, electrolyte balance, and TNSc in the second round.

Fig 1 presents the percentages of abnormal cases of major health indicators for Group H and Group L in 2009–2011. The results show a consistent positive relationship between abnormal test results and intensity of pesticide use. This indicates that participants more frequently exposed to pesticides may suffer from CBC, hepatic, renal, glucose metabolic, peripheral, and central neurological ailments.

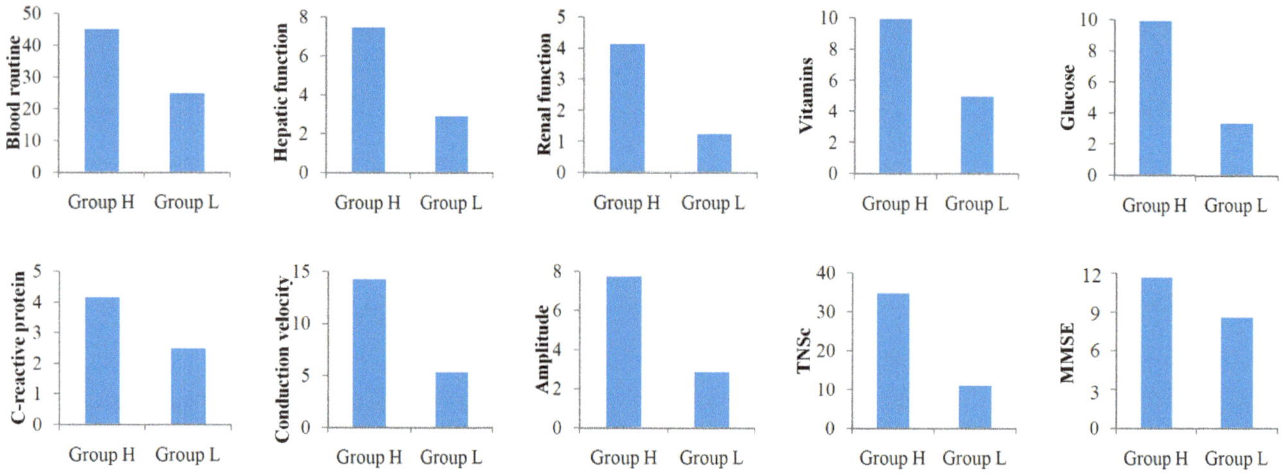

Figure 1. Percentage of abnormal cases by Group H and Group L in 2009–2011. The eight bar charts showed that the percentages of abnormal cases of major health indicators, including blood routine, hepatic function, renal function, vitamins, glucose, C-reactive protein, TNSc, MMSE, conduction velocity and amplitude, were higher in Group H than in Group L.

Multivariate regression analysis was conducted to control for the impact of other factors. Results showed significant differences in five health indicators between Group H and Group L (Table 3), while no difference was found in the other indicators. Specifically, participants with greater pesticide exposure had higher white blood cell counts, but lower creatinine and potassium levels. In addition, Group H demonstrated lower tibial nerve proximal and distal CMAP amplitude.

Table 3. Estimated Long-term Health Consequences of Pesticide Exposure among Farmers in China, 2009–2011 (OLS Estimation).

Variables	WBC	Cr	K	TNPCMAPA	TNDCMAPA
Pesticide application:					
Group H	0.64 (0.03, 1.26)*	-5.32 (-8.91, -1.73)**	-0.28 (-0.48, -0.08)**	-1.35 (-2.61, -0.10)**	-1.53(-3.04, -0.02)*
Control variables:					
Female	0.35 (-0.41, 1.12)	-18.00 (-22.43, -13.56)**	-0.02 (-0.27, 0.23)	-0.59 (-2.14, 0.95)	0.07 (-1.79, 1.93)
Age (year)	-0.01 (-0.04, 0.01)	0.17 (0.04, 0.31*	0.00 (-0.01, 0.00)	-0.12 (-0.17, -0.07)**	-0.11 (-0.17, -0.05)**
Height (cm)	-0.01 (-0.05, 0.04)	0.03 (-0.25, 0.31)	-0.01 (-0.02, 0.01)	0.03 (-0.07, 0.13)	0.12 (0.00, 0.24)*
Weight (kg)	0.01 (-0.02, 0.04)	0.16 (0.01, 0.31*	0.01 (-0.00, 0.01)	-0.06 (-0.11, -0.01)*	-0.06 (-0.13, 0.00)
Current smoker	0.59 (0.03, 1.16)*	-3.91 (-7.19, -0.63)*	0.16 (-0.02, 0.34)	-0.12 (-1.26, 1.03)	-0.37 (-1.75, 1.01)
Current drinker	-0.07(-0.58, 0.44)	-0.65 (-3.60, 2.31)	0.20 (0.04, 0.37)*	-0.16 (-1.20, 0.88)	-0.61 (-1.87, 0.64)
Constant	6.83 (-0.94, 14.60)	50.29 (5.19, 95.39)	5.47 (2.94, 7.99)**	17.40 (1.35, 33.44)*	4.19 (-15.12, 23.50)
Observations	242	242	242	246	246
R-squared	0.11	0.46	0.19	0.14	0.10

Abbreviations: CI, confidence interval; Cr, creatinine; K, potassium; TNPCMAPA, tibial nerve proximal compound muscle action potential amplitude; TNDCMAPA, tibial nerve distal compound muscle action potential amplitude; WBC, white blood cell.

* and ** represent the levels of significance at 5% and 1%, respectively. Province dummy variables were included, but not reported.

doi:10.1371/journal.pone.0128766.t003

However, while increases or decreases in the above health indicators were observed in Group H, some scores were still within the normal range. To address this issue, each of the health indicators was measured as a dummy (or binary) variable (abnormal = 1, normal = 0). A Probit model was used for estimation. The results showed a positive association between

Group H and abnormalities in nerve conduction velocities (CVs), CMAP, and SNAP amplitudes; however, no relationship was found with other health indicators. A significant increase in sensory nerve abnormalities relative to motor nerve abnormalities was observed in peripheral nervous system (Table 4).

Table 4. Estimated Long-term Neurological Effects of Pesticide Exposure among Farmers in China, 2009–2011 (Dprobit Estimation).

Variables	Abnormal conduction Velocity			Abnormal amplitude		
	All nerves	Motor nerves	Sensory nerves	All nerves	Motor nerves	Sensory nerves
Pesticide application:						
Group H	0.16 (0.06, 0.25)**	0.05 (-0.01, 0.10)	0.10 (0.03, 0.18)**	0.05 (0.00, 0.10)*	0.02 (-0.03, 0.06)	0.03 (0.01, 0.06) *
Control variables:						
Female	-0.03 (-0.17, 0.12)	-0.02 (-0.10, 0.06)	0.00 (-0.11, 0.11)	0.00 (-0.08, 0.08)	0.03 (-0.06, 0.12)	-0.01 (-0.03, 0.01)
Age (year)	0.02 (0.01, 0.02)**	0.01 (0.00, 0.01)**	0.01 (0.01, 0.01)**	0.01 (0.00, 0.01)**	0.00 (0.00, 0.01)**	0.00 (-0.00, 0.00)*
Height (cm)	0.00 (-0.01, 0.01)	0.00 (-0.01, 0.01)	-0.00 (-0.01, 0.01)	0.00(-0.00, 0.01)	0.00 (-0.00, 0.01)	0.00 (-0.00, 0.00)
Weight (kg)	0.00 (-0.00, 0.01)	0.00 (-0.00, 0.00)	0.00 (-0.00, 0.00)	0.00 (-0.00, 0.00)	0.00 (-0.00, 0.00)	-0.00 (-0.00, 0.00)
Current smoker	-0.04 (-0.14, 0.07)	0.01 (-0.05, 0.07)	-0.05 (-0.13, 0.03)	0.02 (-0.03, 0.08)	0.05 (-0.00, 0.10)	-0.01 (-0.04, 0.01)
Current drinker	0.05 (-0.05, 0.16)	0.04 (-0.03, 0.11)	0.03 (-0.06, 0.11)	0.07 (0.00, 0.13)*	0.04 (-0.01, 0.10)	0.02 (-0.01, 0.05)
Observations	246	246	246	246	246	246

Abbreviations: CI, confidence interval

* and ** represent the levels of significance at 5% and 1%, respectively. Province dummy variables were included, but not reported.

doi:10.1371/journal.pone.0128766.t004

Short-Term Effects

The two rounds of health investigations in the present study allowed for the analysis of the short-term effects of pesticide exposure (frequency of pesticide application in the previous three days and in the previous 4–10 days) on health indicators. To accomplish this, a fixed effect model was used to control for individual differences, including the accumulated health effects of pesticide exposure prior to the first round of investigation.

CBC analysis showed that pesticide application during the previous three days had significant effects on the majority of selected health indicators, such as monocytes, monocyte percentage, red blood cell, hemoglobin, hematocrit, mean corpuscular volume, mean corpuscular hemoglobin, mean corpuscular hemoglobin concentration, red cell distribution width coefficient of variation, platelet count, and platelet distribution width (Table 5). However, the effect of prior pesticide exposure on most indicators was absent (red blood cell, hemoglobin, platelet count, etc.) after 3 days.

Table 5. Estimated Short-term Health Effects of Pesticide Exposure on Blood Tests among Farmers in China, 2009–2011 (FE Estimation).

Dependent variables: ΔIndicator	Independent variables: Δ frequencies of pesticide application		
	In past 3 days	In past 4–10 days	Constant
Mon	-0.11 (-0.15, -0.06)**	-0.05 (-0.09, 0.02)**	0.37 (0.34, 0.40)**
Monp	-1.98 (-2.73, -1.23)**	-0.94 (-1.50, -0.38)**	6.14 (5.66, 6.62)**
RBC	-0.17 (-0.25, -0.09) **	-0.03 (-0.09, 0.03)	4.69 (4.64, 4.75)**
Hb	-2.01 (-3.87, -0.16)*	-1.26 (-2.64, 0.12)	144.30 (143.1, 145.5)**
Hct	-0.88 (-1.63, -0.14)*	0.08 (-0.47, 0.64)	42.70 (42.22, 43.17)**
MCV	1.66 (0.72, 2.60)**	0.83 (-0.13, 1.52)*	91.62 (91.02, 92.22)**
MCH	0.70 (0.33, 1.06)**	-0.08 (-0.35, 0.19)	30.96 (30.72, 31.19)**
MCHC	2.27 (-0.25, 4.80)	-3.76 (-5.63, -1.88) **	337.17 (335.55, 338.78)**
RDW_CV	-0.16 (-0.33, 0.01)*	0.13 (0.01, 0.25)*	12.56 (12.45, 12.66)**
PLT	-8.67 (-16.53, -0.81)*	-3.47 (-9.30, 2.36)	215.62 (210.59, 220.65)**
PDW	0.75 (0.36, 1.14)**	0.21 (-0.07, 0.50)	14.65 (14.40, 14.90)**
ALT	4.54 (2.64, 6.43)**	-0.45 (-1.86, 0.95)	21.28 (20.06, 22.49)**
AST	3.59 (1.76, 5.43)**	-0.82 (-2.18, 0.53)	24.01 (22.84, 25.18)**
CHE	-355.24(-534.1, -176.4) **	-237.37(-370.1, -104.7)**	8641.03(8526.6, 8755.5)**
TP	-3.56 (-4.79, -2.32)**	-0.48 (-1.40, 0.43)	74.58 (73.78, 75.37)**
BUN	0.22 (0.02, 0.43)*	0.02 (-0.13, 0.18)	5.25 (5.12, 5.38)**
Na	-0.78 (-1.29, -0.27)**	-0.42 (-0.79, -0.04)*	141.83 (141.51, 142.16)**
P	0.07 (0.02, 0.12)**	-0.04 (-0.07, -0.005)*	1.24 (1.21, 1.27)**
VB_{12}	-4.14 (-60.85, 52.56)	57.52 (15.46, 99.59) **	459.25 (422.98, 495.52) **
FPG	-0.25 (-0.40, -0.10)**	-0.17 (-0.28, -0.06)**	5.35 (5.25, 5.44)**

Abbreviations: ALT, alanine aminotransaminase; AST, aspartate aminotransferase; BUN, blood urea nitrogen; CHE, cholinesterase; CI, confidence interval; FPG, fasting plasma glucose; Hb, haemoglobin; Hct, hematocrit; Mon, monocytes; Monp, monocyte percentage; MCH, mean corpuscular haemoglobin; MCHC, mean corpuscular haemoglobin concentration; MCV, mean corpuscular volume; Na, Sodium; P, inorganic phosphorus; PDW, platelet distribution width; PLT, platelet count; RBC, red blood cell; RDW_CV, red cell distribution width coefficient of variation; TP, total protein; VB_{12}, Vitamin B_{12}.

* and ** represent the levels of significance at 5% and 1%, respectively.

doi:10.1371/journal.pone.0128766.t005

Similar results were observed in the analysis of short-term health effects of pesticide exposure on blood chemistry tests (Table 5) and nerve conduction studies (Table 6). Blood chemistry results from three days post-exposure showed a significant increase in liver enzyme, blood urea nitrogen, and inorganic phosphorus levels; and a decrease in plasma cholinesterase, total protein, sodium, and fasting plasma glucose levels. However, these effects were not observable at 4–10 days post-exposure. Three CV indicators (median motor, ulnar motor, and ulnar sensory) were significantly increased at three days post-exposure, but not at 4–10 days post-exposure. The effects of pesticide exposure on health indicators related to DMLs (ulnar nerve) and amplitudes of CMAP (median, ulnar, peroneal) and SNAP (median, ulnar) were negative, which differs from the positive effects on CVs; however, the impact patterns over time are similar.

Table 6. Estimated Short-term Health Effects of Pesticide Exposure on Nerve Conduction among Farmers in China, 2009–2011 (FE Estimation).

Dependent variables: ΔIndicator	Independent variables: Δ frequencies of pesticide application		
	In past 3 days	In past 4–10 days	Constant
Median MCV	0.74 (0.10, 1.38)*	-0.15 (-0.66, 0.37)	59.01 (58.55, 59.46)**
Ulnar MCV	0.62 (0.03, 1.21)*	0.28 (-0.20, 0.75)	57.18 (56.76, 57.60)**
Ulnar SCV	0.89 (0.14, 1.64)*	0.32 (-0.29, 0.92)	53.46 (52.92, 53.99)**
Median DML	-0.06 (-0.15, 0.03)	-0.07 (-0.15, -0.001)*	3.46 (3.40, 3.53)**
Ulnar DML	-0.16 (-0.25, -0.07)**	-0.04 (-0.12, 0.04)	2.83 (2.76, 2.89)**
Median proximal CMAP	-0.70 (-1.17, -0.23)**	-0.12 (-0.50, 0.26)	13.27 (12.94, 13.61)**
Median distal CMAP	-0.61 (-1.07, -0.14)*	0.02 (-0.35, 0.39)	13.76 (13.43, 14.09)**
Ulnar proximal CMAP	-0.74 (-1.06, -0.42)**	-0.20 (-0.46, 0.06)	12.25 (12.03, 12.48)**
Ulnar distal CMAP	-0.45 (-0.74, -0.16)**	-0.20 (-0.44, 0.04)	12.89 (12.68, 13.10)**
Peroneal proximal CMAP	-0.12 (-0.39, 0.14)	-0.23 (-0.44, -0.01)*	6.79 (6.60, 6.98)**
Median SNAP	-0.75 (-1.06, -0.45)**	0.15 (-0.09, 0.40)	8.20 (7.98, 8.41)**
Ulnar SNAP	-0.33 (-0.57, -0.09)**	-0.07 (-0.26, 0.12)	6.19 (6.02, 6.35)**

Abbreviations: CI, confidence interval; CMAP, compound muscle action potential; DML, distal motor latency; MCV, motor conduction velocity; SCV, sensory conduction velocity; SNAP, sensory nerve action potential.

* and ** represent the levels of significance at 5% and 1%, respectively.

doi:10.1371/journal.pone.0128766.t006

DISCUSSION

Nervous System

This study was the first to comprehensively examine the human health effects of frequent pesticide exposure in China. Organophosphate-induced delayed polyneuropathy (OPIDN) might develop after acute organophosphates exposure [12], but there was no evidence of such dysfunction after prolonged low-level exposure to organophosphates or other pesticide species [23]. To detect clinical evidence of polyneuropathy, we used the total neuropathy score, which was widely used in studies of toxic neuropathies [22] and other distal length-dependent neuropathies [24]. Its reduced version (the TNSc) included only items of symptoms and signs revealed by clinical neurological examinations. This score graded and combined the severity of motor, sensory, and autonomic symptoms and neurological signs, and prevented the diluting effect of other non-distal symptoms such as radiculopathy and entrapment syndromes. We found that sensory abnormalities were much more common than motor abnormalities (Table 4). However, the score was not sensitive enough to detect neurological function changes among different exposure levels and over different periods.

Nerve conduction studies were used as objective measurements in the present research. We tested nerves on the non-dominant side to minimize the potential confound of nerve injury due to work activities rather than pesticides. In our estimation of long-term effects, the proximal and distal CMAP of the tibial nerve, which reflects axonal function in the lower limbs, decreased in participants with a higher exposure level, but most of them did not meet the abnormal criteria. Abnormal electrophysiological profiles (CV, CMAP, and SNAP amplitude) were seen in the higher exposure group, and mostly in sensory nerves. These findings suggest that prolonged, low-level pesticide exposure can also cause peripheral neurotoxicity, especially in sensory nerves, which was consistent with the TNSc findings. This form of sensory neuropathy is different from OPIDN, in which prominent sensory loss is not a main feature [25]. Although OP compound exposure has been associated with sensory neuropathy based on neurological symptoms, neurological examinations or quantitative sensory testing [26], some studies did not find evidence of sensory neuropathy among workers with long-term occupational exposure to Chlorpyrifos, an OP [27]. Our results are positive and robust for two reasons. First, the present study examined a larger sample of participants that were exposed to higher pesticide levels; secondly, we measured neurological variables at consistent times following exposure. Since nerve conduction measurements merely reflect abnormal large fiber populations, we need to further evaluate small nerve fibers, which mediate pain, thermal sensation, and autonomic function.

The present study also contributes to better understanding of the short-term health effects of pesticide exposure in crop production. We found that most effects were evident within 3 days post-exposure, but significantly attenuated after 4 days. It is presumed that these short-term reversible alterations in both amplitude and CV might be due to transient actions of pesticides on ion channels in peripheral nerves, ultimately resulting in altered intracellular ionic levels. Such alterations may trigger secondary changes that induce cytotoxicity in axons and neuronal cell bodies, similar to chemotherapy-induced peripheral neurotoxicity [28]. However, the precise mechanisms require further study.

We used the MMSE, the most common cognitive screening assessment tool, to assess changes in the central nervous system. As shown in Fig 1, frequent pesticide exposure tended to increase the likelihood of an abnormal MMSE score; however, this association was not significant. The lack of significance may have been due to the smaller sample size of our study compared with other studies, particularly one that indicated an increased risk of Alzheimer disease among pesticide-exposed individuals [29]. Nevertheless, this preliminary finding necessitates further investigation into this problem.

Hematological System

In the hematological system, our study is consistent with some recent findings—alterations in CBC parameters are related to pesticides exposure [30]. Yet, it provides better understanding of pesticide health effects in long- and short-term alterations. Specifically, only white blood cell count remains significantly increased over time. A meta-analysis showed a positive relationship between pesticide exposure and hematopoietic cancers, including non-Hodgkin's lymphoma [31]. Although the increase in the number of white blood cells in peripheral blood was still within the normal range, it might suggest that pesticides affect the hematopoietic function of bone marrow. On the contrary, short-term pesticide exposure decreased monocytes, hemoglobin, and platelets, which suggests a direct toxic effect of pesticides on peripheral blood cells.

With regard to the metabolic panel, our results showed that pesticide exposure led to acute hepatic dysfunction and FPG reduction, which was opposite to previous reports of pesticide-associated hyperglycemia [15]. However, there were some reports of hypoglycemia following pesticide exposure [32]. This paradoxical phenomenon might suggest that different types of pesticides exert different effects. Electrolytes, including potassium, sodium, and inorganic phosphorus, were also affected either in long-term or in short-term. To date, only one study has shown electrolyte alterations following pesticide (acetamiprid) exposure [33], and the mechanism and significance of this has not yet been explored. Finally, renal function was also both acutely and chronically affected, and these abnormalities have been shown in previous studies [34].

Strengths and Limitations

We selected three provinces that representing high (Guangdong), middle (Jiangxi), and low (Hebei) levels of pesticide application provinces in China in order to have large variations on pesticide used based on farmers' average pesticide use by province from the surveys conducted by the National Development and Reform Commission, but after we conducted our surveys in the selected counties in the above three provinces, we found that farmers used less pesticide in Jiangxi than that in Hebei. However, this does not affect our analysis because our data show that there were large variations on pesticide uses among our samples among provinces. In contrast to earlier studies, the present study thoroughly evaluated various health indicators, used novel methodology for analyzing the duration and extent of pesticide exposure, contained a large sample size, and had a high response rate. Based on our survey, the major crops are: rice, vegetables, and orchards in Guangdong; rice, cotton, and rapeseed in Jiangxi; and wheat, maize, and cotton in Hebei. Moreover, the control was unique. Farmers with lower level pesticides exposure were used as control for long-term health effects estimation, while self-control was used when short-term effects were evaluated. Yet, several limitations are noticed. First, the base for pesticide application was established based on farmers' recall of their pesticide application in 2009–2011. Such measure is subject to recall errors and may not represent participants' lifetime exposure of pesticide. For example, participants who sprayed pesticides frequently before 2009 might have sprayed less than 50 times in 2009 to 2011, and vice versa. Secondly, to estimate the chronic health effects of pesticide exposure, such as cancer, Parkinson's disease, and Alzheimer's disease, it is necessary to follow subjects for an even longer time period. Finally, the pesticide-associated changes in the health indicators measured in the present study demonstrate the general effects of multiple pesticides, but other indicators should additionally be examined to yield insight into specific disorders. Similarly, the effects of each kind of pesticide, particularly high toxic pesticides, should be further analyzed.

CONCLUSIONS

In summary, we found that there were extensive long-term and short-term health effects in farmers exposed to pesticides. The former involved peripheral nervous system, white blood cells, liver, electrolytes, and the latter involved blood cells, liver, kidney, electrolytes and peripheral nervous system. Future studies are necessary to elucidate the mechanisms of these effects. It should be indicated that the high toxic pesticides, such as omethoate and other organophosphates are still used extensively in the crops production. Pyrethroid and some types of median toxic pesticides are more widely used. The high and median toxic pesticides are most harmful to the farmers. The government should revise the policy to reduce the pesticides production and application.

AUTHOR CONTRIBUTIONS

Conceived and designed the experiments: RH JH XH. Performed the experiments: CZ YL YY ZC JC FC RH XH.

Analyzed the data: RH CZ YL YJ. Contributed reagents/materials/analysis tools: YL CZ. Wrote the paper: RH JH YL.

REFERENCES

1. Oerke EC. Crop losses to pests. J Agric Sci, 144, pp 31–43. doi: 10. 1017/s0021859605005708
2. FAOSTAT. http: //faostat. fao. org/ 2012.
3. Peshin R, Bandral RS, Zhang W, Wilson L, Dhawan AK. Integrated Pest Management: A Global Overview of History, Programs and Adoption. In: Peshin R., Dhawan A. K., editors. Integrated Pest Management: Innovation-Development Process. Springer Netherlands; 2009. pp. 1–49.
4. Zhang W, Jiang F, Qu J. Global pesticide consumption and pollution: with China as a focus. Proceedings of the International Academy of Ecology and Environmental Sciences. 2011; 1(2): 125–144.
5. Ekstrom G, Akerblom M. Pesticide management in food and water safety: international contributions and national approaches. Rev Environ Contam Toxicol. 1990; 114: 23–55. pmid: 2403690 doi: 10. 1007/978-1-4612-3368-8_2
6. Naylor RL, Ehrlich PR. Natural pest control services and agriculture. In: Daily GC, editor. Natue's Services: Societal Dependence on Natural Ecosystems. Washintong, DC: Island Press; 1997. p. 151–176.
7. Pimentel D, Stachow U, Takacs DA, Brubaker HW, Dumas AR, Meaney JJ, et al. Conserving biological diversity in agricultural/forestry systems. Bioscience. 1992; 42: 354–362. doi: 10. 2307/1311782
8. Huang J, Hu R, Rozelle S, Qiao F, Pray CE. Transgenic varieties and productivity of smallholder cotton farmers in China. Aust J Agric Resour Econ. 2002; 46(3): 367–387. doi: 10. 1111/1467-8489. 00184
9. Pingali PL, Roger PA. Impact of Pesticides on Farmer Health and the Rice Environment. In: Pingali PL, Roger PA, editors. Boston, MA: Kluwer Academic Publishers; 1995.
10. Rosenstock L, Keifer M, Daniell WE, McConnell R, Claypoole K. Chronic central nervous system effects of acute organophosphate pesticide intoxication. The Pesticide Health Effects Study Group. Lancet. 1991; 338(8761): 223–227. pmid: 1676786 doi: 10. 1016/0140-6736(91)90356-t
11. Steenland K, Dick RB, Howell RJ, Chrislip DW, Hines CJ, Reid TM, et al. Neurologic function among termiticide applicators exposed to chlorpyrifos. Environ Health Perspect. 2000; 108(4): 293–300. pmid: 10753086 doi: 10. 1289/ehp. 00108293
12. Jayasinghe SS, Pathirana KD, Buckley NA. Effects of acute organophosphorus poisoning on function of peripheral nerves: a cohort study. PloS One. 2012; 7(11): e49405. doi: 10. 1371/journal. pone. 0049405. pmid: 23185328
13. Qiao F, Huang J, Zhang L, Rozelle S. Pesticide use and farmers' health in China's rice production. China Agricultural Economic Review. 2012; 4(4): 468–484. doi: 10. 1108/17561371211284821
14. Starks SE, Hoppin JA, Kamel F, Lynch CF, Jones MP, Alavanja MC, et al. Peripheral nervous system function and organophosphate pesticide use among licensed pesticide applicators in the Agricultural Health Study. Environ Health Perspect. 2012; 120(4): 515–520. doi: 10. 1289/ehp. 1103944. pmid: 22262687
15. Karami-Mohajeri S, Abdollahi M. Toxic influence of organophosphate, carbamate, and organochlorine pesticides on cellular metabolism of lipids, proteins, and carbohydrates: a systematic review. Hum Exp Toxicol. 2011; 30(9): 1119–1140. doi: 10. 1177/0960327110388959. pmid: 21071550
16. Stokes L, Stark A, Marshall E, Narang A. Neurotoxicity among pesticide applicators exposed to organophosphates. Occup Environ Med. 1995; 52(10): 648–653. pmid: 7489054 doi: 10. 1136/oem. 52. 10. 648
17. London L, Nell V, Thompson ML, Myers JE. Effects of long-term organophosphate exposures on neurological symptoms, vibration sense and tremor among South African farm workers. Scand J Work Environ Health. 1998; 24(1): 18–29. pmid: 9562397 doi: 10. 5271/sjweh. 274
18. Kamel F, Engel LS, Gladen BC, Hoppin JA, Alavanja MC, Sandler DP. Neurologic symptoms in licensed private pesticide applicators in the agricultural health study. Environ Health Perspect. 2005; 113(7): 877–882. pmid: 16002376 doi: 10. 1289/ehp. 7645
19. Alavanja MC, Hoppin JA, Kamel F. Health effects of chronic pesticide exposure: cancer and neurotoxicity. Annu Rev Public Health. 2004; 25: 155–197. pmid: 15015917 doi: 10. 1146/annurev. publhealth. 25. 101802. 123020
20. Hossain F, Pray CE, Lu Y, Huang J, Fan C, Hu R. GM cotton and farmer's health in China: an econometric analysis of the relationship between pesticide poisoning and GM cotton use in China. Int J Occup Environ Health. 2004; 10: 296–303. pmid: 15473084
21. Engel LS, Keifer MC, Checkoway H, Robinson LR, Vaughan TL. Neurophysiological function in farm workers exposed to organophosphate pesticides. Arch Environ Health. 1998; 53(1): 7–14. pmid: 9570304 doi: 10. 1080/00039899809605684
22. Cavaletti G, Frigeni B, Lanzani F, Piatti M, Rota S, Briani C, et al. The Total Neuropathy Score as an assessment tool for grading the course of chemotherapy-induced peripheral neurotoxicity: comparison with the National Cancer Institute-Common Toxicity Scale. J Peripher Nerv Syst. 2007; 12(3): 210–215. pmid: 17868248 doi: 10. 1111/j. 1529-8027. 2007. 00141. x
23. Lotti M. Low-level exposures to organophosphorus esters and peripheral nerve function. Muscle Nerve. 2002; 25(4): 492–504. pmid: 11932966 doi: 10. 1002/mus. 10086
24. Cornblath DR, Chaudhry V, Carter K, Lee D, Seysedadr M, Miernicki M, et al. Total neuropathy score: validation and reliability study. Neurology. 1999; 53(8): 1660–1664. pmid: 10563609 doi: 10. 1212/wnl. 53. 8. 1660
25. Moretto A, Lotti M. Poisoning by organophosphorus insecticides and sensory neuropathy. J Neurol Neurosurg

Psychiatry. 1998; 64(4): 463–468. pmid: 9576536 doi: 10. 1136/jnnp. 64. 4. 463

26. Jamal GA, Hansen S, Pilkington A, Buchanan D, Gillham RA, Abdel-Azis M, et al. A clinical neurological, neurophysiological, and neuropsychological study of sheep farmers and dippers exposed to organophosphate pesticides. Occup Environ Med. 2002; 59(7): 434–441. pmid: 12107290 doi: 10. 1136/oem. 59. 7. 434
27. Albers JW, Garabrant DH, Schweitzer S, Garrison RP, Richardson RJ, Berent S. Absence of sensory neuropathy among workers with occupational exposure to chlorpyrifos. Muscle Nerve. 2004; 29(5): 677–686. pmid: 15116371 doi: 10. 1002/mus. 20007
28. Jaggi AS, Singh N. Mechanisms in cancer-chemotherapeutic drugs-induced peripheral neuropathy. Toxicology. 2012; 291(1–3): 1–9. doi: 10. 1016/j. tox. 2011. 11. 005. pmid: 22120539
29. Hayden KM, Norton MC, Darcey D, Ostbye T, Zandi PP, Breitner JC, et al. Occupational exposure to pesticides increases the risk of incident AD: the Cache County study. Neurology. 2010; 74(19): 1524–1530. doi: 10. 1212/WNL. 0b013e3181dd4423. pmid: 20458069
30. Fareed M, Pathak MK, Bihari V, Kamal R, Srivastava AK, Kesavachandran CN. Adverse respiratory health and hematological alterations among agricultural workers occupationally exposed to organophosphate pesticides: a cross-sectional study in North India. PloS One. 2013; 8(7): e69755. doi: 10. 1371/journal. pone. 0069755. pmid: 23936093
31. Merhi M, Raynal H, Cahuzac E, Vinson F, Cravedi JP, Gamet-Payrastre L. Occupational exposure to pesticides and risk of hematopoietic cancers: meta-analysis of case-control studies. Cancer Causes Control 2007; 18(10): 1209–1226. pmid: 17874193 doi: 10. 1007/s10552-007-9061-1
32. Eissa FI, Zidan NA. Haematological, biochemical and histopathological alterations induced by abamectin and Bacillus thuringiensis in male albino rats. Actabiol Hung. 2010; 61(1): 33–44. doi: 10. 1556/ABiol. 61. 2010. 1. 4. pmid: 20194097
33. Mondal S, Ghosh RC, Mukhopadhyaya SK. Studies on the electrolytes and microelements in Wistar rat following multiple exposures to acetamiprid. Toxicol Ind Health. 2012; 28(5): 422–427. doi: 10. 1177/0748233711413800. pmid: 21937532
34. Payan-Renteria R, Garibay-Chavez G, Rangel-Ascencio R, Preciado-Martinez V, Munoz-Islas L, Beltran-Miranda C, et al. Effect of chronic pesticide exposure in farm workers of a Mexico community. Arch Environ Occup Health. 2012; 67(1): 22–30. doi: 10. 1080/19338244. 2011. 564230. pmid: 22315932

CHAPTER 6

OCCUPATIONAL EXPOSURE OF WORKERS TO PESTICIDES: TOXICOGENETICS AND SUSCEPTIBILITY GENE POLYMORPHISMS

Lúcia Maria de Miranda Adad, Heloísa Helena Rodrigues de Andrade, Kátia Kvitko, Mauricio Lehmann, Ana Amélia de Carvalho Melo Cavalcante, and Rafael Rodrigues Dihl

Keywords: Human monitoring, buccal micronucleus test, pesticides, *PON 1*, DNA repair genes

INTRODUCTION

The extent and complexity of health hazards inherent to workers who live with injuries caused by some types of work activities have highlighted the need to develop a model study of the future of occupational health (Higashi, 2006). In this scenario, agriculture is one of the main activities deserving special attention with regard to workers' health, since most farmers handle or manipulate highly hazardous compounds such as pesticides.

Pesticides are chemicals used in agriculture to control weeds and pests and to fight plant diseases. Most pesticides are a mixture of several chemical components, characterizing them as a complex mixture (Bolognesi, 2003).

Pesticides are associated with the increase of various types of cancer at specific sites, such as lip, skin, prostate and brain tumors, non-Hodgkin lymphoma, Hodgkin's disease (Blair and Zahm, 1991; Georgellis *et al.*, 1999), leukemia (Purdue *et al.*, 2007), multiple myeloma (Khuder and Mutgi, 1997), cancers of the immune, nervous, reproductive and hematological systems (Beck, 1991; Mourad, 2005) and lipid discords (Remor *et al.*, 2009; Sharma *et al.*, 2010). Many studies have pointed to the action of these chemicals as inducers of chromosomal aberrations (CA) (Au *et al.*, 1999; Zeljezic and Garaj-Vrhovac, 2001), sister chromatid exchange (SCE) (Shaham *et al.*, 2001) and the formation of micronuclei (MN) (da Silva *et al.*, 2008).

Citation: Adad LM de M, de Andrade HHR, Kvitko K, Lehmann M, Cavalcante AA de CM, Dihl RR. Occupational exposure of workers to pesticides: Toxicogenetics and susceptibility gene polymorphisms. Genetics and Molecular Biology. 2015; 38(3):308-315. DOI:10.1590/S1415-475738320140336.

Among the techniques used in human biomonitoring toxicogenetics, the micronucleus assay is universally considered a validated methodology for assessing the genetic instability induced by genotoxic agents (Zalacain *et al.*, 2005). Micronuclei are corpuscles present in the cytoplasm that resemble the core in its structure and color. They are the result of loss of chromosomal fragments or whole chromosomes due to clastogenic and/or aneugenic events, respectively (Maluf and Erdtmann, 2000).

Individual susceptibility plays a critical role in the response to pesticide exposure, determining the onset or absence of clinical symptoms, as well as acute poisoning. This sensitivity is directly associated with polymorphisms of key metabolism enzymes such as paraoxonase 1 (PON 1) (Hernández-Jerez, 2006). Organophosphates are activated by cytochrome P450 and hydrolyzed by PON 1, present in the liver (Costa *et al.*, 2003). Two polymorphisms in *PON1* are common, affecting the efficiency of the protein and its catalytic function for different substrates. A consequence of these polymorphisms is the differential response regarding the incidence of DNA damage in individuals exposed to organophosphates (Morahan *et al.*, 2007). Additionally, polymorphisms that reduce the DNA repair capacity can be associated with enhanced mutagenic effects and should be investigated in order to understand the differences in susceptibility to pesticide exposure (Winkinsonz and Clapper, 1997).

The aim of this study was to assess the effects of human exposure to complex mixtures of pesticides. To evaluate the mutagenic effects of occupational exposure we used the buccal micronucleus assay in exfoliated cells of oral mucosa, besides investigating the polymorphisms of the *PON 1* and DNA repair

OGG1 (8-Oxoguanine glycosylase), *XRCC1* (X Ray Repair Cross-Complementing 1) and *XRCC4* (X Ray Repair Cross-Complementing 4) genes, biochemical parameters such as butyrylcholinesterase (BchE) enzyme, and hematological and lipid parameter activities.

MATERIALS AND METHODS

Study Population

This study was conducted in a setting of agricultural enterprises in the municipalities of Teresina, Nazária and José de Freitas, in the state of Piaui, Brazil. It was approved by the Research Ethics Committee of the Universidade Luterana do Brasil - ULBRA), and individual written informed consent was obtained from all participants. All subjects included completed a detailed questionnaire (a Portuguese version of the International Commission for Protection against Environment Mutagens and Carcinogens) (Carrano and Natarajan, 1988).

The sample used in this study comprised 100 male workers (out of a total of 150 registered workers found in the selected municipalities) exposed to pesticides, who were divided into two groups: group l: 80 individuals registered with the State Occupational Health Reference Center (Association), working with maize, bean and watermelon crops; group 2: 20 workers of a private company that grows lemons and mangoes. The control group was matched for age, sex, ethnic group, smoking and drinking habits. Individuals with a history of exposure to clastogens and aneugens were excluded, according to the protocol of Minozzo *et al.* (2004). A 10 mL blood sample was collected from each individual by venipuncture, using vacutainers with heparine and EDTA. Each blood sample was divided into two 5 mL aliquots, one to be used for DNA extraction and the other for analyzing biochemical, hematological and lipid parameter activities. The buccal samples for the micronucleus assay were obtained by rubbing the inside of the cheeks with a cytobrush. The samples were kept at 4 °C until processing.

Buccal Micronucleus Assay

An MN test in exfoliated epithelial cells of oral mucosa was performed according to the method described by Salaija *et al.* (2006). The buccal cells were collected from the inner cheeks of the subjects with a cytobrush, washed with 5 mL of cold 1% phosphate-buffered saline solution (PBS) (pH 7.4), centrifuged at 1,500 rpm for 8 min, and fixed with a 3:1 methanol-acetic acid solution. The cell suspension was dropped onto a slide and air-dried at room temperature. The slides were then stained with 2% Giemsa solution for 10 min, rinsed in distilled water, and air-dried. The nuclear abnormalities were evaluated as described by Holland *et al.* (2008). Biomarkers of DNA damage (micronuclei), cytokinetic defects (binucleated cells) and cell death (karyorrhectic and karyolytic cells) were scored in a total of 3,000 buccal cells of each subject. The analysis was performed by light microscopy with a magnification of 1000×.

DNA Extraction and Genotyping

Genomic DNA was isolated from whole blood according to the method described byLahiri and Nurnberger Jr (1991), and stored at −20 °C. Four polymorphic markers were investigated by genotyping, using the PCR-RFLP (polymerase chain reaction - restriction fragment length polymorphism) method.

PON1Gln192Arg polymorphism

The *PON1Gln192Arg* polymorphism (location: 7q21.3) was genotyped by PCR/RFLP, as described by Humbert *et al.* (1993). An aliquot of the PCR product was digested with AlwI, and the genotypes were resolved on a 3% agarose gel stained with ethidium bromide.

OGG1Ser326Cys Polymorphism

The *OGG Ser326Cys* polymorphism (location: 3p26.2) was genotyped using the primers and PCR conditions described by De Ruyck *et al.* (2005). An aliquot of the PCR product was digested with Alw I, and the genotypes were resolved using on a 3% agarose gel stained with ethidium bromide.

XRCC1Arg194Trp Polymorphism

The XRCC1Arg194Trp polymorphism (location: 19q13.2) was genotyped by PCR/RFLP according to Lunn *et al.*(1999). The *XRCC1194Arg* and *XRCC1 194Trp* alleles were detected after digestion with enzyme PvuII, and the genotypes were resolved on a 3% agarose gel stained with ethidium bromide.

XRCC4Ile 401Thr Polymorphism

The *XRCC4*Ile 401Thr* polymorphism was genotyped using the primers and PCR conditions described by Relton *et al.* (2004). An aliquot of the PCR product was digested with BstNI, and the genotypes were resolved on a 3% agarose gel stained with ethidium bromide.

Determination of Plasmatic BchE Activity

The level of BchE activity was measured using the Dietz methodology with a Hitachi/Roche P800 automatic equipment to read absorbance in a spectrometer at 410 nm, in accordance with the protocol of Chaves (Chaves TVS, 2007, Master Thesis, Universidade Federal do Ceará. Master Thesis).

Hematological and Lipid Parameters

The hematological study was performed using an automatic analyzer (Sysmed KX21) to measure the following hematological parameters: leucocytes (granulocytes, lymphocytes and monocytes), erythrocytes, hematocrit, hemoglobin and platelets. The lipid parameters analyzed were LDL (low density lipoprotein) and HDL (high density lipoprotein) cholesterol, and triglycerides, using Labtest® methods.

Statistical Analysis

The normality of the variables was evaluated by the Kolmogorov-Smirnov test, and Student's *t* test was used to compare the characteristics of the study population. Since the values of MN, karyorrhexis and karyolysis were not normally distributed, even after transformation, the non-parametric Kruskal-Wallis test was used to compare them. The gene frequencies were estimated by gene counting, and the Hardy-

Weinberg equilibrium was evaluated using the chi-square test, adjusted for small samples. Differences between the genotypes in the exposed groups were tested by the non-parametric Mann-Whitney U-test. All analyses were made using the SPSS 17.0 / PC statistical software.

RESULTS AND DISCUSSION

Previous studies have been conducted to investigate the genotoxicity of pesticides in human biomonitoring assessment (Bolognesi, 2003; Rohr *et al.*, 2011). However, there is little information in the scientific literature regarding the assessment of mutagenic occupational exposure using the buccal micronucleus assay associated with *PON 1* polymorphisms and DNA repair genes.

The characteristics of the study population of workers exposed to pesticides and the controls are summarized in Table 1. There were no statistically significant differences between groups 1 and 2 regarding age, smoking and drinking habits. However, the time of exposure was significantly longer in group 1 (Association), whereas group 2 (Private) showed significant differences only in the use of full personal protective equipment (PPE).

Table 1. Characteristics of the study population.

Characteristics	Control (n = 62)	Association (n = 80)	Private (n = 20)	p
Age (years)	38.68 ± 12.2	43.1 ± 14.9	35.6 ± 10.2	0.246
Exposure time (years)	-	17.9 ± 12.6	10.5 ± 5.0	0.021
Application				
Hand pump, n (%)	-	80 (100)	0	-
Tractor, n (%)	-	0	20 (100)	-
Full PPE, n (%)[a]	-	26 (32.5)	20 (100)	0.016
Smoking, n (%)[b]	13 (20.9)	20 (25)	4 (20)	0.363
Drinking, n (%)[c]	25 (40.3)	32 (40)	7 (35)	0.328

Data are presented as mean ± standard deviation. Mann-Whitney test.

[a]At least two kinds of PPE during pesticide application.

[b]Individuals who smoke more than 5 cigarettes per day.

[c]Individuals who drink more than two times/week.

The two groups of workers included in this study were exposed to different types of pesticides. Those belonging to group 1 dealt basically with two organophosphates, considered extremely toxic, according to the World Health Organization (2005) (Table 2). The workers of group 2 were exposed to a greater number of pesticides, including two organophosphates, two triazines, a biological insecticide, and organic copper (Table 2). Of these compounds, only one - dimethoate - is classified as severely toxic (World Health Organization (WHO), 2009). Also concerning Table 2, only triazine is not listed as being carcinogenic in humans (International Agency for Research on Cancer (IARC), 2007).

Table 2. List of pesticides used by the employees of the association (Group1) and of the private company (Group 2).

Group	Pesticides	Compound	Chemical Class
Association	Insecticides	Methyl parathion	Organophosphorus
		Dimethoate	Organophosphorus
Private	Fungicide	Copper oxychloride	Inorganic copper
	Insecticide	Abamectin	Biological insecticide
		Dimethoate	Organophosphate
	Herbicide	Glyphosate	Organophosphate
		Atrasine	Triazine
		Ametrine	Triazine

Regarding the hematological markers (erythrocytes, leukocytes and platelets), both study groups presented normal values, according to Karazawa and Jamra (1989) and Remor *et al.* (2009), and no significant differences were observed between both exposed groups and the controls (Table 3).

Table 3. Hematological parameters in the study groups (mean ± SD).

	Control group (n = 100)	Farm workers (n = 100)
Leukocyte count (×10^3µL)	7.07 ± 1.66	6.86 ± 0.84
Granulocytes (%)	61.50 ± 8.01	52.62 ± 7.07
Lymphocytes (%)	32.55 ± 3.21	43.67 ± 6.65
Monocytes (%)	1.05 ± 0.83	1.27 ± 0.71
Erythrocyte count (×10^3/µL)	5.25 ± 1.02	4.80 ± 0.26
Hematocrit (%)	47.15 ± 8.18	44.18 ± 2.58
Hemoglobin (g/dL)	15.75 ± 2.73	13.99 ± 0.93
Platelets (×10^3/µL)	307.42 ± 76.13	285.80 ± 17.6

Table 4 shows the lipid profiles of farm workers and controls. No significant difference was found between the two groups, and both presented normal lipid profiles, according to reference values of the Brazilian Consensus on Dyslipidemia. Additionally, no significant difference was found either in BChE activity values between non-exposed (9.530 ± 1.713 UL^{-1}) and exposed subjects (8.203 ± 2.865 UL^{-1}), which were within the normal range. Similar results were reported by Shadnia *et al.* (2005) and Benedetti *et al.*(2013), who did not find any association between chronic exposure to organophosphates and cholinesterase inhibition.

Table 4. Lipid profile of study groups (mean ± SD).

	Controls (n:100)	Farm workers (n:100)	Reference values*
			< 200 Desirable
Total cholesterol (mg/dL)	185.08 ± 27.92	193.18 ± 40.37	200–239 Borderline
			> 240 Increased
			< 130 Desirable
LDL (mg/dL)	116.75 ± 18.33	113.31 ± 25.41	130–159 Borderline
			> 160 Increased
HDL (mg/dL)	40.41 ± 4.05	49.43 ± 6.15	< 35 Desirable
			< 200 Desirable
Triglycerides (mg/dL)	145.51 ± 26.51	119.68 ± 23.65	
			> 200 Increased

*According to the Brazilian Consensus on Dyslipidemia.

To assess the occupational risk of DNA damage, the buccal micronucleus test was performed in oral mucosa, using the following markers: MNi (chromosomal mutations), karyorrhexis (nuclear fragmentation), karyolysis (nuclear breakup) and presence of binucleated cells (defective cytokinesis). Group 1 (Association) and Group 2 (Private) were also compared with their respective negative controls (Figure 1 and 2), and the frequencies of MNi, karyorrhexis, karyolysis and binucleated cells were found to be significantly increased in groups 1 and 2. However, no significant differences were observed in the frequencies of cytogenetic parameters when group 1 was compared with group 2. In contrast to the frequencies of MNi and binucleated cells, there was an increase, although not significant, in the frequencies of karyorrhexis and karyolysis in group 1 compared to group 2 (Figure 2).

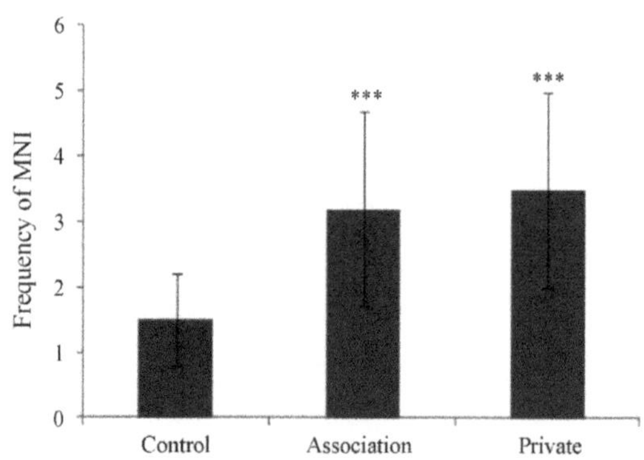

Figure 1. Comparison of the MNi frequencies between groups of workers exposed to pesticides and the control group (in 1,000 cells). ***$p < 0.001$ (Kruskal-Wallis test).

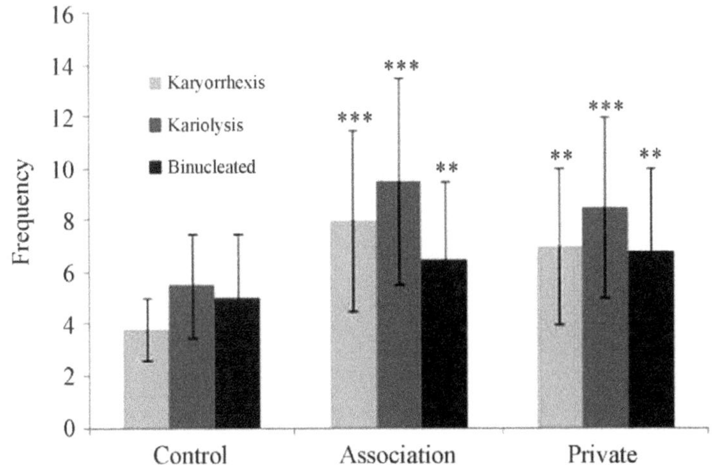

Figure 2. Comparison between nuclear abnormality frequencies in groups of workers exposed to pesticides and the control group (in 1,000 cells). $^{**}p < 0.01$; $^{***}p < 0.001$ (Kruskal-Wallis test).

Some researchers have argued that workers exposed to pesticides can present genetic damage, which can be minimized if proper personal protective equipment is used (Bull *et al.*, 2006). Another factor that complicates comparisons between exposed groups is related to the different pesticides employed (da Silva *et al.*, 2008). In fact, the two groups of workers included in this study were exposed to different pesticides, and there were also significant differences in the use of PPE. Group 1 handled two organophosphates, which are applied from top to bottom, leading to a lower risk of spreading, but only 32.5% used PPE. The subjects of Group 2, exposed to two triazines and two organophosphates, used PPE (100%) and applied pesticides with the pump turned on. All these observations may explain the absence of significant differences in the induction of MN between the two groups.

Both groups 1 and 2 exhibited higher frequencies of cytogenetic alterations compared to their (non-exposed) controls. It is well documented in the scientific literature that organophosphates can induce oxidative damage as well as reduce the levels of cellular antioxidants and antioxidant enzyme activity (Norppa, 2004). Therefore, considering that subjects of groups 1 and 2 may carry an enhanced body burden of reactive genotoxic agents, such as that induced by organophosphates, increased frequencies of cytogenetic alterations should be expected. Even in group 2, in which 100% of workers used full PPE, an increased frequency of events related to DNA damage and cell death was observed. In fact, studies have demonstrated that the use of pesticides such as dithiocarbamates, atrazine and malathion for extended periods of time can cause chromosomal breaks, acentric fragments, dicentrics, sister chromatid exchange and micronucleus frequency (Battershill, 2005).

The presence of binucleated cells is indicative of events of aneuploidy and could lead to failure in cytokinesis after the last nuclear division (Bonassi *et al.*, 2011). Karyorrhexis and karyolysis in turn are related to cell death (Thomas *et al.*, 2008), the first being characterized by the appearance of chromatin aggregation (on the condensation of chromatin), due to fragmentation and nuclear disintegration. Yet, cells in karyolysis show only the shadow of the nucleus (the nucleus "ghost") and are at a late stage of the cell death process.

In order to investigate whether individual genetic variations in xenobiotic metabolizing and DNA damage repair could influence the individual susceptibility to different DNA damage effects of pesticides, the exposed subjects were genotyped for *PON1*, *OGG1*,*XRCC1* and *XRCC4* genes. PON1 is the serum enzyme responsible for the metabolism of organophosphates, and genotoxic metabolites could be increased due to unfavorable genotypes of the exposed individuals (Au *et al.*, 1999). While OGG1 and XRCC1 proteins act in the base excision repair (BER) pathway for the correction of oxidized bases and DNA adducts (Goode *et al.*, 2002; Marsin *et al.*, 2003), XRCC4 acts at the non-homologous end, joining (NHEJ) the repair pathway for the correction of DNA double strand breaks (DSBs).

Genotype distribution and variant allele frequencies in the subjects studied are presented in Table 5. No deviations from the Hardy-Weinberg equilibrium were observed. Table 6 shows the individual effect of genetic polymorphisms in xenobiotic metabolizing and DNA repair enzyme genes on different results of micronucleus, karyorrhexis and karyolysis findings in exfoliated epithelial cells of oral mucosa. No significant difference was observed for the parameters of DNA damage or nuclear fragmentation in relation to the genes investigated. Previous studies have provided evidence that significant DNA damage induced by pesticide exposure was not associated with metabolizing or DNA repair enzyme genes (Sözmen *et al.*, 2007; Silva *et al.*, 2012). However, other studies have demonstrated that polymorphism in the *PON1* and BER pathways could modulate the susceptibility to DNA damage caused by pesticide exposure in the cytokinesis-blocked micronucleus (CBMN) assay in human lymphocytes (da Silva *et al.*, 2008, Rohr*et al.*, 2011). Further efforts should be made to characterize the combined effects of metabolizing and DNA repair variants on the response of individuals to pesticide exposure.

Table 5. Distribution of genotypes and variant allele frequencies in the exposed groups.

Gene	Variant alleles	N (%)	p
OGG1 Ser326Cys	*Ser/Ser*	77 (68.1)	
	Ser/Cys	36 (31.9)	0.370
	Cys/Cys	0 (0.0)	
XRCC1 Arg194Trp	*Arg/Arg*	97 (85.8)	
	Arg/Trp	16 (14.2)	0.367
	Trp/Trp	0 (0.0)	
XRCC4 Ile401Thr	*Ille/Ille*	90 (91.2)	
	Thr/Ille	10 (8.8)	0.369
	Thr/Thr	0 (0.0)	
PON1 Gln192Arg	*Gln/Gln*	37 (37)	
	GLn/Arg	46 (46)	0.156
	Arg/Arg	17 (17)	

p-value calculated using the Kruskal-Wallis test for comparing allelic frequencies.

Table 6. Effects of individual genotypes for metabolism (*PON1 Gln192Arg*) and DNA repair genes (*OGG1 Ser326Cys, XRCC1 Arg194Trp,XRCC4 Ile 401Thr*) polymorphism on the frequencies of the biomarkers evaluated (micronuclei, karyorrhexis and karyolysis) in the exposed groups (mean ± S.D).

Biomarkers	Genotypes	Mean ± SD	p
Micronuclei	*PON1 Gln/Gln*	1.77 ± 1.72 (n = 36)	0.240
	PON1 Arg/–	2.89 ± 2.62 (n = 64)	
	OGG1 Ser/Ser	2.75 ± 2.76 (n = 58)	0.221
	OGG1 Cys/–	2.06 ± 1.68 (n = 29)	
	XRCC1 Arg/Arg	2.70 ± 2.61 (n = 74)	0.117
	XRCC1 Arg/Trp	1.60 ± 1.50 (n = 15)	
	XRCC4 Ille/Ille	2.60 ± 2.32 (n = 66)	0.186
	XRCC4 Thr/–	1.63 ± 1.50 (n = 11)	
Karyorrhexis	*PON1 Gln/Gln*	19.11 ± 11.52 (n = 36)	0.563
	PON1 Arg/–	21.09 ± 18.56 (n = 64)	
	OGG1 Ser/Ser	21.74 ± 17.83 (n = 58)	0.249
	OGG1 Cys/–	20.90 ± 14.60 (n = 29)	
	XRCC1 Arg/Arg	21.19 ± 17.31 (n = 74)	0.349
	XRCC1 Arg/–	16.73 ± 13.32 (n = 15)	
	XRCC4 Ille/Ille	19.14 ± 13.02 (n = 66)	0.400
	XRCC4 Thr/–	23.27 ± 24.25 (n = 11)	
Karyolysis	*PON1 Gln/Gln*	23.42 ± 12.35 (n = 36)	0.986
	PON1 Arg/–	23.47 ± 15.59 (n = 64)	
	OGG1 Ser/Ser	24.41 ± 16.17 (n = 58)	0.888
	OGG1 Cys/–	23.93 ± 12.60 (n = 29)	
	XRCC1 Arg/Arg	23.62 ± 14.91 (n = 74)	0.782
	XRCC1 Arg/Trp	22.47 ± 13.79 (n = 15)	
	XRCC4 Ille/Ille	22.68 ± 11.67 (n = 66)	0.489
	XRCC4 Thr/–	25.73 ± 21.73 (n = 11)	

p value calculated using the Mann-Whitney U-test in relation to a variant (heterozygous and/or homozygous) genotype from the same group.

In this study, we determined the mutagenic and cell death potential of pesticides through the increased frequency of MNi and other nuclear abnormalities in exposed individuals. The individual metabolism and DNA repair genotypes of the exposed subjects did not show any effect on the investigated biomarkers.

These results also indicate that the buccal mucosa micronucleus test is useful for assessing genotoxic effects of pesticides. Hence, the association with genetic polymorphisms of enzymes involved in xenobiotic metabolism and DNA repair in populations exposed to genotoxic agents such as pesticides are part of the evaluation of individual gene modulation of the cytological response to chemical exposure. This integration is relevant to prevent the harmful effects of pesticides on workers.

ACKNOWLEDGMENTS

This work was supported by grants from the Instituto Federal de Educação, Ciência e Tecnologia do Piauí (IFPI) and Fundação de Amparo à Pesquisa do Estado do Piauí (FAPEPI).

REFERENCES

1. Au WW, Sierra-Torres CH, Cajas-Salazar N, Shipp BK, Legator MS. Cytogenetic effects from exposure to mixed pesticides and the influence from genetic susceptibility. Environ Health Perspect. 1999; 107: 501–515.
2. Battershill JM. The multiple chemicals and actions model of carcinogenesis. A possible new approach to developing prevention strategies for environmental carcinogenesis. Hum Exp Toxicol. 2005; 24: 547–558.
3. Beck WS. Hematology. 5th edition. Cambridge University Press; Cambridge: 1991. p. 667.
4. Benedetti D, Nunes E, Sarmento M, Porto C, Santos CE, Dias JF, da Silva J. Genetic damage in soybean workers exposed to pesticides: Evaluation with the comet and buccal micronucleus cytome assays. Mutat Res. 2013; 752: 28–33.
5. Blair A, Zahm SH. Cancer among farmers. J Occup Med. 1991; 6: 335–354.
6. Bolognesi C. Genotoxicity of pesticides: A review of human biomonitoring studies. Mutat Res. 2003; 543: 251–272.
7. Bonassi S, Coskun E, Ceppi M, Lando C, Bolognesi C, Burgaz S, Holland N, KIrsh-Volders M, Knasmueller S, Zeiger E, et al. The Human Micronucleus project on eXfoLiated buccal cells (HUMN(XL)): The role of life-style, host factors, occupational exposures, health status, and assay protocol. Mutat Res.2011; 728: 88–97.
8. Bull S, Fletcher K, Boobis AR, Battershill JM. Evidence for genotoxicity of pesticides in pesticide appliers: A review. Mutagenesis. 2006; 21: 93–103.
9. Carrano AV, Natarajan AT. Considerations for population monitoring using cytogenetic techniques. International Commission for Protection against Environmental Mutagens and Carcinogens (ICPEMC publication 14) Mutat Res.1988; 204: 379–406.
10. Costa LG, Cole TB, Furlong CE. Polymorphisms of paraoxonase (PON1) and their significance in clinical toxicology of organophosphates. J Toxicol Clin.2003; 41: 37–45.
11. da Silva J, Moraes CR, Heuser VD, Andrade VM, Silva FR, Kvitro K, Emmel V, Rohr P, Bordin DL, Andreazza AC, et al. Evaluation of genetic damage in a Brazilian population occupationally exposed to pesticides and its correlation with polymorphisms in metabolizing genes. Mutagenesis. 2008; 35: 415–422.
12. De Ruyck K, Van Eijkeren M, Claes K, Morthier R, De Paepe A, Vral A, De Ridder L, Thierens H. Radiation-induced damage to normal tissues after radiotherapy in patients treated for gynecologic tumors: Association with single nucleotide polymorphisms in XRCC1, XRCC3, and OGG1 genes and in vitro chromosomal radiosensitivity in lymphocytes. Int J Radiat Oncol Biol Phys.2005; 62: 1140–9.
13. Georgellis A, Kolmodin-Hedman B, Kourestas D. Can traditional epidemiology detect cancer risk caused by occupation exposure to pesticides? J Exp Clin Cancer.1999; 18: 159–166.
14. Goode REL, Ulrich CM, Potter JD. Polymorphisms in DNA repair genes and associations with cancer risk. Cancer Epidemiol Biomarkers Prev. 2002; 11: 1513–1530.
15. Hernández-Jerez A. Polimorfismos de paraoxonasa (PON 1). Un ejemplo de biomarcadores de susceptibilid a plaquicidas. Acta Toxicol Argent. 2006; 14: 29–30.
16. Higashi T. Study on a model for future occupational health. Ind Health.2006; 44: 541–555.
17. Holland N, Bolognesi C, Kirsch-Volders M, Bonassi S, Zeiger E, Knasmueller S, Fenech M. The micronucleus assay in human buccal cells as a tool for biomonitoring DNA damage: The HUMN project perspective on current status and knowledge gaps. Mutat Res. 2008; 659: 93–108.
18. Humbert R, Adler DA, Disteche CM, Hasset C, Omiecinski CJ, Furlong CE. The molecular basis of the human serum paraoxonase activity polymorphism. Nat Genet. 1993; 3: 73–76.
19. Karazawa EH, Jamra M. Normal hematological parameters. Rev Saúde Publ.1989; 23: 58–66. [in Portuguese with Abstract in English]
20. Khuder SA, Mutgi AB. Meta-analyses of multiple myeloma and farming. Am J Ind Med. 1997; 32: 510–551.
21. Lahiri DK, Nurnberger JI., Jr A rapid non-enzymatic method for the preparation of HMW DNA from blood for RFLP studies. Nucleic Acids Res. 1991; 19: 5444.
22. Lunn RM, Langlois RG, Hsieh LL, Thompson CL, Bell DA. XRCC1 polymorphisms: Effects on Aflotoxin B1-DNA adducts and Glycophorin A variant frequency. Cancer Res. 1999; 59: 2557–2561.
23. Maluf SW, Erdtmann B. Follow up study of the genetic damage in lymphocites of pharmacists and nurses handling antineoplastic drugs evaluated by cytokinesis block micronucleus; analysis and single cell gel electrophoresis assay. Mutat Res.2000; 471: 21–27.
24. Marsin S, Vidal AE, Sossou M, Menissier JM, Le Page

F, Boiteux S, de Murcia G, Radicella JP. Role of XRCC1 in the coordination and stimulation of oxidative DNA damage repair initiated by the DNA glycosylase hOGG1. J Biol Chem.2003; 278: 44068–44074.

25. Minozzo R, Deimling LI, Gigante LP, Santos-Melo R. Micronuclei in peripheral blood lymphocytes of workers exposed to lead. Mutat Res. 2004; 565: 53–60.

26. Morahan JM, Yu B, Trent RJ, Pamphlett R. A gene-environment study of paraoxonose 1 gene pesticides in amyotropic lateral sclerosis. Neurotoxicology.2007; 28: 532–540.

27. Mourad TA. Adverse impact of insecticides on the health of Palestinian farm workers in the Gaza Strip: A hematologic biomarker study. Int J Occup Environ Health. 2005; 11: 144–149.

28. Norppa H. Cytogenetic biomarkers and genetic polymorphisms. Toxicol Lett.2004; 149: 309–334.

29. Purdue MP, Hoppin JA, Blair A, Dosemeci M, Alavanja MC. Occupational exposure to organochlorine insecticides and cancer incidence in the agricultural health study. Int J Cancer. 2007; 120: 642–649.

30. Relton CL, Daniel CP, Hammal DM, Parker L, Janet ET, Burn J. DNA repair gene polymorphisms, pre-natal factors and the frequency of somatic mutations in the glycophorin-A gene among healthy newborns. Mutat Res. 2004; 545: 49–57.

31. Remor AP, Totti CC, Moreira DA, Dutra GP, Heuser VD, Boeira JM. Occupation exposure of workers to pesticides: Biochemical parameters and evaluation of genotoxicity. Environ Int. 2009; 35: 273–278.

32. Rohr P, da Silva J, Erdtmann B, Saffi J, Guecheva TN, Henriques JAP, Kvitko K. BER gene polymorphisms (*OGG1 Ser326Cys* and *XRCC1 Arg194Trp*) and modulation of DNA damage due to pesticides exposure. Environ Mol Mutagen.2011; 52: 20–27.

33. Salaija N, Chandrasekhar M, Rekhadevi PV, Mahboob M, Rahman MF, Vuyyuri SB, Danadevi K, Hussain SA, Grover P. Genotoxic evaluation of workers employed in pesticide production. Mutat Res. 2006; 609: 74–80.

34. Shadnia S, Azizi E, Hosseini R, Khoei S, Fouladdel S, Pajoumand A, Jalali N, Abdollahi M. Evaluation of oxidative stress and genotoxicity in organophosphorus insecticide formulators. Hum Exp Toxicol. 2005; 24: 439–445.

35. Shaham J, Kaufman Z, Gurvich R, Levi Z. Frequency of sister-chromatid exchange among greenhouse farmers exposed to pesticides. Mutat Res.2001; 491: 71–80.

36. Sharma P, Shankar S, Singh R. Variation in serum lipids & liver function markers in lindane exposed female wister rats: Alternating effect of curcumin, vitamin C & vitamin E. Asian J Exp Biol Sci. 2010; 1: 440–444.

37. Silva FR, da Silva J, Nunes E, Benedetti D, Kahl V, Rohr P, Abreu MB, Thiessen FV, Kvitko K. Application of the buccal micronucleus cytome assay and analysis of PON1Gln192Arg and CYP2A6*9(−48T > G) polymorphisms in tobacco farmers. Environ Mol Mutagen. 2012; 53: 525–534.

38. Sözmen B, Peker S, Kaya Ü, Erkan M, Sözmen EY. Markers of long-term exposure to organophoshorus pesticides in farmers who work in viniculture and tobacco production in Turkey. Toxicol Mech Methods. 2007; 17: 379–384.

39. Thomas P, Harvey S, Gruner T, Fenech M. The buccal cytome and micronucleus frequency is substantially altered in Down's syndrome and normal ageing compared to young healthy controls. Mutat Res. 2008; 638: 37–47.

40. Winkinsonz J, Clapper ML. Detoxication enzymes and chemoprevention. Proc Soc Exp Biol Med. 1997; 216: 192–200.

41. World Health Organization (WHO) The WHO Recommended Classification of Pesticides and Guidelines to Classification. WHO; Geneva: 2009. p. 78.

42. Zalacain M, Sierrasesumagal L, Patino A. The cytogenetic assay as a measure of genetic instability induced by genotic agents. An Sist San de Navarra.2005; 28: 227–236.

43. Zeljezic D, Garaj-Vrhovac V. Chromosomal aberration and single-cell electrophoresis (comet) assay in the longitudinal risk assessment of occupational exposure to pesticides. Mutagenesis. 2001; 16: 359–363.

CHAPTER 7

MIXTURE EFFECTS OF ESTROGENIC PESTICIDES AT THE HUMAN ESTROGEN RECEPTOR α AND β

Bettina Seeger, Frank Klawonn, Boris Nguema Bekale, and Pablo Steinberg

INTRODUCTION

Many substances used as crop protection products possess hormonal activity, which may influence human health by imitating or disrupting endogenous hormones [1]. Exposure is barely avoidable considering the widespread occurrence in conventionally grown fruits, vegetables and other crops. Up to 2013 there were about 800 substances with known hormonal activity [2]; their total number still remains unknown, since many substances have not been tested for that type of activity [2]. Nowadays, high-throughput bioassays for screening purposes, which are suited to evaluate the potential of pesticides with endocrine activity, are needed. Besides the identification of effects emanating from single substances there is scientific evidence that chemical mixtures of substances sharing the same mode of action elicit predominately additive effects *in vitro* as well as *in vivo* [3–6]. Pesticide residues of substances acting in a similar way on the same cellular targets are found in/on one food sample caused by simultaneous application of various pesticides, by cross-contamination due to common storage or by application of pesticide formulations containing mixtures of pesticides sharing the same mode of action [7]. The individual residues are usually present in low concentrations, mostly below their individual maximum residue levels, but have been shown to act additively, thereby eliciting remarkable effects, even when applied in combination with the individual compounds at concentrations below their individual No Observed Adverse Effect Levels (NOAELs) [4,5,8]. A recent cumulative risk assessment approach considers evaluating pesticides in mixtures, grouped by organ-specific toxicity, in addition to evaluating individual substances [9]. The tested pesticides (pirimicarb, propamocarb, fenhexamid, fludioxonil, chlorpyrifos, fenarimol) were selected based on their occurrence as residues listed in the 2013 European Union report on pesticide residues in food [10] and their estrogenic activity known from the literature [1,11–14]. We included pesticides frequently used, like fenhexamid and fludioxonil, as well as 2,4'-DDT and 4,4'-DDT, which were banned a number of years ago and are not detected in plant-derived foodstuffs anymore [10], but are well-characterized estrogenic substances. Therefore, they were used to test whether the test systems are suited to detect compounds capable of activating the hERα and hERβ, but were not included in the mixture studies, since their occurrence in plant-derived foodstuffs, even in low concentrations, is unlikely. Unfortunately, data on human exposure to hormonally active pesticides is rarely available [15,16]. In this context, an analysis by Kortenkamp et al. [16] showed that anti-androgenic environmental contaminants are present in human serum in picomolar to nanomolar concentrations. At such concentration levels one would not expect a significant effect by individual chemicals, but mixtures of substances being present at low concentrations and sharing the same mode of action could influence the human endocrine system [4,5,8].

We investigated the effects of single pesticides (pirimicarb, propamocarb, fenhexamid, fludioxonil, chlorpyrifos, fenarimol, 2,4'-DDT and 4,4'-DDT) as well as selected binary

Citation: Seeger B, Klawonn F, Nguema Bekale B, Steinberg P (2016) Mixture Effects of Estrogenic Pesticides at the Human Estrogen Receptor α and β. PLoS ONE 11(1): e0147490. DOI:10.1371/journal.pone.0147490.

and ternary mixtures of them at low effect concentrations in a β-galactosidase reporter gene assay, the broadly used Yeast-based Estrogen Screen (YES) assay, as well as in the human U2-OS cell-based ERα chemical-activated luciferase gene expression (ERα CALUX) assay. Full concentration-response curves were evaluated for the mathematical modeling, but the assessment of additivity was restricted to low effect concentrations (EC01 and EC10) in the range of human-relevant concentrations. Furthermore, the substances were screened in combination with a saturating concentration of 17β-estradiol (E2) to test for an E2 potentiating or an anti-estrogenic activity in the YES assay, and the anti-estrogenic substances were also tested for anti-estrogenic activity in the ERα CALUX assay.

Most studies have analyzed the mixture effects of pesticides at the hERα, while only a few reports have dealt with the effects of individual pesticides on the human estrogen receptor β (hERβ) and to our knowledge no study has investigated pesticide mixture effects at the hERβ. While the hERα frequently occurs in tissues related to reproductive activity (uterus, mammary gland), the hERβ is more widely distributed, and the ligand binding domains of the two isoforms slightly differ (59% homology) (reviewed by Gustafsson [17]), thereby indicating differing effects of the substances at the receptor isoforms. Since the hERβ is mostly regarded as a negative regulator of the hERα [18] and the risk emanating from an estrogenic substance depends on its activity at both receptor subtypes, we additionally investigated the effects of the individual test substances as well as combinations of them on the hERβ in the ERβ CALUX assay.

The well-known concept of concentration addition (CA), based on the work of Loewe and Muischnek [19], was used for the prediction of the outcome of the mixture experiments, supposing that additive effects of the pesticides occurred at the hERα and hERβ level.

The aim of this study was to evaluate the suitability of the established YES assay and the more recent ERα CALUX assay to identify estrogenic or anti-estrogenic effects at the hERα and to investigate mixture effects of estrogenic pesticides at low concentrations via CA. Additionally, the effects of the pesticides and mixtures at the hERβ, mostly acting as a counterpart of the hERα *in vivo* [18], were analyzed. The data generated for pesticide residue mixtures at the level of the hERα and hERβ support the assumption of additive effects of pesticides sharing the same mode of action, again emphasizing the importance of a cumulative risk assessment of pesticides.

MATERIAL AND METHODS

Tested Chemicals

17ß-estradiol (E2; CAS# 521-18-6; ≥ 98% purity), 17α-methyltestosterone (CAS# 58-18-4; 99.5% purity), 2,4'-dichlorodiphenyltrichloroethane (2,4'-DDT CAS# 789-02-6; 99.5% purity), 4,4'- dichlorodiphenyltrichloroethane (4,4'-DDT CAS# 50-29-3; 99.8% purity), 4-hydroxytamoxifen (4-HT; CAS# 68047-06-3; ≥ 98% purity), chlorpyrifos (CAS# 2921-88-2; 99,7% purity), corticosterone (CAS# 50-22-6; ≥ 98.5% purity), fenarimol (CAS# 60168-88-9; 99.9% purity), fenhexamid (CAS# 126833-17-8; 99,7% purity), fludioxonil (CAS# 131341-86-1; 99.9% purity), ICI 182,780 (ICI; CAS# 129453-61-8; > 98% purity), pirimicarb (CAS# 23103-98-2; 98.5% purity), propamocarb (CAS# 24579-73-5; 99.3% purity), resveratrol (CAS# 501-36-0, ≥ 99% purity), and tamoxifen (CAS# 10540-29-1; ≥99% purity) were purchased from Sigma Aldrich (Schnelldorf, Germany). By dissolving the chemicals in dimethyl sulfoxide (DMSO; CAS# 67-68-5; ≥ 99.5% purity; Carl Roth, Karlsruhe, Germany), dilution series were stored at -20°C in glass vials freeze-thawed for each experiment or batched in polypropylene vials and only thawed once.

Yeast (Anti)-estrogenicity Screen (YES) Assay

The YES assay developed by Routledge and Sumpter [14] was used as described by Kolle et al. [20] with slight modifications: The maximal concentration of DMSO in the medium was 3% v/v and the E2 concentrations ranged from 0.01 pM to 10 nM. Each pesticide and pesticide mixture was tested in quadruplicate in at least five experiments. To asses anti-estrogenic, estrogenic or potentiating effects in combination with E2, the substances were initially tested with and without a saturating concentration of E2 (1 nM) in increasing concentrations. 4-HT at a concentration of 1 μM combined with 1 nM E2 was used as anti-estrogenic control in a complete concentration-response curve in each plate.

A cytotoxic effect was defined as a decrease in optical density of the cell suspension at 690 nm by more than 30% when compared to the solvent control [20], and such concentrations were excluded from the tests.

Substances that could not be fully solubilized in the YES assay were tested in a cell-free control plate, and the turbidity, as a benchmark for insolubility, was measured at a wavelength of 690 nm. This was necessary to distinguish between a real growth induction of the yeast cells by the tested substance and an artifact resulting from the insolubility of the compound at higher concentrations, since in both cases an increase of turbidity in the cell-containing assay would be observed. It should be taken into account that an increased turbidity could also mask cytotoxicity.

ERα/ERβ CALUX

The ERα and ERβ CALUX assay were performed to evaluate the effect of the test substances on the hERα and hERβ as described by van der Burg et al. [21] with slight modifications. The pesticides were tested in three independent experiments, individually and as mixtures, in triplicates in 96-well plates. DMSO was used as solvent and was present in the test medium at a maximal concentration of 0.2% v/v. Solvent controls were included in each plate. Controls were selected based on the Organisation for Economic Co-operation and Development (OECD) Test Guideline 455 [22]. A full dose-response curve for E2 was included in one 96-well plate per experiment (E2 concentration range: 0.1 pM to 0.1 nM in the ERα CALUX assay, 0.3 pM to 30 nM in the ERβ CALUX assay). The same plate also contained 3 μM 17α-methyltestosterone as additional estrogenic control as well as 10 nM corticosterone as negative control. On each subsequent plate per experiment a middle concentration of E2 (3 pM in the ERα CALUX assay, 0.1 nM in the ERβ CALUX assay) and a high concentration of E2 (0.1 nM in the ERα CALUX assay, 30 nM in the ERβ CALUX assay) were tested, and the effects were compared to the respective controls on the first plate. A prescreen, in which all test substances were tested for estrogenicity between 0.01 µM and 60 µM, was performed. The three highest test substance concentrations and controls were checked for their ability to affect cell membrane integrity by quantification of the lactate dehydrogenase leakage (LDH) with the CytoTox-ONE™ Homogeneous Membrane Integrity Assay (Promega, Madison, WI, USA), which was performed as recommended by the manufacturer. Those test substance concentrations leading to a decline of the effect after a maximum was reached as well as those leading to insolubility were excluded from further experiments.

In the case of anti-estrogenicity assessment in the ERα CALUX assay, test substances that showed an anti-estrogenic activity in the YES assay, fenhexamid and fludioxonil, were tested in combination with 3 pM E2 (~ EC50). One plate per experiment included a full dose-response curve of the competitive hERα antagonist tamoxifen (3 nM to 10 µM), 1 nM 4-HT as additional anti-estrogenic control and 10 µM resveratrol as negative control, all also combined with 3 pM E2.

Data Normalization and Analysis

The raw optical density values obtained at 690 nm in the YES assay were subtracted as background from the values obtained at 540 nm and thereafter the corresponding solvent control values were also subtracted. All data points were normalized to the effect of 1 nM E2 (~ EC100), and mean values were generated in the YES assay. The data from the ERα and ERβ CALUX assays were evaluated in a similar manner: Solvent controls were subtracted from the raw luminescence values and all mean values were normalized to 0.1 nM E2 (~ EC100) in the ERα CALUX assay and 30 nM (~ EC100) in the ERβ CALUX assay, with one exception: When fludioxonil and fenhexamid were tested for anti-estrogenic effects in the ERα CALUX assay, the values were normalized to 3 pM E2.

Application of Nonlinear Regression in the Evaluation of Single Substances and Mixtures

Normalized concentration-response data from the YES, ERα and ERβ CALUX assays were analyzed by nonlinear regression after the x-axis with the test substance concentrations was log-transformed. The best-fit approach established by Scholze et al. [23] was used with few modifications: The regression model selection was performed using the Akaike Information Criterion with a correction for finite sample sizes, whereby logit, probit, Weibull, generalized logit I and II, as well as Aranda-Ordanz were the applicable models, while the 95% confidence intervals (CIs) were determined by application of the bootstrap method with 1,000 bootstrap simulations. All mathematical/statistical analyses were performed using R version 3.1.2 (R Foundation for Statistical Computing, Vienna, Austria).

Mixture Ratios and Analysis of Mixture Effects by use of Concentration Addition (CA)

Fixed mixtures ratios were calculated to be iso-effective [24], i.e. in proportion to their individual concentration that caused a 1 or 10% effect, as calculated from the concentration-response curve of the individual compounds (effect concentrations EC01 and EC10, as listed in Table 1 for the corresponding test systems). CA was applied as described by Rajapakse et al. [25]. A worked out example of the mixture ratio calculation, production of the mixtures and interpretation of the data can be found.

Table 1. EC01 and EC10 values of the single substances in the corresponding test systems.

	YES EC01		ERα CALUX EC01		ERβ CALUX EC01		YES EC10		ERα CALUX EC10		ERβ CALUX EC10	
substance	M	[CI]	M	[CI]	M	[CI]	M	[CI]	M	[CI]	M	[CI]
propamocarb	-	-	2.09E-07	[1.66E-07-3.72E-07]	1.15E-05	[3.55E-06-2.69E-05]	-	-	7.24E-07	[6.31E-07-9.77E-07]	2.63E-05	[1.41E-05-2.95E-05]
chlorpyrifos	7.94E-08	[3.09E-08-4.57E-06]	6.17E-07	[2.14E-07-1.07E-06]	-	-	-	-	3.47E-06	[2.63E-06-4.27E-06]	-	-
fenarimol	1.17E-05	[3.89E-06-2.95E-05]	7.94E-07	[6.61E-07-1.15E-06]	3.63E-06	[1.35E-06-7.94E-06]	2.19E-05	[1.17E-05-3.72E-05]	2.82E-06	[2.63E-06-3.31E-06]	1.35E-05	[9.12E-06-2.57E-05]
fludioxonil	8.32E-07	[4.42E-08-9.22E-07]	3.63E-07	[3.16E-07-5.50E-07]	8.91E-07	[1.00E-07-1.23E-06]	2.63E-05	[6.92E-06-3.80E-05]	1.26E-06	[1.12E-06-1.48E-06]	2.09E-06	[5.89E-07-2.40E-06]
fenhexamid	2.19E-07	[1.51E-08-6.92E-07]	6.31E-07	[2.95E-07-8.71E-07]	1.17E-06	[5.89E-07-2.88E-06]	2.14E-05	[2.04E-06-2.63E-05]	2.63E-06	[1.82E-06-3.02E-06]	5.01E-06	[3.98E-06-5.62E-06]
4,4'-DDT	3.47E-07	[1.51E-07-6.76E-07]	4.57E-07	[3.63E-07-5.13E-07]	1.02E-06	[5.50E-07-1.82E-06]	8.32E-06	[6.61E-06-1.12E-05]	1.07E-06	[9.55E-07-1.15E-06]	-	-
2,4'-DDT	3.31E-08	[1.82E-08-6.76E-08]	4.07E-08	[2.45E-08-6.46E-08]	5.50E-08	[8.32E-09-1.86E-07]	4.68E-07	[3.55E-07-6.92E-07]	1.02E-07	[7.41E-08-1.45E-07]	6.03E-07	[4.27E-07-1.23E-06]

Comparison of the EC01 and EC10 values, the effect concentration needed to elicit a 1 or 10% effect of 1 nM E2 with the approximate 95% confidence interval [CI] of the single substances in the corresponding test systems. Pirimicarb was excluded, since it did not show an effect in any test system.

doi:10.1371/journal.pone.0147490.t001

RESULTS

Effect of E2 in the YES, ERα CALUX and ERβ CALUX Assays

The hERα was trans-activated concentration-dependently by E2 in the YES assay (Fig 1A) and the ERα CALUX assay (Fig 1B). The maximally induced hERα activity was reached at 1 nM E2 in the YES assay and at 0.1 nM in the ERα CALUX assay, while the EC50 values were 0.16 nM and 4.9 pM, respectively. Trans-activation of the hERβ in the ERβ CALUX assay (Fig 1C) was at its maximum at 30 nM E2 with an EC50 of 0.21 nM.

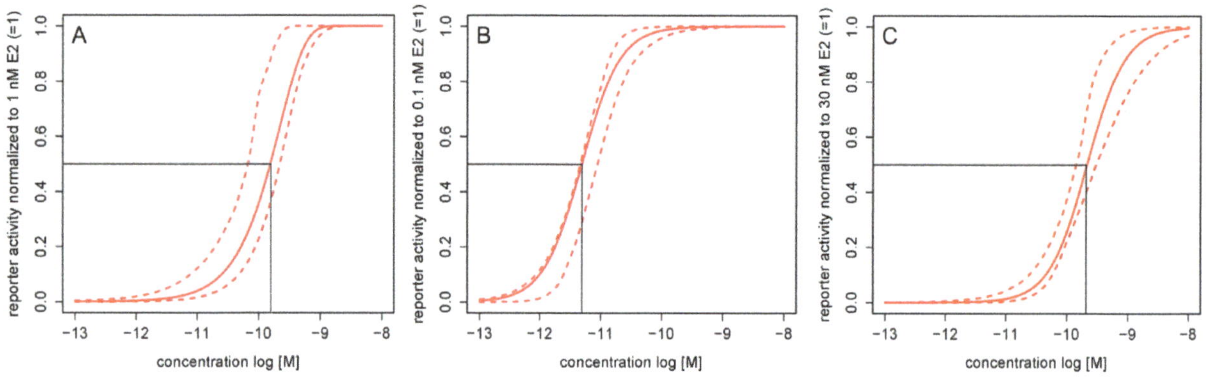

Figure 1. E2 in the YES, ERα CALUX and ERβ CALUX assays. Regression models with the indicated EC50 concentrations and the 95% confidence bands for E2 in the (A) YES, (B) ERα CALUX and (C) ERβ CALUX assays.

Dose-response Analysis of Single Substances in the Presence or Absence of E2

All tested pesticides showed estrogenic activity with full concentration-response curves in the YES and ERα CALUX assays with the exception of pirimicarb, which was negative in both assays, and propamocarb, which was only active in the ERα CALUX assay (Fig 2A and 2B, Table 1). Most substances were in the effect range of E2 in the YES assay, except for fenarimol with a maximum of 2.69 normalized to the effect of 1 nM E2 (Fig 2A). Fenarimol (maximum of 1.17) also exceeded the effect of 0.1 nM E2 in the ERα CALUX assays, as did fenhexamid (maximum of 1.37) and 2,4'-DDT (maximum of 1.23) (Fig 2B). Fenarimol, chlorpyrifos and 2,4'-DDT enhanced the effect of 1 nM E2 in the YES assay to a maximum of 4.36, 1.84 and 1.22, respectively (Fig 3A).

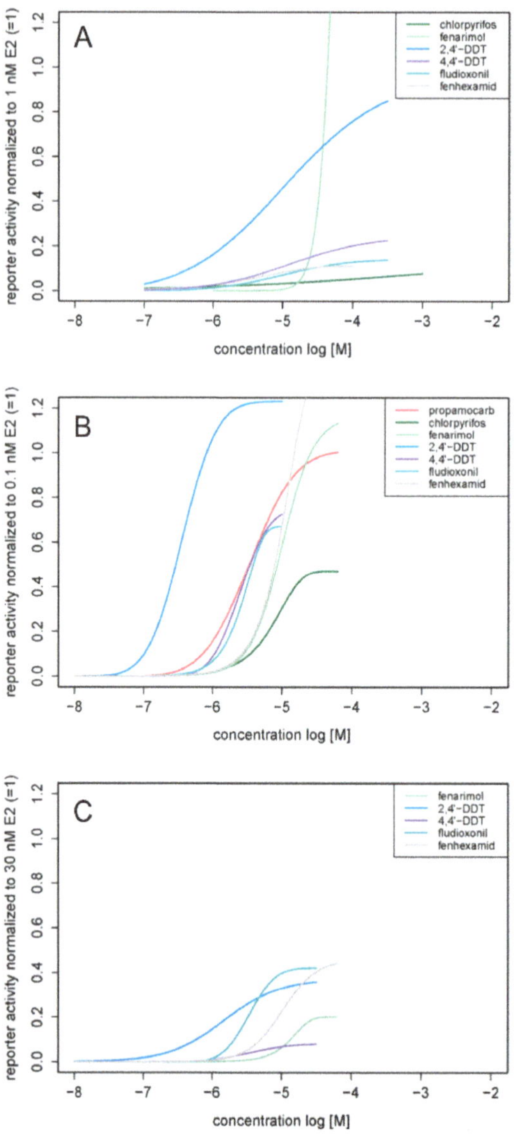

Figure 2. Regression models of the test substances. Regression models of the test substances in the (A) YES, (B) ERα CALUX and (C) ERβ CALUX assays. Substances not eliciting an effect are not shown.

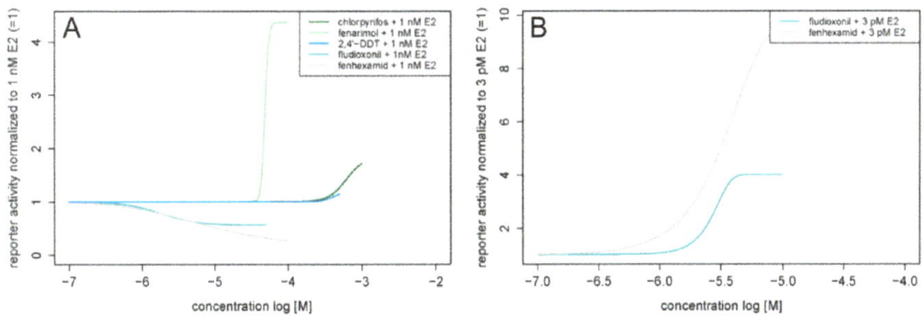

Figure 3. Regression models of the test substances applied in combination with E2. Regression models of the test substances in the (A) YES (with 1 nM E2) and the (B) ERα CALUX assay (with 3 pM E2). Substances not eliciting an effect are not shown.

Fenhexamid and fludioxonil showed concentration-dependent anti-estrogenic effects when combined with 1 nM E2 in the YES assay (Fig 3A). The two above-mentioned compounds were also tested in combination with 3 pM E2 in the ERα CALUX assay, but in both cases an anti-estrogenic effect was not observed (Fig 3B).

In the ERβ CALUX propamocarb, fenarimol, fludioxonil, fenhexamid, 4,4'-DDT and 2,4'-DDT showed a dose-dependent estrogenic activity at the hERβ (Fig 2C, Table 1).

The concentrations of the test substances to be tested were limited depending on their solubility. A decreased solubility led to an increase of turbidity/optical density measured at 690 nm in the yeast-based assay or was determined by microscopic examination in the CALUX assays. Concentrations leading to insolubility were excluded in the CALUX assays, but not in the YES assay. In the CALUX assays it was possible to generate full concentration-response curves at lower concentrations. It should be pointed out that fenarimol and chlorpyrifos only showed reporter gene activity in the YES assay when applied in concentrations leading to insolubility; nevertheless, these were tested in the YES assay (100 μM fenarimol, 100–1000 μM chlorpyrifos, 100–500 μM fludioxonil, 100 μM fenhexamid, 50–500 μM 2,4'- and 4,4'-DDT). Cytotoxicity, measured by performing the LDH leakage assay, was not observed in the case of the U2-OS cells (data not shown). A decrease of turbidity in the YES assay was assumed to be equivalent to an inhibition of yeast cell growth. Fenhexamid decreased turbidity in the YES assay by more than 30% in comparison to the solvent control at concentrations ≥ 500 μM and the concentrations were therefore excluded (data not shown).

Concentration-response Analysis of Mixtures

Iso-effective combinations of different estrogenic pesticides resulted in a concentration-dependent increase of reporter gene activity in the YES assay, the ERα CALUX assay as well as in the ERβ CALUX assay. In the case of the ERα CALUX assay, 95% CIs of the EC01 and EC10 were very narrow (Table 2); therefore, deviations from additivity were identified, mostly towards subadditivity (seven out of twelve predicted EC01/EC10 values were below the experimentally observed 95% CIs. Five out of ten comparisons between the experimental and the predicted data sets showed slight synergism in the YES assay, the predictions being higher than the experimentally obtained 95% CIs; the others were clearly additive (Table 3). The predicted values for the EC01 and EC10 mostly correlated with the 95% CIs of the experiments in the ERβ CALUX assay (Table 4).

Table 2. Regression model parameters of the mixtures in the ERα CALUX assay and comparison of the observed and predicted EC01 and EC10 values.

	Concentration-response function					EC01		EC10	
mixture	RM	$\hat{\theta}_1$	$\hat{\theta}_2$	$\hat{\theta}_{min}$	$\hat{\theta}_{max}$	predicted M	observed M [CI]	predicted M	observed M [CI]
fludioxonil fenhexamid EC01	Weibull	23.39	4.46	0	0.81	4.90E-07	5.89E-07 [5.01E-07-7.41E-07]	1.86E-06	2.00E-06[1.86E-06-2.19E-06]
fludioxonil fenhexamid EC10	Weibull	23.18	4.41	0	0.84	5.01E-07	5.50E-07 [4.90E-07-7.76E-07]	1.95E-06	1.86E-06 [1.86E-06-2.24E-06]
chlorpyrifos fludioxonil fenhexamid EC01	Weibull	20.96	3.99	0	0.68	5.37E-07	4.90E-07 [4.07E-07-6.31E-07]	2.29E-06	1.95E-06 [1.82E-06-2.14E-06]
chlorpyrifos fludioxonil fenhexamid EC10	Weibull	27.65	5.30	0	0.43	5.50E-07	1.20E-06 [8.51E-07-1.86E-06]	2.45E-06	3.39E-06 [3.09E-06-4.17E-06]
propamocarb fludioxonil fenhexamid EC01	Weibull	24.47	4.66	0	0.76	3.98E-07	6.76E-07 [5.25E-07-7.76E-07]	1.45E-06	2.14E-06 [1.95E-06-2.29E-06]
propamocarb fludioxonil fenhexamid EC10	Weibull	26.43	5.09	0	0.76	4.07E-07	9.12E-07 [7.94E-07-1.10E-06]	1.51E-06	2.63E-06 [2.51E-06-2.82E-06]

RM, the selected regression model; $\hat{\theta}_1$, $\hat{\theta}_2$ the estimated model parameters, $\hat{\theta}_{min}$, set 0; $\hat{\theta}_{max}$, the mean of the highest effect gained in the assay; **EC01/EC10** the effect concentration needed to elicit a 1 or 10% effect of 0.1 nM E2; **[CI]**, the approximate 95% confidence interval.

doi:10.1371/journal.pone.0147490.t002

Table 3. Regression model parameters of the mixtures in the YES assay and comparison of the observed and predicted EC01/10values.

mixture	Concentration-response function RM	$\hat{\theta}_1$	$\hat{\theta}_2$	$\hat{\theta}_3$	$\hat{\theta}_{min}$	$\hat{\theta}_{max}$	EC01 predicted M	EC01 observed M [CI]	EC10 predicted M	EC10 observed M [CI]
fludioxonil fenhexamid EC01	logit	18.07	3.42	-	0	0.15	2.69E-07	8.71E-07 [1.12E-08-7.24E-06]	2.69E-05	8.32E-06 [2.63E-06-7.08E-05]
fludioxonil fenhexamid EC10	probit	11.05	2.14	-	0	0.15	3.89E-07	1.35E-06 [2.57E-08-5.89E-06]	3.89E-05	1.07E-05 [3.80E-06-1.10E-04]
chlorpyrifos fludioxonil fenhexamid EC01	logit	16.80	3.19	-	0	0.15	5.62E-07	7.94E-07 [2.63E-09-5.75E-06]	1.10E-04	8.91E-06 [3.98E-06-3.72E-05]

RM, the selected regression model, **glogitl**, generalized logit I; $\hat{\theta}_1$, $\hat{\theta}_2$, $\hat{\theta}_3$, the estimated model parameters, $\hat{\theta}_{min}$, set 0; $\hat{\theta}_{max}$, the mean of the highest effect gained in the assay; **EC01/EC10**, the effect concentration needed to elicit a 1 or 10% effect of 1 nM E2; **[CI]**, the approximate 95% confidence interval.

doi:10.1371/journal.pone.0147490.t003

Table 4. Regression model parameters of the mixtures in the ERβ CALUX assay and comparison of the observed and predicted EC01 and EC10 values.

mixture	Concentration-response function RM	$\hat{\theta}_1$	$\hat{\theta}_2$	$\hat{\theta}_{min}$	$\hat{\theta}_{max}$	EC01 Predicted M	EC01 Observed M [CI]	EC10 Predicted M	EC10 Observed M [CI]
fludioxonil fenhexamid EC01	logit	19.33	3.80	0	0.59	1.02E-06	6.92E-07 [5.37E-07-1.74E-06]	3.09E-06	3.09E-06 [2.75E-06-5.62E-06]
fludioxonil fenhexamid EC10	logit	16.88	3.36	0	0.58	1.07E-06	5.89E-07 [3.31E-07-1.62E-06]	3.55E-06	3.09E-06 [2.45E-06-5.62E-06]
propamocarb fludioxonil fenhexamid EC01	Weibull	26.92	5.72	0	0.19	4.47E-06	6.03E-06 [3.80E-06-8.13E-06]	1.23E-05	1.74E-05 [1.17E-05-2.29E-05]
propamocarb fludioxonil fenhexamid EC10	Weibull	79.24	16.21	0	0.22	3.72E-06	8.32E-06 [4.37E-06-9.12E-06]	1.12E-05	1.20E-05 [1.15E-05-1.91E-05]

RM, the selected regression model; $\hat{\theta}_1$, $\hat{\theta}_2$ the estimated model parameters, $\hat{\theta}_{min}$, set 0; $\hat{\theta}_{max}$, the mean of the highest effect gained in the assay; **EC01/EC10** the effect concentration needed to elicit a 1 or 10% effect of 30 nM E2; **[CI]**, the approximate 95% confidence interval.

doi:10.1371/journal.pone.0147490.t004

DISCUSSION

In the present study mostly additive effects of pesticide mixtures at low effect concentrations in the YES, the ERα CALUX as well as the ERβ CALUX assays were observed. The predicted EC01/10 values not lying in the range of the 95% CIs of the experiments in the ERα CALUX (Table 2) can be explained by the fact that the 95% CIs were very narrow. The concentrations of the prediction are maximally 1.32-fold higher or 0.45-fold lower than the experimentally obtained data (Table 2), which does not represent a biologically relevant deviation from additivity. Those findings underpin the applicability of CA to predict mixture effects of pesticides sharing a common mode of action in the human cell-based ERα CALUX and ERβ CALUX assays as well as in the yeast-based YES assay. This additive behavior of estrogenic substances was previously shown in studies with hERα reporter gene bioassays [4] as well as in a MCF-7 proliferation assay [13] and *in vivo* in the immature mouse uterotrophic assay [26]. To our knowledge, this is the first study that has evaluated mixture effects of pesticides at the hERα and hERβ. Since the hERβ may acts as counterpart of the hERα, it could be of great relevance for the risk assessment process not to limit the analysis to the mixture effects of low concentrations of pesticides to the hERα level, since most hERα ligands also bind to the hERβ, which is also valid for the test substances in this study. The extent of the estrogenic effect in the different tissues largely depends on their hERα:hERβ ratio (reviewed by Böttner et al. [18]). In the present study, it has been shown that the pesticides did not elicit the same maximal response at the hERβ as compared to the effects at the hERα (e.g. Fig 2), but the potencies of the substances to elicit a 1 or 10% effect (Table 1, EC01 and EC10 in comparison in the corresponding test systems) are in the same order of magnitude when compared to the ECs in the ERα CALUX. These results, in turn, lead to the conclusion that most pesticides tested at low concentrations as well as their mixtures induce stronger effects at the hERα when directly compared to those at the hERβ, but the same substances elicit estrogenic effects in an additive manner at the hERβ. Additive effects of pesticides at the hERα activating the corresponding receptors should be considered in a cumulative risk assessment, and it is an important finding that the substances eliciting additive effects at the hERα in most cases also influence the hERβ additively, albeit with a lower efficacy.

There was a high level of concordance between the ERα CALUX and the YES assay regarding the identification of estrogenic effects of single substances. With the exception of pirimicarb and propamocarb, all tested pesticides trans-activated the hERα in both test systems. Pirimicarb was negative in both bioassays. There were certain differences in the identification of estrogenic and anti-estrogenic activity in the selected reporter gene assays: Propamocarb was identified as an estrogenic compound with an effect at the hERα in the ERα CALUX assay but not in the YES assay. Both substances were applied alone in the YES and in the ERα CALUX assay and in combination with a saturating concentration of 1 nM E2 in the YES assay. As known from the scientific literature, pirimicarb [11,27] and propamocarb [11] did not stimulate the proliferation of MCF-7 cells in the E-Screen with or without E2, but they induced a stronger effect in a β-galactosidase and luciferase-based transactivation assay performed with MCF-

7 BOS cells, when added in combination with a saturating concentration of E2 [11]. In this case, it was suggested that the substances could influence the interaction of E2 or cofactors with the estrogen receptor or the binding of the ligand-receptor-complex to its response element [11]. Bonefeld-Jorgensen et al. [28] also showed that propamocarb did not elicit an effect alone and in combination with an E2 concentration at the EC50 level in a stably transfected reporter gene assay based on MVLN cells and speculated that propamocarb might interact with the reporter vector and/or internal standard vector in the MCF-7 BOS cell-based assay used in the study of Andersen et al. [11], therefore leading to an artifact. Pirimicarb was negative in both reporter gene assays used in this study, even when combined with 1 nM E2 (saturating effect) in the YES assay. Interestingly, propamocarb, which did not induce any effects in the YES assay, was able to elicit concentration-dependent effects in the ERα and ERβ CALUX assays when applied alone. As far as we know, there is no *in vivo* data concerning the estrogenic activity of these two substances, and it is not possible to assess their estrogenicity based on the outcome of the above-mentioned studies, but it is most likely that the described potentiation the E2 effect in the transiently transfected MCF-7 BOS cell assay used in the study of Andersen et al. [11] depends on a particular interaction in this assay. An alternative explanation for the positive outcome in the CALUX assays could be that propamocarb is unable to cross the yeast cell wall or that it could be metabolized to compounds with a certain estrogenicity in the U2-OS cells used in the CALUX assays. Although U2-OS cells do not express any cytochrome P450 enzymes, they possess a certain enzymatic capacity, as shown in a proteomic analysis by Niforou et al. [29], which could impact the estrogenicity of propamocarb in the ERα CALUX assay, but not in the YES assay.

We detected an anti-estrogenic activity of fenhexamid and fludioxonil when applied in combination with E2 in the YES assay (Fig 3A), which could not be confirmed in the ERα CALUX assay (Fig 3B). Fenhexamid and fludioxonil led to an inhibition of the E2 effect in the YES assay by 79 and 44%, which is in line with a study by Mertl et al. [30] on food contact materials: Some of these substances showed an anti-estrogenic effect in the YES assay, but not in the ERα CALUX assay. Teng et al. [31] reported an increase of miR-21 expression in breast cancer cells and an inhibition of E2-induced cell proliferation in MCF-7 cells when applying fenhexamid and fludioxonil, which are associated with downstream anti-estrogenic effects.

Supramaximal estrogenic effects of substances eliciting higher putative effects than E2 itself were observed (Fig 2A and 2B) and were also shown to act additively as a mixture in the YES assay. Those supramaximal effects are receptor-dependent, mostly occur at high concentrations not relevant for the risk assessment of pesticide residues at low concentrations and are most likely to be a post-transcriptional artifact in hERα reporter gene bioassays [32–34] and do not deviate from the assumption of additivity, consequently not influencing the results of the present mixture study. Furthermore, fenarimol and chlorpyrifos were tested in a MCF-7 proliferation assay by Vinggaard et al. [1] without exceeding the effect of E2, thereby underlining that the biological relevance of the supramaximal effects in the YES assay is questionable.

CONCLUSIONS

To our knowledge, this is the first study to test mixtures of pesticides for additive effects not only at the hERα but also at the hERβ level. Most pesticides being active at the hERα were also active at the hERβ, and additive effects were observed when applied as mixtures. Although the pesticide mixtures were less efficient at the hERβ, one should analyze their effects at both receptor subtypes, since after an exposure one would expect that the compounds interact with both receptors in humans, the receptor subtype ratio influencing the estrogenic outcome. Additive effects of the tested pesticides at the hERα were mostly observed in the YES assay as well as in the ERα CALUX assay, an observation that supports the assumption of additivity of pesticides sharing a mode of action. The used *in vitro* assays are suitable for the identification of estrogenic effects at the hERα and hERβ level and, moreover, for the hazard identification step in the risk assessment process. The shown additive effects implicate that pesticide mixtures sharing a mode of action should be taken into account, besides a single substance assessment, in a cumulative risk assessment approach.

ACKNOWLEDGMENTS

This study was financially supported by the Federal Institute for Risk Assessment in Berlin, Germany (grant FK.3 1329–484 6433898).

AUTHOR CONTRIBUTIONS

Conceived and designed the experiments: BS PS. Performed the experiments: BS. Analyzed the data: BS PS. Contributed reagents/materials/analysis tools: FK BNB PS. Wrote the paper: BS PS.

REFERENCES

1. Vinggaard AM, Breinholt V, Larsen JC (1999) Screening of selected pesticides for oestrogen receptor activation in vitro. Food Addit Contam 16: 533–542. pmid:10789375 doi: 10. 1080/026520399283678
2. Bergman A, Heindel JJ, Kasten T, Kidd KA, Jobling S, Neira M, et al. (2013) The impact of endocrine disruption: a consensus statement on the state of the science. Environ Health Perspect 121: A104–106. doi: 10. 1289/ehp. 1205448. pmid:23548368
3. Payne J, Rajapakse N, Wilkins M, Kortenkamp A (2000) Prediction and assessment of the effects of mixtures of four xenoestrogens. Environ Health Perspect 108: 983–987. pmid:11049820 doi: 10. 1289/ehp. 00108983
4. Silva E, Rajapakse N, Kortenkamp A (2002) Something from "nothing"—eight weak estrogenic chemicals combined at concentrations below NOECs produce significant mixture effects. Environ Sci Technol 36: 1751–1756. pmid:11993873 doi: 10. 1021/es0101227
5. Rajapakse N, Silva E, Kortenkamp A (2002) Combining xenoestrogens at levels below individual no-observed-effect concentrations dramatically enhances steroid hormone action. Environ Health Perspect 110: 917–921. pmid:12204827 doi: 10. 1289/ehp. 02110917
6. Hass U, Scholze M, Christiansen S, Dalgaard M, Vinggaard AM, Axelstad M, et al. (2007) Combined exposure to anti-androgens exacerbates disruption

of sexual differentiation in the rat. Environ Health Perspect 115 Suppl 1: 122–128. doi: 10. 1289/ehp. 9360. pmid:18174960

7. Seeger B, Klawonn F, Nguema Bekale B, Steinberg P (2016) The ability of the YAS and AR CALUX assays to detect the additive effects of anti-androgenic fungicide mixtures. Toxicol Lett 241: 193–199. doi: 10. 1016/j. toxlet. 2015. 11. 015. pmid:26602169
8. Hass U, Boberg J, Christiansen S, Jacobsen PR, Vinggaard AM, Taxvig C, et al. (2012) Adverse effects on sexual development in rat offspring after low dose exposure to a mixture of endocrine disrupting pesticides. Reprod Toxicol 34: 261–274. doi: 10. 1016/j. reprotox. 2012. 05. 090. pmid:22659286
9. EFSA Panel on Plant Protection Products and their Residues (2013) Scientific Opinion on the identification of pesticides to be included in cumulative assessment groups on the basis of their toxicological profile. EFSA J 11.
10. European Food Safety Authority (EFSA) (2015) The 2013 European Union report on pesticide residues in food. EFSA J 2015 13(3):4038: 169 pp. doi: 10. 4135/9781412963855. n432
11. Andersen HR, Vinggaard AM, Rasmussen TH, Gjermandsen IM, Bonefeld-Jorgensen EC (2002) Effects of currently used pesticides in assays for estrogenicity, androgenicity, and aromatase activity in vitro. Toxicol Appl Pharm 179: 1–12. doi: 10. 1006/taap. 2001. 9347
12. Medjakovic S, Zoechling A, Gerster P, Ivanova MM, Teng Y, Klinge CM, et al. (2014) Effect of nonpersistent pesticides on estrogen receptor, androgen receptor, and aryl hydrocarbon receptor. Environ Toxicol 29: 1201–1216. doi: 10. 1002/tox. 21852. pmid:23436777
13. Payne J, Scholze M, Kortenkamp A (2001) Mixtures of four organochlorines enhance human breast cancer cell proliferation. Environ Health Perspect 109: 391–397. pmid:11335188 doi: 10. 1289/ehp. 01109391
14. Routledge EJ, Sumpter JP (1996) Estrogenic activity of surfactants and some of their degradation products assessed using a recombinant yeast screen. Environ Toxicol Chem 15: 241–248. doi: 10. 1002/etc. 5620150303
15. Kortenkamp A (2007) Ten years of mixing cocktails: a review of combination effects of endocrine-disrupting chemicals. Environ Health Perspect 115 Suppl 1: 98–105. doi: 10. 1289/ehp. 9357. pmid:18174957
16. Kortenkamp A, Scholze M, Ermler S (2014) Mind the gap: can we explain declining male reproductive health with known antiandrogens? Reproduction 147: 515–527. doi: 10. 1530/REP-13-0440. pmid:24435164
17. Gustafsson JA (1999) Estrogen receptor beta-a new dimension in estrogen mechanism of action. J Endocrinol 163: 379–383. pmid:10588810 doi: 10. 1677/joe. 0. 1630379
18. Böttner M, Thelen P, Jarry H (2014) Estrogen receptor beta: Tissue distribution and the still largely enigmatic physiological function. J Steroid Biochem 139: 245–251. doi: 10. 1016/j. jsbmb. 2013. 03. 003
19. Loewe S, Muischnek H (1926) Über Kombinationswirkungen. N-S Arch Pharmacol 114: 313–326. doi: 10. 1007/bf01952257
20. Kolle SN, Kamp HG, Huener HA, Knickel J, Verlohner A, Woitkowiak C, et al. (2010) In house validation of recombinant yeast estrogen and androgen receptor agonist and antagonist screening assays. Toxicol in Vitro 24: 2030–2040. doi: 10. 1016/j. tiv. 2010. 08. 008. pmid:20732407
21. van der Burg B, Winter R, Weimer M, Berckmans P, Suzuki G, Gijsbers L, et al. (2010) Optimization and prevalidation of the in vitro ERα CALUX method to test estrogenic and antiestrogenic activity of compounds. Reprod Toxicol 30: 73–80. doi: 10. 1016/j. reprotox. 2010. 04. 007. pmid:20435135
22. OECD (2015) Test No. 455: Draft Performance-Based Test Guideline for Stably Transfected Transactivation In Vitro Assays to Detect Estrogen Receptor Agonists and Antagonists: OECD Publishing.
23. Scholze M, Boedeker W, Faust M, Backhaus T, Altenburger R, Grimme LH (2001) A general best-fit method for concentration-response curves and the estimation of low-effect concentrations. Environ Toxicol and Chem 20: 448–457. doi: 10. 1002/etc. 5620200228
24. Kortenkamp A, Altenburger R (1998) Synergisms with mixtures of xenoestrogens: a reevaluation using the method of isoboles. Sci Total Environ 221: 59–73. pmid:9810734 doi: 10. 1016/s0048-9697(98)00261-7
25. Rajapakse N, Silva E, Scholze M, Kortenkamp A (2004) Deviation from additivity with estrogenic mixtures containing 4-nonylphenol and 4-tert-octylphenol detected in the E-SCREEN assay. Environ Sci Technol 38: 6343–6352. pmid:15597891 doi: 10. 1021/es049681e
26. Charles GD, Gennings C, Zacharewski TR, Gollapudi BB, Carney EW (2002) An approach for assessing estrogen receptor-mediated interactions in mixtures of three chemicals: a pilot study. Toxicol Sci 68: 349–360. pmid:12151631 doi: 10. 1093/toxsci/68. 2. 349
27. Bitsch N, Körner W, Failing K, Brunn H (2002) In vitro Screening von Pflanzenschutzmittel-Wirkstoffen auf östrogene Aktivität. Umweltwissenschaften und Schadstoff-Forschung 14: 76–84. doi: 10. 1065/uwsf2002. 04. 016
28. Bonefeld-Jorgensen EC, Grünfeld HT, Gjermandsen IM (2005) Effect of pesticides on estrogen receptor transactivation in vitro: A comparison of stable transfected MVLN and transient transfected MCF-7 cells. Mol Cell Endocrinol 244: 20–30. pmid:16219411 doi: 10. 1016/j. mce. 2005. 01. 017
29. Niforou KM, Anagnostopoulos AK, Vougas K, Kittas C, Gorgoulis VG, Tsangaris GT (2008) The proteome profile of the human osteosarcoma U2OS cell line. Cancer Genomics Proteomics 5: 63–78. pmid:18359981
30. Mertl J, Kirchnawy C, Osorio V, Grininger A, Richter A, Bergmair J, et al. (2014) Characterization of estrogen and androgen activity of food contact materials by different in vitro bioassays (YES, YAS, ERalpha and AR CALUX) and chromatographic analysis (GC-MS,

HPLC-MS). PLoS One 9: e100952. doi: 10. 1371/ journal. pone. 0100952. pmid:25000404

31. Teng Y, Manavalan TT, Hu C, Medjakovic S, Jungbauer A, Klinge CM (2013) Endocrine disruptors fludioxonil and fenhexamid stimulate miR-21 expression in breast cancer cells. Toxicol Sci 131: 71–83. doi: 10. 1093/ toxsci/kfs290. pmid:23052036
32. Auld DS, Thorne N, Nguyen DT, Inglese J (2008) A specific mechanism for nonspecific activation in reporter-gene assays. ACS Chem Biol 3: 463–470. doi: 10. 1021/cb8000793. pmid:18590332
33. Montaño M, Bakker EJ, Murk AJ (2010) Meta-analysis of supramaximal effects in in vitro estrogenicity assays. Toxicol Sci 115: 462–474. doi: 10. 1093/toxsci/kfq056. pmid:20176621
34. Sotoca AM, Bovee TFH, Brand W, Velikova N, Boeren S, Murk AJ, et al. (2010) Superinduction of estrogen receptor mediated gene expression in luciferase based reporter gene assays is mediated by a post-transcriptional mechanism. J Steroid Biochem122: 204–211. doi: 10. 1016/j. jsbmb. 2010. 06. 010

CHAPTER 8 BIOCONCENTRATION AND ACUTE INTOXICATION OF FRESHWATER FISHES BY THE METHYL PARATHION ORGANOPHOSPHATE PESTICIDE

João Bosco de Salles, Renato Matos Lopes, Cristiane M. C. de Salles, Vicente P. F. Cassano, Manildo Marcião de Oliveira, Vera L. F. Cunha Bastos, and Jayme Cunha Bastos

INTRODUCTION

Pesticides are a group of toxic compounds with a deep effect on aquatic life and water quality. Organophosphates (OP) are a group of pesticides widely used in Latin America. For instance, in Brazil, methyl parathion (O,O-dimethyl O-p-nitrophenylphosphorothioate) has extensively been applied in agriculture, food storage shelters, pest control programs, and aquaculture ponds to control aquatic insect larvae, which are predators of fingerling [1, 2]. The development of the Brazilian freshwater fish industry was accompanied by the extensive use of methyl parathion to control ectoparasite infestations in fish. This practice resulted in the discharge of large amounts of methyl parathion-treated waters into the nearby area. Since OP has no selectivity for any specific target organism, discharged waters might cause the intoxication of natural populations of aquatic organisms [3, 4].

An important mechanism of acute toxicity by methyl parathion is the inhibition of acetylcholinesterase (AChE) activity. Methyl parathion is a weak acetylcholinesterase inhibitor, but it can be activated into a more potent metabolite, its oxon derivative, methyl paraoxon (O,O-dimethyl O-p-nitrophenyl phosphate), by a desulfuration reaction catalyzed by cytochrome P-450 [5]. Methyl parathion is metabolized by both plants and animals and it is not expected to persist or bioconcentrate. However, studies have reported that Girardinichthys multiradiatus captured in the Ignacio Ramirez Dam (Mexico) accumulated methyl parathion more than 13,000 times in relation to water levels of this compound [6, 7]. Methyl parathion can be detoxified by dealkylation by glutathione S-transferases (GST) [8], whereas methyl paraoxon can be removed from blood by scavenger enzymes, such as carboxylesterase and cholinesterases, or degraded by paraoxonase, yielding 4-nitrophenol and dimethylphosphoric acid [9].

Intoxication mechanisms of aquatic organisms by OP are not completely understood. Dembéle and coworkers have attributed fish death to asphyxia resulting from gill irritation, not dependent on brain AChE inhibition [10]. Patil and David [11] reported that sublethal OP levels could induce oxidative stress and inferred that oxidative damages might be related to the death of exposed animals. Nevertheless, OP toxicity can be attributed to AChE inhibition. In neuromuscular junctions and the central nervous system, AChE inhibition induces excessive cholinergic stimulation, producing in coordination, fatigue, involuntary muscle contractions, and eventually, paralysis of the body extremities and the respiratory muscles [12]. Accordingly, kinetic characterization of AChE and butyrylcholinesterase (BChE) (another cholinesterase form) allows the determination of sensitivity differences between both enzymes to organophosphate pesticides, which is essential for environmental monitoring programs [4, 13]. Therefore, the aim of this work was to study the effects of organophosphate methyl parathion on ChE activities from different tissues and organs from three freshwater fish species that dwell in Brazilian waters.

Citation: João Bosco de Salles, Renato Matos Lopes, Cristiane M. C. de Salles, et al., "Bioconcentration and Acute Intoxication of Freshwater Fishes by the Methyl Parathion Organophosphate Pesticide," BioMed Research International, vol. 2015, Article ID 197196, 9 pages, 2015. DOI:10.1155/2015/197196.

MATERIAL AND METHODS

Chemicals

Acetylthiocholine iodide (ASCho), butyrylthiocholine iodide (BSCho), propionylthiocholine iodide (PrSCho),p-nitrophenol (PNP), 5,5′-dithiobis (2-nitrobenzoic acid) (DTNB), and methyl paraoxon (O,O-dimethyl O-p-nitrophenyl phosphate) were obtained from Sigma (St. Louis, MO, USA). Triton X-100 was acquired from Riedel of Haën AG (Hannover). Isooctane HPLC grade was acquired from Merck. Folidol 600 (methyl parathion 600 g L^{-1}, Bayer-Brazil) was purchased at the local market. All other reagents were of analytical grade.

Animals and Experimental Procedures

Common and scientific names, geographical origin, and size of fishes examined in this work are shown in Table1. Specimens of Piaractus mesopotamicus (Holmberg, 1887), commonly named pacu, and Leporinus macrocephalus Garavello & Britski, 1988 (piavussu), were supplied by the Morro Grande Fish Farm, RJ, Brazil. Specimens of curimbatá, Prochilodus lineatus (Valenciennes, 1836), were supplied by the Sol Nascente Fish Farm, RJ, Brazil. Experimental procedures were carried out according to the ethical principles of animal experimentation elaborated by the Brazilian College for Animal Experimentation (COBEA), which is in agreement with the uniform requirements for manuscript submissions to biomedical journals.

Table 1: Common and scientific name, suppliers, and average size of fishes examined

Common name	Scientific name	Supplier[a]	Size and SD[b] (cm)
Curimbatá	Prochilodus lineatus (Valenciennes, 1836)	Sol Nascente Fish Farm	18 ± 5
Pacu	Piaractus mesopotamicus (Holmberg, 1887)	Morro Grande Fish Farm	18 ± 4
Piavussu	Leporinus macrocephalus Garavello and Britski, 1988	Morro Grande Fish Farm	20 ± 5

[a]All suppliers are located in the state of Rio de Janeiro, Brazil.

[b]SD: standard deviation.

Prior to the assays, the fish were separated by species and kept into 500 L tanks with dechlorinated water at 25 ± 2°C for 15 days. Water was constantly aerated through a biological filtering system pushed by common pumps to produce 5 mg O_2 L^{-1} (measured with an oximeter). Fish were always on a 14:10-h light/dark cycle and fed with a commercial pellet once a day at 9 o'clock AM. Water pH was 6.4 ± 0.2.

In order to be exposed fish were removed from the larger tanks and individually kept in 40 L aquaria filled with water obtained from the 500 L tanks. Methyl parathion (as Folidol) was added once to a final 5 ppm concentration. All fish were not fed 24 h prior to and during the experiments in order to limit organic matter in the water. Moreover, the water during fish exposure was not filtered to avoid removal of the pesticide. Control fish were submitted to the same conditions, without Folidol.

The concentration of 5 ppm methyl parathion was chosen to elicit a quick response and to correlate the inhibition of cholinesterase with the bioconcentration of pesticide in several tissues and organs of the fish.

The intervals of 30 min, 24 hours, and 78 hours were chosen to carry out laboratory analyses. Curimbatás showed more sensitivity to methyl parathion. This was noticed when they stopped moving their opercula. Once this happened, they were immediately removed and laboratory analyses were immediately carried out.

Tissue Samples

Blood was collected by puncture of the dorsal aorta. Then, the fish were euthanized by quickly sectioning their spinal cord. Liver, brain, and heart were removed using scissors and tongs. Also, a portion of epaxial muscle (1 g wet tissue) was excised. All tissues were separately washed quickly with 50 mL of ice-cold saline (0.9% NaCl) and placed into cryogenic vials that were dropped into liquid nitrogen for storage. They were thawed separately by suspension in four volumes of ice-cold 0.1 mol L^{-1} potassium phosphate buffer, pH 7.0. They were minced with scissors and homogenized with 20 strokes in a Potter-Elvehjem apparatus while being maintained in an ice bath at 5-6°C. ChE assay in brain and liver tissue were carried out using this crude homogenate as sample. Samples for assaying heart and muscle ChE were produced by mixing their homogenates with three volumes of 10 mmol L^{-1} Tris-HCl buffer, pH 7.0, containing 3% Triton X-100 and 1 mol L^{-1} sodium chloride. After centrifuging these mixtures at 3,000 ×g for 10 min at 5°C enzyme assays were carried out in the resulting heart and muscle Triton X-100 soluble supernatant fractions.

Methyl Parathion Determinations

Serum and homogenate samples (200 μL) were extracted by using a mixture of 6 mL of isooctane, 200 μL methanol, and 200 μL saturated sodium chloride solution. The extracts were centrifuged at 3,000 ×g for 10 min at 5°C. Three milliliters from each supernatant fraction was collected and evaporated under a gentle nitrogen stream. The residue was reconstituted in 200 μL of acetonitrile. From this reconstituted sample, 50 μL were injected onto a 200 mm × 4.6 mm ODS Hypersil RP-18, 5 μm particle size HPLC column, using (50:50 v/v) acetonitrile/ultrapure water as mobile phase with a flow-rate of 1 mL min^{-1}, and examined under UV light with the detector set at 270 nm. Under these conditions, methyl parathion recovery was estimated at approximately 92%, following quantification based on a standard prepared with a 98% methyl parathion pure sample previously obtained by thin-layer chromatography.

Cholinesterase Assays

Serum and tissue homogenate samples were placed in a medium containing 1.8 mmol L^{-1} from one of the three substrates (acetylthiocholine, propionylthiocholine, or butyrylthiocholine), with 0.32 mmol L^{-1} DTNB. Enzyme activity was continuously recorded up to 90 s, at 412 nm, using a Shimadzu spectrophotometer, model UV-160A, according to the Ellman method [14]. The reaction was carried out in microcuvettes. All reagents were dissolved to 200 µL final volume with a 0.1 mmol L^{-1} sodium phosphate buffer, pH 7.5, at 25°C, and the thionitrobenzoate ion concentration was estimated using an extinction coefficient of 14,150 $M^{-1}cm^{-1}$. One unit (1 U) of enzyme activity was defined as the amount that hydrolyzes 1.0 µmol of substrate per minute.

RESULTS

Tissue Distribution of ChE Activity

The studied fish species (control groups) exhibited remarkable differences in tissue-specific cholinesterase activity (Figure 1). The ChE activity measured in the serum of curimbatá and pacu specimens was lower than the activity measured in brain, liver, heart, and muscle. By contrast, in serum of piavussu the ChE activity was similar to that measured in liver, brain, heart, and muscle.

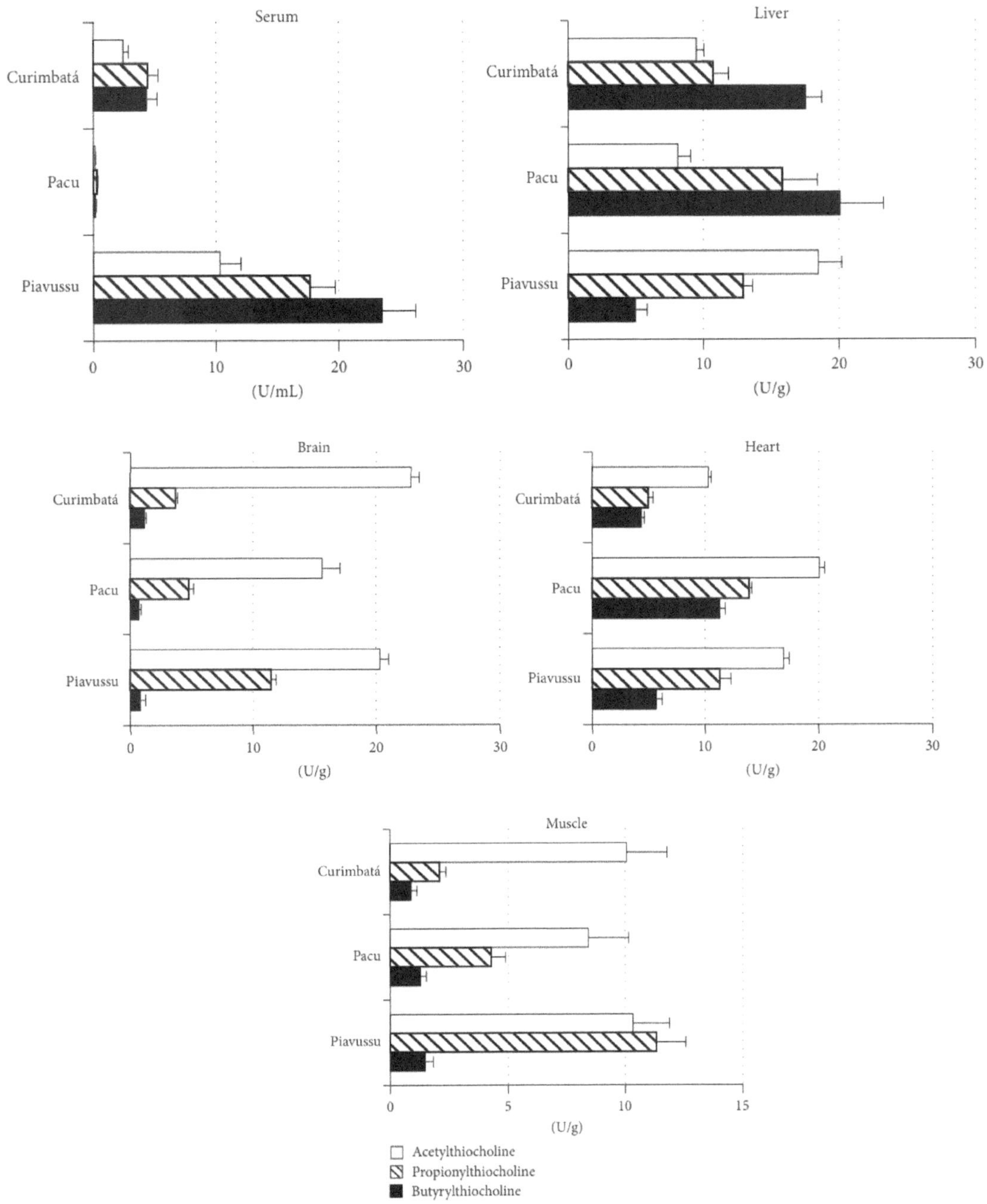

Figure 1: Tissue distribution of cholinesterase activity in pacu, piavussu and curimbatá (control fish). The box shows the substrates used in enzyme assays. Results are expressed asµmol of products formed per minute per gram of wet tissue (U g^{-1}) or per mL of serum (U mL^{-1}). Liver and brain ChE activities were assayed in homogenates, while heart and muscle ChE activities were assayed in the enzymatic fraction solubilized with Triton and NaCl. Results are average values with the corresponding SEM of assays carried out in five individuals of each species.

Methyl Parathion Bioconcentration

Muscle, heart, brain, liver, and serum from pacu, piavussu, and curimbatá exhibited similar methyl parathion bioconcentration patterns (Figure 2). Brain tissue showed the highest capacity for methyl parathion bioconcentration, reaching 80 ppm (16-fold increase) after 30 minutes exposure to methyl parathion in water (Figure 2). The maximum methyl parathion concentration in muscle, heart, brain, and serum was achieved after 30 minutes, with one exception: methyl parathion concentrations in piavussu showed an increase after 24 hours exposure, reaching a maximum of 120 ppm in liver, corresponding to a 24-fold increase (Figure 3). Methyl parathion was not found in samples from control animals.

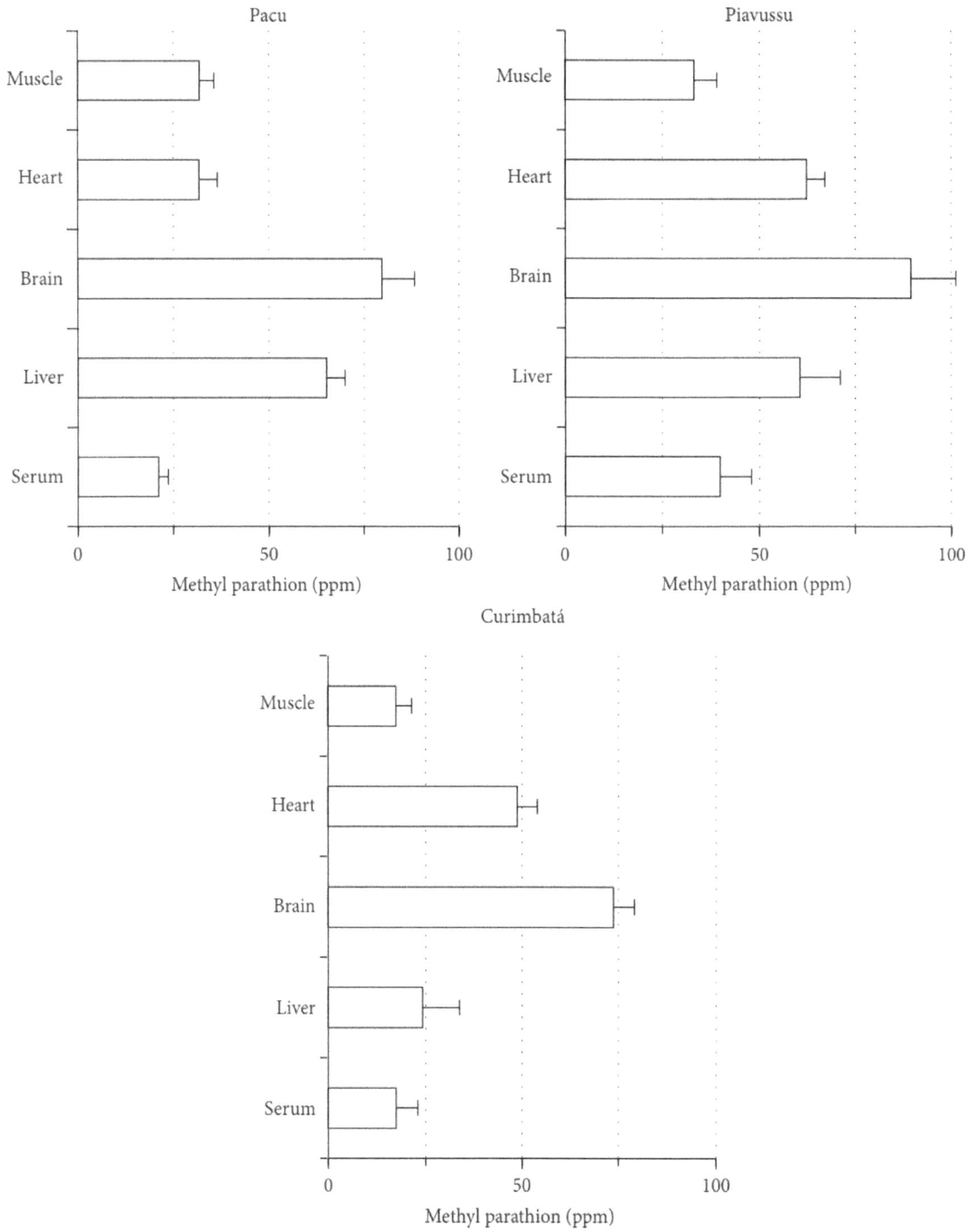

Figure 2: Concentration of methyl parathion in muscle, heart, brain, and liver and serum from pacu, piavussu, and curimbatá. Each one of four animals per species was placed in separated tanks containing 40 L of water with Folidol to produce 5 ppm methyl parathion. After 30 minutes exposure, blood samples were collected and the fishes were euthanized. Tissues were dissected and homogenized. Methyl parathion was extracted from serum and tissue homogenates and quantified by HPLC.

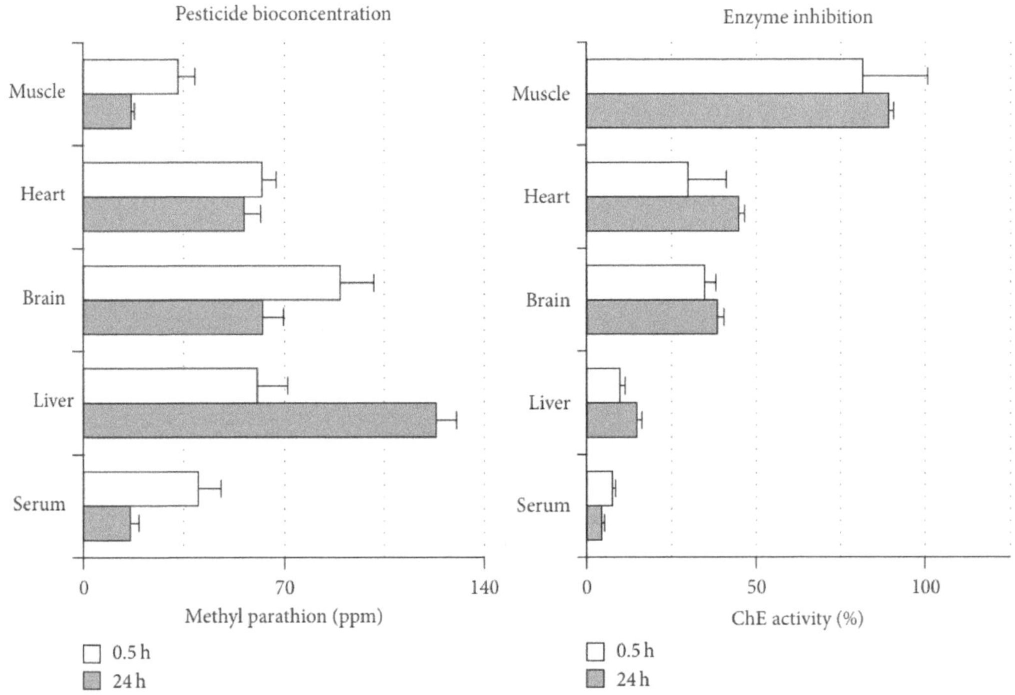

Figure 3: Piavussu methyl parathion concentrations and cholinesterase inhibition after exposure. Each one of eight piavussus was individually placed in separated tanks containing 40 L of water with Folidol to produce 5 ppm methyl parathion. Incubation happened for the showed times. After exposure, blood was collected, fishes were submitted to euthanasia and tissues were dissected and homogenized. Cholinesterase activity was assayed in serum and in homogenates tissues with acetylthiocholine and expressed as percentage of control activity. Methyl parathion was extracted from serum and tissue homogenates and quantified by HPLC.

Cholinesterase Inhibition

Heart, liver, and serum cholinesterase activities for pacu, piavussu, and curimbatá were greatly inhibited after 30 minutes exposure to Folidol. By contrast, skeletal muscle ChE activity for the three species and brain AChE activity for pacu showed no inhibition within this time period (Figure 4). Exposure of fishes to water with Folidol for 78 h resulted in 70% brain AChE activity inhibition in pacu and 60% brain AChE activity inhibition in piavussu.

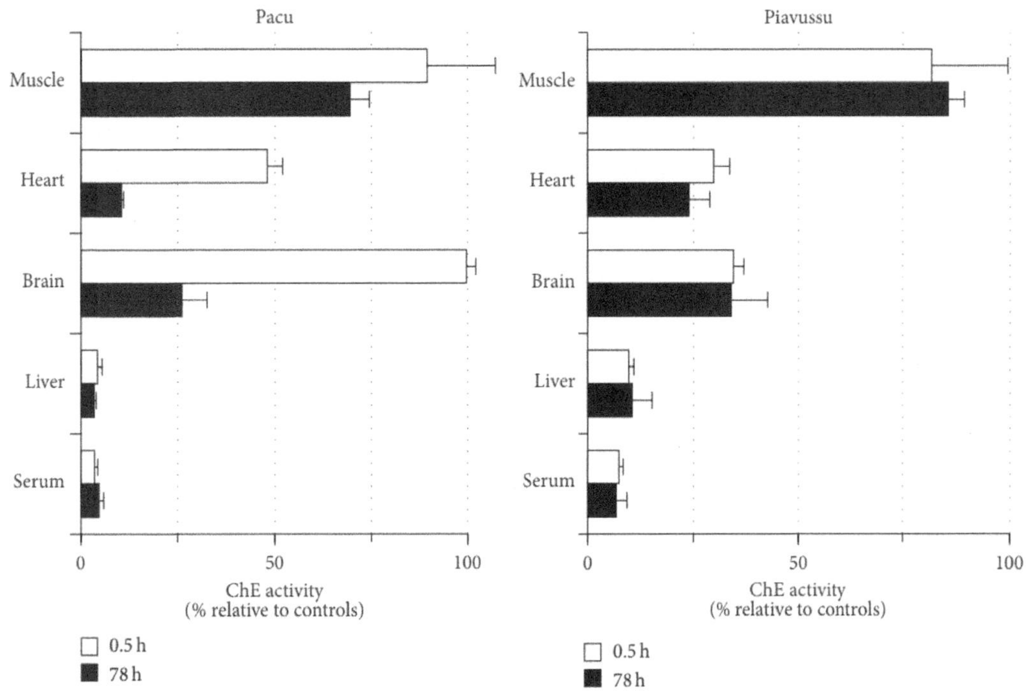

Curimbatá

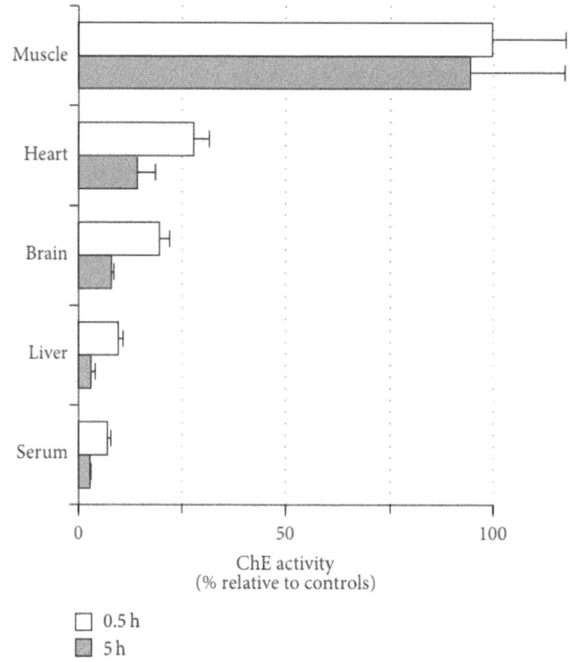

Figure 4: Cholinesterase activity in tissues of pacu, piavussu, and curimbatá exposed to 5 ppm methyl parathion. Each one of eight animals by species was individually treated with Folidol 600 for the indicated time in a 40 L aquarium. After 30 min of exposure blood was collected from four fishes by species and these fishes were submitted to euthanasia. Then, of the remaining fish, four pacus and four piavussus were euthanized after 78 h exposure, while four curimbatás were collected immediately after they showed no movement of opercula (from 3 to 5 h of exposure). Tissues were dissected and homogenized. Cholinesterase was assayed in serum and homogenates with acetylthiocholine and expressed as percentage of the activity found in controls.

All curimbatás stopped moving their opercula between 3 and 5 hours of exposure to 5 ppm Folidol, showing more than 90% inhibition of brain AChE activity and no inhibition of skeletal muscle ChE activity (Figure 4). Increasing the time of exposure from 30 min to 78 hours did not considerably modify muscle, heart, brain, liver, or serum ChE activities in piavussu (Figure 4).

DISCUSSION

Brain tissue showed the highest pesticide bioconcentration, reaching 80 ppm after 30 minutes of exposure to methyl parathion (Figure 2). The three fish species selected for this study present different brain cholinesterase sensitivities to intoxication by methyl paraoxon [15]. In these fishes, the sensitivity of brain ChE to intoxication by methyl paraoxon is inversely related to the activity of the brain enzyme for ASCho [15], in contrast to what was described for trout and rat by Kemp and Wallace [16].

The results demonstrate that the fishes exposed to Folidol rapidly absorbed methyl parathion, since concentrations of methyl parathion in the tissues and serum peaked within 30 min of exposure to Folidol. This concentration continued to increase only in liver, in which the amount of methyl parathion doubled after 24 hours (Figure 3). Such findings can be ascribed to the promptness with which methyl parathion undergoes biotransformation into methyl paraoxon inside liver cells, as compared to other organs. Since P-450 can be inhibited by the sulphur removed from parathion [17, 18] it is plausible that metabolism of methyl parathion molecules occurred in liver up to 30 min and then a consequent inhibition of liver P-450 allowed that more methyl parathion could accumulate and be extracted. When rainbow trout was exposed to the 75 ng mL^{-1} ethyl parathion Abbas and coworkers [19] observed a peak concentration of this pesticide in plasma up to 4.5 hours. In the experiments conducted in the present study, pesticide bioconcentration occurred over a shorter period, probably due to the significantly higher pesticide concentration used. Our results indicate that methyl parathion can bioconcentrate in brain of the tested fish up to approximately 80 ppm (4 times its solubility in water) (Figure 2). This bioconcentration capacity is probably due to high lipid solubility of this OP compound. The half-life of parathion in fish has been reported as 5 times longer than in rats [19]. This suggests that this pesticide could persist for a relatively long time in fish, raising concerns regarding human consumption of fish previously exposed to OP.

The results reported herein demonstrated that ChE inhibition in fish exposed to Folidol does not seem to depend on the methyl parathion bioconcentration in their tissues. No ChE inhibition in pacu brain was observed up to 30 min of exposure, despite the fact that this organ presented 80 ppm of methyl parathion at that time. Therefore, this data is important for further investigations. Tissues in which ChE inhibition occurred more quickly, such as pacu liver (Figure 4) showing higher BChE activity in comparison to AChE activity (Figure 1). BChE has been shown to be more sensitive to methyl paraoxon and is probably more quickly inhibited [20]. In addition, activation of OP compounds occurs mainly in the liver because of the high concentrations of P-450 in that tissue. Thus, liver ChE undergoes inhibition by methyl paraoxon locally generated in tissue before this metabolite leaks out to general blood circulation and reaches other target tissues.

In their work, de Aguiar and coworkers [2] described an 87% brain AChE inhibition in matrinxã (Brycon cephalus) exposed to water with 2 ppm Folidol 600 for 96 hours. The present study indicates that piavussu and pacu might be more tolerant to Folidol than matrinxã, since piavussu presented 66% and pacu 74% brain AChE inhibition when exposed to water with 5 ppm methyl parathion for 78 hours. On the other hand, curimbatá showed a 92% brain AChE inhibition with only 5 hours of exposure to 5 ppm methyl parathion, probably due to the higher sensitivity of this species AChE to inhibition by methyl paraoxon [15].

Mammals poisoned by OP usually die by asphyxia. However, fishes have been reported to be more resistant to poisoning by high levels of organophosphates compounds than rats [21]. The actual causes and mechanisms of fish poisoning by OP are not fully understood. Our findings reinforce that interspecific differences in AChE inhibition by their oxon derivatives should be considered. Physiological and behavioral disturbances start at 50% AChE inhibition and death usually follows when inhibition exceeds 80% in mammals and birds [22]. We found here that curimbatás also perished when brain AChE inhibition reached above 80%.

This study suggests that a major contributing factor to acute fish toxicity by Folidol is brain AChE sensitivity to methyl paraoxon, the oxon derivative of methyl parathion. The same explanation, that is, the sensitivity of brain AChE to oxon derivatives, has been suggested for chlorpyrifos, parathion, and methyl parathion toxicity in mosquito fish (Gambusia affinis) [23]. Although it was clearly established that curimbatás possess the most sensitive brain AChE among the studied species, the death of curimbatás exposed to 5 ppm methyl parathion cannot be solely credited to brain AChE inhibition. Other possible causes of this fish death, such as Na^+ K^+ATPase inhibition [24] and tissue hypoxia, which compromises heart function, need to be examined.

CONCLUSION

Prochilodus lineatus (curimbatá), Piaractus mesopotamicus (pacu), and Leporinus macrocephalus (piavussu) studied here showed similar capacities to bioconcentrate methyl parathion in their tissues after exposure to 5 ppm in water. However, only curimbatá, with the highest brain AChE sensitivity to methyl paraoxon, died after 5 hours of exposure to Folidol. Pacu and piavussu are more resistant to methyl paraoxon and were alive up to 78 hours of exposure to 5 ppm of methyl parathion. Brain AChE sensitivity to methyl paraoxon might be a decisive factor for determining the sensitivity of these species to poisoning by high concentrations of organophosphate compounds. The present study indicates that fishes whose brain acetylcholinesterase activity is more sensitive to oxon derivatives will suffer more severe impacts from environmental contamination by organophosphate pesticides.

Measures reducing the use of organophosphate pesticides in fish culture should be adopted in order to minimize the discharge and consequent impact of these chemicals to natural communities inhabiting rivers and lakes, of which the sensitivity to intoxication by pesticides is largely unknown.

ACKNOWLEDGMENTS

The authors are grateful to Morro Grande Fish Farm and Sol Nascente Fish Farm for supplying the fishes. This paper is dedicated to Dr. Moacelio Veranio Silva Filho (in memoriam).

REFERENCES

1. E. Barbieri and L. A. A. Ferreira, "Effects of the organophosphate pesticide Folidol 600 on the freshwater fish, Nile Tilapia (Oreochromis niloticus)," Pesticide Biochemistry and Physiology, vol. 99, no. 3, pp. 209–214, 2011.
2. L. H. de Aguiar, G. Moraes, I. M. Avilez, A. E. Altran, and C. F. Corrêa, "Metabolical effects of Folidol 600 on the neotropical freshwater fish matrinxã, Brycon cephalus," Environmental Research, vol. 95, no. 2, pp. 224–230, 2004.
3. D. Hernández-Moreno, M. Pérez-López, F. Soler, C. Gravato, and L. Guilhermino, "Effects of carbofuran on the sea bass (Dicentrarchus labrax L.): study of biomarkers and behaviour alterations," Ecotoxicology and Environmental Safety, vol. 74, no. 7, pp. 1905–1912, 2011.
4. C. R. D. Assis, A. G. Linhares, V. M. Oliveira et al., "Comparative effect of pesticides on brain acetylcholinesterase in tropical fish," Science of the Total Environment, vol. 441, pp. 141–150, 2012.
5. L. G. Sultatos, "Mammalian toxicology of organophosphorus pesticides," Journal of Toxicology and Environmental Health, vol. 43, no. 3, pp. 271–289, 1994.
6. O. Arellano-Aguilar and C. M. Garcia, "Effects of methyl parathion exposure on development and reproduction in the viviparous fish girardinichthys multiradiatus," Environmental Toxicology, vol. 24, no. 2, pp. 178–186, 2009.
7. M. Y. de la Vega Salazar, L. Martínez Tabche, and C. Macías García, "Bioaccumulation of methyl parathion and its toxicology in several species of the freshwater community in Ignacio Ryamirez dam in Mexico," Ecotoxicology and Environmental Safety, vol. 38, no. 1, pp. 53–62, 1997.
8. J. E. Chambers and H. W. Chambers, "Biotransformation of organophosphorous insecticides in mammals," in Pesticide Transformation Products: Fate and Significance in the Environment, L. Somasundaram and J. R. Coats, Eds., pp. 32–42, American Chemistry Society, Washington, DC, USA, 1991.
9. B. N. LaDu, "Human serum paraoxonase/arylesterase," in Pharmacogenetics of Drug Metabolism. Pergamon, W. Kalow, Ed., pp. 51–91, Pergamon, New York, NY, USA, 1992.
10. K. Dembéle, E. Haubruge, and C. Gaspar, "Concentration effects of selected insecticides on brain acetylcholinesterase in the common carp (Cyprinus carpio L.)," Ecotoxicology and Environmental Safety, vol. 45, no. 1, pp. 49–54, 2000.

11. V. K. Patil and M. David, "Oxidative stress in freshwater fish, Labeo rohita as a biomarker of malathion exposure," Environmental Monitoring and Assessment, vol. 185, no. 12, pp. 10191–10199, 2013.
12. M. Jokanović, "Current understanding of the mechanisms involved in metabolic detoxification of warfare nerve agents," Toxicology Letters, vol. 188, no. 1, pp. 1–10, 2009.
13. R. M. Lopes, M. V. S. Filho, J. B. de Salles, V. L. F. C. Bastos, and J. C. Bastos, "Cholinesterase activity of muscle tissue from freshwater fishes: characterization and sensitivity analysis to the organophosphate methyl-paraoxon," Environmental Toxicology and Chemistry, vol. 33, no. 6, pp. 1331–1336, 2014.
14. G. L. Ellman, K. D. Courtney, V. Andres Jr., and R. M. Featherstone, "A new and rapid colorimetric determination of acetylcholinesterase activity," Biochemical Pharmacology, vol. 7, no. 2, pp. 88–95, 1961.
15. M. V. S. Filho, M. M. Oliveira, J. B. Salles, V. L. F. C. Bastos, V. P. F. Cassano, and J. C. Bastos, "Methyl-paraoxon comparative inhibition kinetics for acetylcholinesterases from brain of neotropical fishes,"Toxicology Letters, vol. 153, no. 2, pp. 247–254, 2004.
16. J. R. Kemp and K. B. Wallace, "Molecular determinants of the species-selective inhibition of brain acetylcholinesterase," Toxicology and Applied Pharmacology, vol. 104, no. 2, pp. 246–258, 1990.
17. J. Halpert, D. Hammond, and R. A. Neal, "Inactivation of purified rat liver cytochrome P-450 during the metabolism of parathion (diethyl p-nitrophenyl phosphorothionate)," The Journal of Biological Chemistry, vol. 255, no. 3, pp. 1080–1089, 1980.
18. B. J. Norman, R. E. Poore, and R. A. Neal, "Studies of the binding of sulfur released in the mixed-function oxidase-catalyzed metabolism of diethyl p-nitrophenyl phosphorothionate(parathion) to diethylp-nitrophenyl phosphate (paraoxon)," Biochemical Pharmacology, vol. 23, no. 12, pp. 1733–1744, 1974.
19. R. Abbas, I. R. Schultz, S. Doddapaneni, and W. L. Hayton, "Toxicokinetics of parathion and paraoxon in rainbow trout after intravascular administration and water exposure," Toxicology and Applied Pharmacology, vol. 136, no. 1, pp. 194–199, 1996.
20. V. L. F. Cunha Bastos, A. Rossini, L. F. Ribeiro Pinto et al., "Different sensitivities to paraoxon of brain and serum cholinesterases from pacu, an indigenous Brazilian fish," Bulletin of Environmental Contamination and Toxicology, vol. 60, no. 1, pp. 1–8, 1998.
21. J. S. Boone and U. E. Chambers, "Time course of inhibition of cholinesterase and aliesterase activities, and nonprotein sulfhydryl levels following exposure to organophosphorus insecticides in mosquitofish (Gambusia affinis)," Fundamental and Applied Toxicology, vol. 29, no. 2, pp. 202–207, 1996.
22. C. H. Walker, "The use of biomarkers to measure the interactive effects of chemicals," Ecotoxicology and Environmental Safety, vol. 40, no. 1-2, pp. 65–70, 1998.
23. J. S. Boone and J. E. Chambers, "Biochemical factors contributing to toxicity differences among chlorpyrifos, parthion, and methyl parathion in mosquitofish (Gambusia affinis)," Aquatic Toxicology, vol. 39, no. 3-4, pp. 333–343, 1997.
24. J. Blasiak, "Allosteric inhibition of the (Na^+ + K^+)-ATPase by parathion and methylparathion," Pesticide Biochemistry and Physiology, vol. 54, no. 1, pp. 40–47, 1996.

CHAPTER 9

THE TOXICITY, PERSISTENCE AND MODE OF ACTIONS OF SELECTED BOTANICAL PESTICIDES AGAINST INSECT PESTS IN COMMON BEANS, P. VULGARIS

Nelson Mpumi, Kelvin Mtei, Revocatus Machunda, and Patrick A. Ndakidemi

Keywords: Toxicity, Persistence, Botanical Pesticides, Mode of Actions, Half Life

INTRODUCTION

Currently, synthetic pesticides are used in controlling crops insect pests and have usually provided strong defense against insect pests [1] . Synthetic pesticides have paid off and contributed to increase in crop yields of up to four times the value of the applied pesticides [2] . Synthetic pesticides work quickly, and are not labour intensive [3] . Besides of being used widely, they are strong in controlling insect pests. However, their availability is unreliable in distant rural areas, because they are diluted to ineffective concentrations by dishonest traders, they are toxic to people and contaminate the environment [4] . Also, synthetic pesticides and their metabolites have high persistence in soil, water and crops themselves and therefore affect environment and the health of human being during preparation, application and the consumption of crops. These constraints of synthetic pesticides have led to increased interest in the application of botanical pesticides for crop protection in the field and during storage [5] .

Botanical pesticides are naturally occurring chemicals extracted from plants [6] [7] . They are also called natural insecticides [6] . Botanical pesticides are cheaply available and easy to prepare and use [8] , environmentally friend, difficult to contaminate farmers during preparation and use and therefore, safer than synthetic pesticides. Also they are friend to the health of human beings because they are less persistent in the environment, less toxicity and less harmful to beneficial insects [8] [9] . The active ingredients of botanical pesticides degrade rapidly in sunlight, air, and moisture and are readily broken down by detoxification enzymes [6] [7] . Commercially, botanical pesticides are cheap and in this case are viable to small-holder farmers and the growing economy of Africa.

In agricultural practices, the insect pests affecting the bean production include: bean stem maggot, Ophiomyia spp, bean foliage beetles, Ootheca spp, aphids, Aphis fabae, flower beetles, Mylabris variabilis, leaf eating caterpillars and legume pod borers [10] . The insect pests affecting the bean production in northern part of Tanzania include; bean stem maggot, Ootheca, beetle, flower beetle, leaf eating caterpillar, pod borer and aphids [11] . Insect pests reduce crop production yield because they consume plant tissues such as pod suckers, accelerating leaf senescence and contaminate the entire crops in the field which lead to poor food quality [12] [13] .

It is estimated that, about 30% - 40% of crop loss in preharvest and postharvest worldwide is due to pests [14] . Therefore, insect pests limit much of the productivity gain made through agricultural innovations. Most smallholder farmers in sub-Saharan Africa in which the use of modern technology is poor, overcome the problem of crop loss through insect pests by the use of traditional methods [15] [16] . This includes mixing more than one crop in the farm field at the same time [17] which is known as intercropping. In northern part of Tanzania, small-holder farmers practice both intercropping and monocroping system [18] . Common beans, for example, are either intercropped with maize or grown alone for nutritional values such as protein, iron and zinc supplements and for

Citation: Nelson Mpumi, Kelvin Mtei, Revocatus Machunda, Patrick A. Ndakidemi (2016). The toxicity, persistence and mode of actions of selected botanical pesticides against insect pests in common beans, P. vulgaris. American Journal of Plant Sciences,07,138-151. doi:10.4236/ajps.2016.71015.

good health including brain development [19] [20] . However, these traditional methods employed by small holder farmers to control insect pests of common beans are less effective. Therefore, the use of botanical pesticides is a promising means for controlling insect pests in common beans. But also there is limited scientific information of the toxicity, persistence and mode of action of botanical pesticides active ingredients for proper use to control insect pests in common beans.

This review paper concentrates on common practices used to control insect pests of common beans, the toxicity, persistence, and mode of action of some of active ingredients of botanical pesticides including Tephrosia vogelii, Vernonia amygdalina, Tithonia diversifolia and Lantana camara, Neem, Azadirachta indica and pyrethrum, Chrysanthemum cinerariifolium as the alternatives of controlling common bean insect pests in the farmers' field, and in storage rooms and increase the possibility of safe food and environmental friendly farming and storage practices [5] .

COMMON PRACTICES USED TO CONTROL INSECT PESTS OF COMMON BEANS IN THE FIELD

Field insect pests infestation is a very serious problem because life stages of insects cause economic damage and deteriorates the crops to be harvested and food products to be stored [21] . Field insect pests infestation cause products lose in the field and some insect life stages are carried up to the storage rooms in which the damage can be estimated up to 9% in developed countries and up to 20% or more in developing countries [22] . There are practices used to control the insect pests of the crop in the field. Those practices used by small-holder farmers include traditional, biological, synthetic and botanical pesticides application.

Traditional Practices

There are many traditional practices, whereby smallholder farmers use to reduce infestations of the insect pests in the farms. Those traditional practices are also called cultural practices. Cultural practices such as site selection, crop rotation, and cultivar and seed selection, proper sowing date can reduce the infestation of certain insect pests [23] . For instance, aphid infestation in wheat and common beans is reduced by early sowing time [24] and also, affect the population of bean stem maggot, Ophiomyia sp, bean foliage beetle, Ootheca, aphids and other arthropods attacking common beans in the field [25] . Apart from that, [26] reported that; planting in either late time or off-season lead to higher infestation of bean stem maggot in the farms grown common beans. In other agronomic studies, row spacing and plant density, weed control and stubble retention have been used to control bean stem maggot [27] . Other studies have reported that, sloping sites and border hedgerows which reduce wind speed promote aphid landing and affects aphids and Ootheca species distribution [28] . Increasing plant density from 22 bean plant/m^2 to 33 bean plant/m^2 was found to decrease common bean virus incidence transmitted by aphids by 10% - 20% [29] . Also [29] reported that; planting cereal border around faba bean field reduces the spread of non-persistently bean transmitted virus. However, cultural practices are not very effective although they are safe and cheap. Therefore, there is a need of conducting detailed study on the use of botanical pesticides for controlling insect pests in the farms of small-holder farmers.

Biological Control Methods

Biological control is defined as the reduction of pest populations (insects, mites, weeds and plant diseases) using other living organisms [30] . Insect pests are suppressed by naturally occurring organisms and environmental factors which are called natural enemies or natural control. Natural enemies of insect pests are known as biological control agents. Natural enemies are organisms which kill, decrease the reproductive potential, or reduce the number of another organism [30] . Those natural enemies include predators, parasitoids, and pathogens.

A predator is an organism that attacks, kills, and feeds on several or many other organisms in life time. Some predators are specialized, which means feeding on only one or few preys while most are generalized meaning that, feeding on variety of organisms [30] . Predators include spiders, lacewings, lady beetles, ground beetles, rove beetles, hover flies, and true bugs [30] . These organisms kill and feed in the insect pests affecting common beans. Ladybird beetles, family Coccinelidea, both adults and larvae feeds on aphids [31] and therefore reduce the population of aphids. Ladybird beetles are stronger, larger and usually more intelligent than the prey and therefore attack several hosts in a short period of time [32] .

Parasitoids are insects which parasitize and kill other invertebrates. Many species of wasps and some flies are parasitoids. Some of species of parasitoids, when are in immature stage develops on or within a single insect host forming mummies and therefore killing the host [31] . Parasitoids are parasitic when are in immature stage and kill their hosts as they reach maturity [30] . Species of entomopathogenic fungi infest aphids through the cuticle and finally killing the host [32] . Apart from that, species in the Braconidae family develop as endopara- sitoids of aphids whereby the larva complete their development in the host [23] . Aphids are also controlled by spinosad. It is a biologically derived insecticide produced by the actinomycete, Saccharopolyspora spinosa, a bacterial organ isolated from soil [34] .

Pathogens are important in biological control of many pests including insect pests, nematodes, mites and weeds [31] . Pathogen such as Bacillus thuringiensis, controls certain caterpillars, beetles and flies but does not affect other arthropods. Biological control is safe and eco-friendly. Therefore, detail studies are needed on toxicity, persistence and mode of actions of active ingredients of botanical pesticides for safety use by small holder farmers and support the present of natural enemies.

Chemical Pesticides

Worldwide, it is estimated that approximately 1.8 billion people engage in agriculture and most of them use approximately 5.6 billion pounds of synthetic pesticides to protect the food and commercial products that they produce [35] . Pesticide use in Africa accounts for only 2% - 4% of the global pesticide market [36] . Synthetic pesticides are reported to be effective, reliable against a wide range of insect pests, quick acting and easily tested for new insect pests [37] . Pesticides chemicals such as endosulfan, diazinon and lindane have been identified by several National bean programs and Research organizations to provide protection to germinating bean plants at a time when they are most vulnerable to attacks especially bean stem maggot [38] . Cypermethion, carbaryl, and Lambda-cyhalothrin have shown efficacy to control the pests in the field and in the storage rooms [39] . However, many of the synthetic pesticides such as endosulfan and lindane are either banned, or are expensive to smallholder farmers in Africa, and are persistence in the environment [40] .

Also, according to the Stockholm Convention, among the 12 Persistence Organic Pollutants (POPs), nine are pesticides including aldrin, chlordane, dichlorodiphenyltrichloroethane (DDT), dieldrin, endrin, heptachlor, hexachlorobenzene (HCB), mirex and toxaphene [40] . Those POPs are associated with human health problems such as cancer. The study done by [40] reported that mothers who are exposed to lindane, can accumulate it into breast milk, o, p-dichlorodiphenyldichloroethane (DDD) can accumulate in maternal serum and total dichlorodiphenyltrichloroethane (DDT) can accumulate in the umbilical serum [40] . This reflects their potential placental and breast milk transfer to her child during pregnancy and lactation [40] . In response to the high costs and the negative side effects of synthetic pesticides to the health of human being [41] , there is a need for studying botanical pesticides in detail which are affordable and have less or no health problems to the applicators, consumers and do not contaminate the environment to replace the synthetic pesticides.

Botanical Pesticides

Botanical pesticides have long been publicized as attractive alternatives to synthetic insecticides for pest management since they pose either little or no threat to the environment, to ecosystems and to human health [42] . In the middle of the 17th century, pyrethrin, nicotine and rotenone form pyrethrum, tobacco and Tephrosia spp respectively were recognized as effective insect control agents [6] . Hence, mankind has used plant extracts for thousands of years to the prevent diseases, treatment of disease, as insecticides to control microbial growth, weeds and many more functions [43] . Therefore, many plants which were used for medicinal purposes locally, also demonstrated potential as insect control agents [44] . Botanical products like tobacco extracts, neem oil and extracts, have found promising and useful for bean pests control [45] . Similarly, Tephrosia vogelii, Azadirachta indica, Annona squamosa, chilli paper, Cupscum frutensces, Allium sativa are reported to control insect pests of beans and cowpeas successfully [46] . Aristolochia ringens and Alium sativum have antifeedants, food poisons, contact poisons and repellents against Sitophilus zeamais [44] . Pesticidal plants such as Tobacco, Nicotiana tabacum, Neem, Azadirachta indica, Garlic, Allium sativum, Eucalyptus, Eucalyptus camaldulemsis and Mehogony, Swietenia mehogany, have been reported in controlling aphids attacking bean plants [47] . Normally, botanical pesticides comprise a mixture of bioactive compounds with many have advantages in terms of efficacy and short life span [48] . Table 1 shows the toxicity of some active ingredients of botanical pesticides

The botanical pesticides are generally pest-specific and are relatively harmless to non-target organisms including man and natural enemies of insect pests [6] [49] , and environmentally ecofriend, degrade rapidly in sunlight, air, and moisture, so they are less persistence in the environment, and are rapid in action to the insect pests, no adverse effect on plant growth, seed viability and cooking quality of the grains and are less expensive and easily available in the farmers natural environment [41] . Now it becomes necessary to search for the alternative means of insect pests control, which can minimize the use of synthetic pesticides. This work reviews the toxicity, persistence and mode of actions of some of botanical pesticides which are locally available plant materials in our environment.

Table 1: Toxicity of certain botanical pesticides active ingredients (mg/kg).

Generic Name	Oral LD_{50}	Dermal LD_{50}	Signal Word
Pyrethrins	1200 - 1500	>1800	Caution
Rotenone	60 - 1500*	940 - 3000	Caution
Sabadilla	4,000	-	Caution
Ryania	750 - 1200	4000	Caution
Nicotine	50 - 60	50	Danger
d-Limonene	>5000	-	Caution
Linalool	2440 - 3180	3578 - 8374	Caution
Neem	13,000	-	Caution

*Toxicity varies greatly depending on type of solvent used as carrier; Source: [6].

TOXICITY AND PERSISTENCE OF SOME OF ACTIVE INGREDIENTS FROM BOTANICAL PESTICIDES

Although botanical pesticides can be used as alternatives to synthetic pesticides but the toxicity of the chemical compounds extracted from botanical pesticides to insect pests and humans, persistence to the environment and mode of actions to insect pests are not clear [7] . Therefore, this part, intends to explain the toxicity and persistence of pyrethrin from pyrethrum, rotenone from T. vogelii, azadirachtin from Azadirachta indica, and some active chemical compounds from V. amygdalina, L. camara, and T. diversifolia. These botanical pesticides are selected, because are commonly found around our homes, along the roads, river banks and bush lands in northern part of Tanzania [11] and may be used as botanical pesticides.

Rotenone

Rotenone, Figure 1 is contained in large amount in plant species especially Tephrosia, Derris, and Lonchocarpus [50] . All these plants are in the family fabaceae in Leguminosae. Rotenone (Figure 1) is used as natural insecticide, piscicide, and pesticide [7] . It is a relatively low toxicity insecticide for use in gardens but is highly toxic to fish and is sometimes used to eliminate unwanted fish from lakes [50] . It occurs naturally in the seeds, stems, leaves and the roots of plants in fabaceae family. It was the first described member of the family of chemical compounds known as rotenoids [51] .

The LD_{50} of rotenone (Figure 1) for rats is 132 - 1500 mg/kg [6] [7] . In human being, rotenone is moderately toxic with an oral LD_{50} ranges from 300 to 1500 mg/kg [7] [51] . This compound (Figure 1) is highly toxic to fish and insects because it is lipophilic in nature [50] . The respiratory mechanism of fish is directly linked to water through the gills and in insect is directly exposed through trachea whereby rotenone is easily taken up through the gills or trachea into the bloodstream of fish, and insects respectively resulting to death [50] .

However, rotenone is less toxic to mammals and birds since the route of ingestion is through the digestive tract whereby the compound is easily broken down to less toxic compounds before toxic quantities can enter the bloodstream [6] [7] [50] [51] . Rotenone is rapidly broken down by sunlight which is both an advantage and disadvantage [50] . Since it breaks down rapidly, it does not accumulate in the environment and less harmful to non-target organisms [6] . However, it must be re-applied at short intervals and is usually applied in the early morning or in the evening to avoid degradation of it by sunlight [7] . In water, the rate of decomposition depends upon several factors, including temperature, pH, turbidity of water and sunlight. The half life of rotenone is four days [6] [7] [50] [51] . The half-life of rotenone in natural waters ranges from half a day at 24°C to 3.5 days at 0°C [6] . However, there are limited scientific information about toxicity of rotenone to organisms and persistence of it in the environment. Therefore, detail studies are needed on the toxicity of rotenone to various animals and persistence of it in the environment for use it sustainably as botanical pesticide.

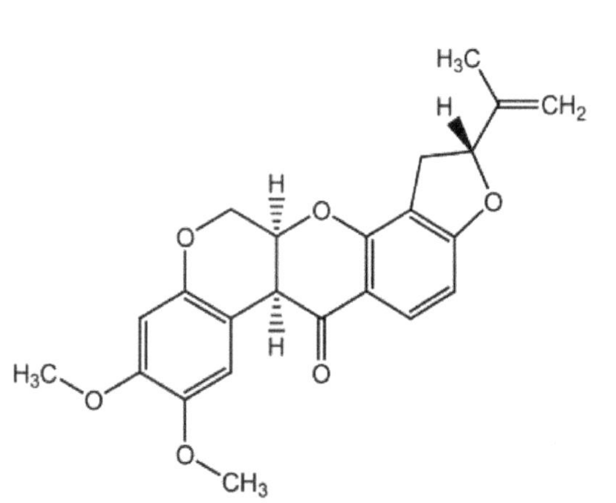

Figure 1: Chemical structure of rotenone.

Azadirachtin from Neem, Azadirachta indica

Neem tree is in Meliaceae family possessing bitter triterpenoids [52] . The active compound in the neem is azadirachtin which is found in the leaves, and also concentrated in the seeds [53] . It is a bitter, complex chemical compound which belongs to the limonoid group and it show strong biological activities among various insect pests [53] . This compound (Figure 2) is a feeding deterrent and growth regulator [51] .

This compound (Figure 2) can affect about 200 species of insects by acting as antifeedant and growth disruptor. Azadirachtin has a toxicity and fascinating effect on insects (LD_{50} (S. littoralis), 15 µg/g) [53] . It has very low toxicity to mammals whereby the LD_{50} in rats is greater than 3540 mg/kg which make it practically non-toxic to mammals [53] and also has been reported to be non mutagenic [52] . Azadirachtin has been found to degrade rapidly under environmental factors such as UV radiation in sunlight, heat, air moisture, acidity and enzymes present in foliar surfaces [53] . The half-life of azadirachtin has been found to be between 48 minutes and 3.98 days under Ultraviolet (UV) light and sunlight and 2.47 days on leaf surface [6] [51] . Therefore, there is a need to use azadirachtin as environmentally compatible insecticides, with selective toxicity to targeted pests, low toxic to plants and mammals and environmental friendly desired stability.

Pyrethrin from Pyrethrum, Tanacetum cinerariifolium (Chrysanthemum cinerariifolium)

Pyrethrum is powdered, dried flower head of the pyrethrum daisy, Tanacetum cinerariaefolium and pyrethrins active compound from pyrethrum with six related insecticidal compounds which occur naturally [6] . There is pyrethrin I and pyrethrin II. The compounds related to pyrethrin I contain methyl group ($-CH_3$) and the compound related to pyrethrin II contain $-CO_2CH_3$ group [51] [52] . The general chemical formula of pyrethrin and the six related pyrethrin compounds are shown inFigure 3 and Figure 4.

Pyrethrins are axonic poisons and have an insect repellent effect when present in little amount [6] . They are harmful to fish, but are less toxic to mammals and birds than many synthetic insecticides. In pure form, the rat oral LD_{50} is 1200 - 1500 mg/kg [51] . The technical grade of pyrethrum is less toxic to rat with the LD_{50} of about 1500 mg/Kg. Pyrethrins degrade easily when is exposed to the environment moisture, air and the sunlight [51] . The half-life of pyrethrins in the environment and field-grown bell pepper fruit is 2 hours or less [6] . However, there are limited scientific information about toxicity of pyrethrins compounds to various organisms and persistence of it in the environment. Therefore, detail studies are needed about the toxicity of pyrethrins compounds to various organisms and persistence of it in the environment for use it sustainably as botanical pesticide.

Sesquiterpene Lactones from T. diversifolia

Many classes of secondary metabolites which are isolated from the Tithonia diversifolia extracts include diterpenoids, flavonoids, sesquiterpene lactones (Figure 5) and chlorogenic acids derivatives [54] [55] . However, the most abundant terpernoids in Tithonia diversifolia are sesquiterpene lactones [55] . But tagitinins compounds (Figure 5) which are in sesquiterpene lactones class are the most studied [56] . The sesquiterpene lactones and diterpenoids have biological activities [16] and contribute to inflammatory activity [55] . T. diversifosia is used as traditional medicine in constipation, stomach pains, indigestion, sore throat, liver pains and to treat malaria [57] . [56] reported that; the extracts of T. diversifolia of 10 mg/kg and 100 mg/kg administered to rats for 90 days were relatively safe with some toxicity observed at 100 mg/kg. However, the later can cause damage of the liver, the kidneys and, to a lesser extent, the heart [56] . The liver damage observed at higher doses of aqueous extracts of Tithonia diversifolia may result from the Chlorogenic acid, while kidney damage results from the Sesquiterpene lactones [55] [56] . However, no clear information about toxicity of these compounds from Tithonia diversifolia.

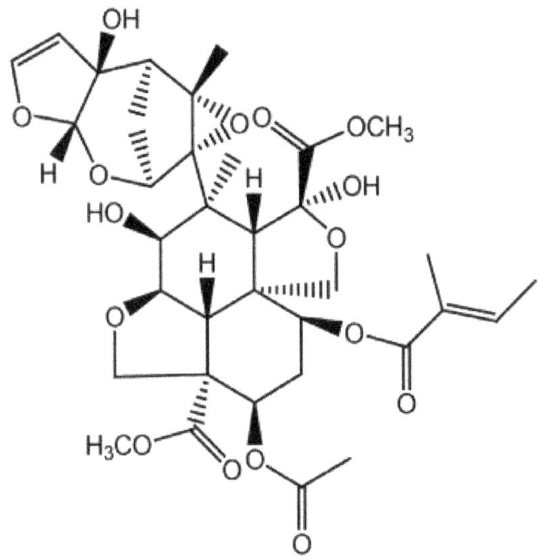

Figure 2: Chemical structure of azadirachtin.

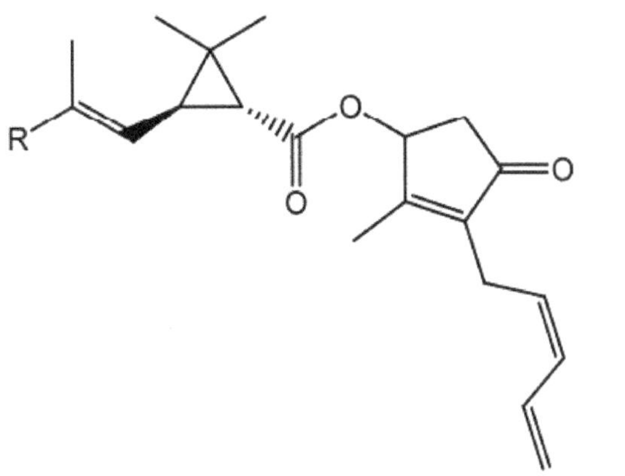

Figure 3: General chemical formula of pyrethrum, Pyrethrum I, R = CH_3, Pyrethrum II, R = CO_2CH_3.

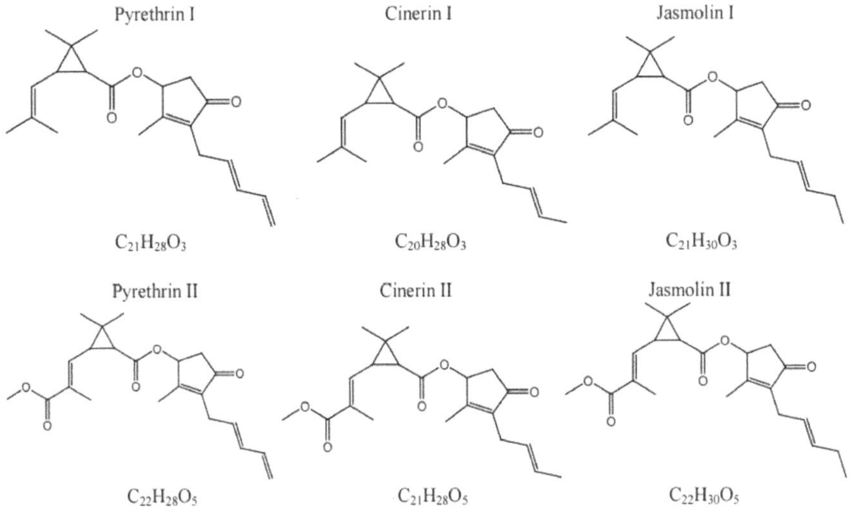

Figure 4: Chemical structures of the six related pyrethrin compounds.

Pentacyclic Triterpenoids from Lantana camara

Lantana camara is recognized to be toxic to cattle, sheep, horses, dogs and goats [58] . The active ingredients causing toxicity of Lantana camara in grazing animals is pentacyclic triterpenoids [47] (Figure 6). It is one of terpernoids, which result in liver damage and photosensitivity [59] . The toxicity of L. camara to human being is undetermined, whereby numerous studies suggest that; ingestion of berries from Lantana camara can be toxic to humans [47] . [60] reported that; leaf extract of L. camara had excellent repellent, moderate toxic and antifeedant activities. However, other studies have found evidence which suggests that ingestion of L. camara fruit, poses no risk to humans and is in fact edible when ripe [47] . Studies conducted in India have found that; Lantana camara leaves can display antimicrobial, fungicidal and insecticidal properties [61] . L. camara has also been used in traditional herbal medicines for treating a variety of ailments, such as cancer, skin itches, leprosy, rabies, chicken pox, measles, asthma and ulcers [61] . Lantana camara has been tested as an alternative to fumigants in stored grains [62] . Therefore, there is a need of finding out more information about the toxicity of Lantana camara to the health of human being.

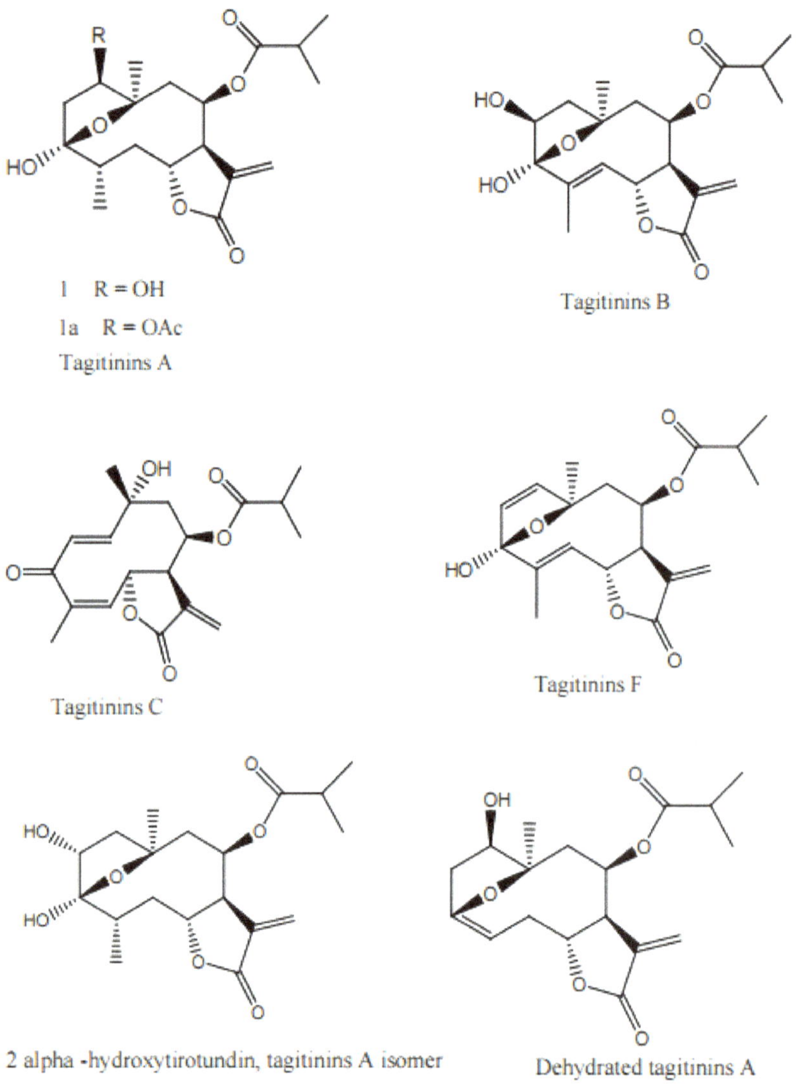

Figure 5: Chemical structures of some sesquiterpene lactones from leaves of T. diversifolia. Source: [57] .

Vernodalin, Vernodalol and Epivernodalol from V. amygdalina

Vernodalin, Vernodalol and Epivernodalol (Figure 7) are sesquiterpene lactone compounds from the members of Asteraceae family. They are the major bioactive constituents isolated from Vernonia species [63] . Phytochemical analysis of leaves of Vernonia amygdalina revealed the presence of important natural products vernolide and vernodalol [64] [65] . The toxicity of these compounds is very low to human being [42] . The LD_{50} was tested in mice and extrapolated to be 1265 mg/kg [63] . The possibility of using nontoxic deterrents and repellents as crop protectants is intuitively attractive [42] since have little effects to health of human being. These sesquiterpene lactones were found to be active on gram positive bacteria and vernolides has reported to show high antifungal activities [66] . Study on the efficacy of V. amygdalina on field insect pests in cowpea (M. sjostedti and B. tabaci) showed effectiveness when it was applied in two weeks interval after crop emergency [67] . These chemicals have deterrent and repellent properties toward insect pests.

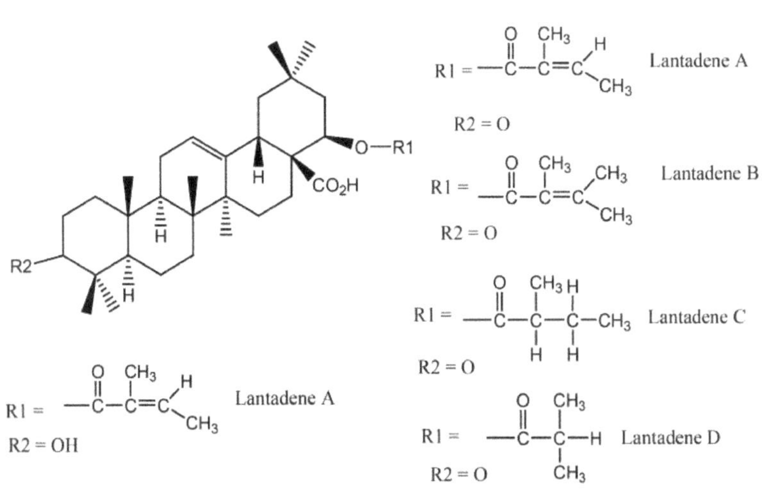

Figure 6: Chemical structures of Lantadenes (Pentacyclic triterpenoids). Source: [47] .

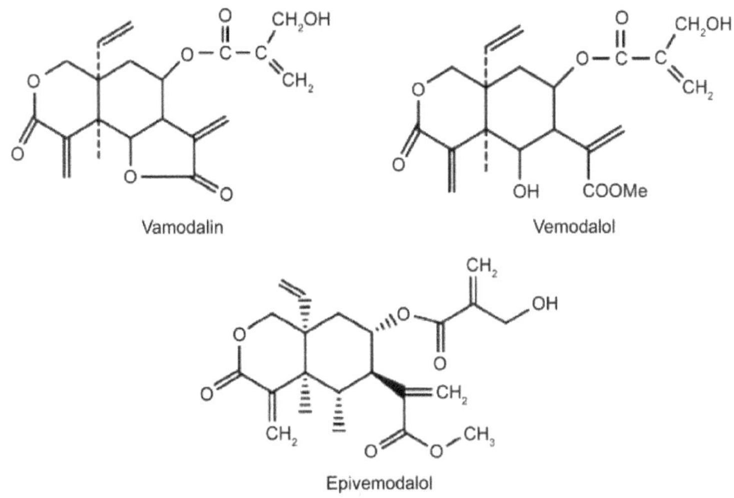

Figure 7: Chemical structures of isolated compounds from V. amygdalina. Source: [64] [65] .

Repellent and feeding deterrent chemicals from pesticidal plants discourage the insects from feeding it. Many compounds or extracts from medicinal plants which demonstrated antifeedant effects lack toxicity if ingested [42] [68] . Sesquiterpene lactone disappears in the environment in 90 days and the disappearance is faster when compared with synthetic pesticides [63] . However, there is little information about the toxicity, persistence and mode of action of V. amygdalina suitable for the protection of the environment from the botanical pesticides metabolites contamination.

MODE OF ACTION OF ACTIVE COMPOUNDS FROM BOTANICAL PESTICIDES

Despite the suggestion, of using botanical pesticides as alternatives to synthetic pesticides but there are limited information about the mode of actions of the chemical compounds to insect pests. Therefore, this part intends to explain the mode of actions of pyrethrin from pyrethrum, rotenone from T. vogelii, azadirachtin from Azadirachta indica, and some active chemical compounds from V. amygdalina, L. camara, and T. diversifolia.

Mode of Action of Rotenone

Rotenone delay the electron transport chain in mitochondria of the insect pests and it is a contact and stomach poison [6] [7] [50] [51] [69] . It inhibits the transfer of electrons from iron-sulfur centers in complex I to ubiquinone and interferes with Nicotinamide adenine dinucleotide hydride (NADH) during the creation of usable cellular energy Adenosine triphosphate (ATP) [69] . In that case, Complex I is unable to pass through its electron to Complex Q, creating a back-up of electrons within the mitochondrial matrix. During this limiting process, cellular oxygen is reduced to the radical which is a reactive species. This reactive species can damage Deoxyribonucleic acid (DNA) and other components of the mitochondria [50] .

Mode of Action of Azadirachtin

Azadirachtin displays strong antifeedant effects on chemoreceptors of the insects and discourage the insect pests to consume the plant [53] . If the insect pest continues in consuming crops sprayed with neem tree extracts, the azadirachtin blocks peptide hormone release, which results in severe growth defects and molting abnormalities [52] [70] . Finally, azadirachtin has a damaging effect on the tissues including the muscles, fat and gut of most of the insect [52] .

Mode of Action of Pyrethrins

Pyrethrins attack the nervous systems of all insects and act like pyrethroids and DDT [6] . Axonic poison substances affect the electrical transmission of the impulses along the axon [51] . During their mode of action pyrethrins upset the sodium and potassium ion exchange process in insect nerve fibers and interfere the normal transmission of nerve impulses [6] . Pyrethrins delay the closure of voltage-gated sodium ion channels in the nerve cells of insects, resulting in repeated and extended nerve firings [51] . This hyperexcitation causes the death of the insect due to loss of motor coordination and paralysis. Sometimes insect pests may develop resistance to pyrethrum [6] . Piperonyl butoxide is paired with pyrethrin to prevent resistance by insect pests. It is synergist compound. Together, these two compounds prevent detoxification in the insect, ensuring insect death. Synergists make pyrethrin more effective, allowing lower doses to be effective [6] . Pyrethrins are effective pesticides because they selectively target insects rather than mammals due to higher insect nerve sensitivity, smaller insect body size, lower mammalian skin absorption, and more efficient mammalian hepatic metabolism [6] .

Mode of Action of Active Ingredients from T. diversifolia to Insect Pests

The phytochemical analysis of T. diversifolia, revealed the presence of non-volatile fractions which are rich in flavonoids and sesquiterpene lactones whereas the essential oil contains mainly monoterpene hydrocarbons, such as b-ocimene, a-pinene and limonene [23] . The plant has been found to have insect feeding deterrent characteristics to insect pests [23] which have effect on chemoreceptors to discourage the insects from consuming bean plants. That is caused by the presence of 6-methoxyapigenin, sesquiterpene lactones, chologenic acids and tagitinins A, B, C and F, with diversiform, tirotundin, tithonine and sulphurein [23] .

Mode of Action of Some Active Ingredients from L. camara

L. camara essential oil and leaves are composed of large amounts of sesquiterpene hydrocarbons, mainly β-caryophyllene [71] , lantadene A, B and C which produce strong hepatotoxic response in rodents and the extracts have fumigant activity against Sitophilus granarius adults [72] [73] . Terpenoids, phenylpropanoids, and flavonoids are thought to be the main components which have biological activities in Lantana camara [60] . The leaves have repellent and antifeedant properties against insects [60] . Apart from that, L. camara active ingredients have acetylcholine inhibition property against insect pests [60] . It is known for its enzyme inhibition and therefore serves as an alternative to synthetic fumigants. However, there is limited information about the effectiveness, toxicity, and persistence of chemical constituents in Lantana camara.

Mode of Action of Active Ingredients from T. diversifolia to Insect Pests

The phytochemical analysis of T. diversifolia, revealed the presence of non-volatile fractions which are rich in flavonoids and sesquiterpene lactones whereas the essential oil contains mainly monoterpene hydrocarbons, such as b-ocimene, a-pinene and limonene [23] . The plant has been found to have insect feeding deterrent characteristics to insect pests [23] which have effect on chemoreceptors to discourage the insects from consuming bean plants. That is caused by the presence of 6-methoxyapigenin, sesquiterpene lactones, chlorogenic acids and tagitinins A, B, C and F, with diversiform, tirotundin, tithonine and sulphurein [23] .

CONCLUSION

Currently, control of bean pests is achieved through the use of synthetic pesticides. But synthetic pesticides have been found to cause problems to the environment, ecology and to the health of human being. The promising alternatives, the botanical pesticides for instance the Tephrosia vogelii, Cupscum species, Vernonia species, Tagetes minuta, Ocimum gratissimum, Pyrethrum, Azadirachta indica, Lantana camara and Tithonia diversifolia can be considered to control the insect pests in the field and in the storage rooms. However, the toxicological and environmental properties of the compounds from botanical pesticides and the mode of action in insects should also be considered for the safety of our environment and the human health and insect resistance. It is known that most toxic mammalian poisons are natural products from plants although they are found to be cheap and easy available and to prepare, short life span and are ecofriendly to the environment and other animals.

ACKNOWLEDGEMENTS

This study was funded by McKnight Foundation through a grand from Bill and Melinda Gates Foundation given to The Nelson Mandela African Institution of Science and Technology (NM-AIST).

REFERENCES

1. Chikukura, L., Mvumi, B., Chikonzo, R., and Chenzara, C. (2011) Evaluation of Selected Indigenous Pesticidal Plant Powders against Stored Maize and Cowpeas Insect Pests. African Crop Science Conference Proceedings, 10, 189-192.
2. Cooper, J. and Dobson, H. (2007) The Benefits of Pesticides to Mankind and the Environment. Crop Protection, 26, 1337-1348. http://dx.doi.org/10.1016/j.cropro.2007.03.022
3. Weinberger, K. and Srinivasan, R. (2009) Farmers' Management of Cabbage and Cauliflower Pests in India and Their Approaches to Crop Protection. Journal of Asia-Pacific Entomology, 12, 253-259. http://dx.doi.org/10.1016/j.aspen.2009.08.003
4. Stevenson, P.C., Kite, G.C., Lewis, G. P., Forest, F., Nyirenda, S.P., Belmain, S.R., Sileshi, G.W. and Veitch, N.C. (2012) Distinct Chemotypes of Tephrosia vogelii and Implications for Their Use in Pest Control and Soil Enrichment. Phytochemistry, 78, 135-146. http://dx.doi.org/10.1016/j.phytochem.2012.02.025
5. Sola, P., Mvumi, B., Ogendo, J., Mponda, O., Kamanula, J., Nyirenda, S., Belmain, S. and Stevenson, P. (2014) Botanical Pesticide Production, Trade and Regulatory

Mechanisms in Sub-Saharan Africa: Making a Case for Plant-Based Pesticidal Products. Food Security, 6, 369-384. http://dx.doi.org/10.1007/s12571-014-0343-7

6. Henn, T. and Weinzierl, R. (1989) Botanical Insecticides and Insecticidal Soaps. University of Illinois at UrbanaChampaign, College of Agriculture, Cooperative Extension Service.
7. El-Wakeil, N.E. (2013) Botanical Pesticides and Their Mode of Action. Gesunde Pflanzen, 65, 125-149. http://dx.doi.org/10.1007/s10343-013-0308-3
8. Amoabeng, B.W., Gurr, G.M., Gitau, C.W. and Stevenson, P.C. (2014) Cost: Benefit Analysis of Botanical Insecticide Use in Cabbage: Implications for Smallholder Farmers in Developing Countries. Crop Protection, 57, 71-76. http://dx.doi.org/10.1016/j.cropro.2013.11.019
9. Cortés, A.J., Monserrate, F.A., Ramírez-Villegas, J., Madriñán, S. and Blair, M.W. (2013) Drought Tolerance in Wild Plant Populations: The Case of Common Beans (Phaseolus vulgaris L.). PLoS ONE, 8, e62898. http://dx.doi.org/10.1371/journal.pone.0062898
10. Maikai Murna, A. (2013) Development of an Integrated Pest Management Strategy for the Control of Maruca Vitrata on Cowpea in Kebbi State of Nigeria. Ph.D. Thesis, University of Greenwich, Greenwich.
11. Mkenda, P.A., Mtei, K. and Ndakidemi, P.A. (2014) Pesticidal Efficacy of Tephrosia vogelii and Tithonia diversifolia against Field Insect Pests of Common Beans
12. Phaseolus vulgaris L.. within African Farming Communities. African Journal of Applied Agricultureal Sciences and Technologies, 2, 9-26.
13. Boote, K., Jones, J., Mishoe, J. and Berger, R. (1983) Coupling Pests to Crop Growth Simulators to Predict Yield Reductions
14. Mathematical Models.. Phytopathology (USA), 11, 1581-1587. http://dx.doi.org/10.1094/Phyto-73-1581
15. Ampofo, J. and Massomo, S. (1998) Some Cultural Strategies for Management of Bean Stem Maggots (Diptera: Agromyzidae) on Beans in Tanzania. African Crop Science Journal, 6, 351-356. http://dx.doi.org/10.4314/acsj.v6i4.27785
16. Oerke, E.-C. and Dehne, H.-W. (2004) Safeguarding Production—Losses in Major Crops and the Role of Crop Protection. Crop Protection, 23, 275-285. http://dx.doi.org/10.1016/j.cropro.2003.10.001
17. Muhammad, B.Y. and Awaisu, A. (2008) The Need for Enhancement of Research, Development, and Commercialization of Natural Medicinal Products in Nigeria: Lessons from the Malaysian Experience. African Journal of Traditional, Complementary, and Alternative Medicines, 5, 120-130.
18. Obembe, D. (2013) Knowledge Sharing, Sustained Relationships and the Habitus. Management Learning, 44, 355-372. http://dx.doi.org/10.1177/1350507612450988
19. Altieri, M.A. (2004) Linking Ecologists and Traditional Farmers in the Search for Sustainable Agriculture. Frontiers in Ecology and the Environment, 2, 35-42. http://dx.doi.org/10.1890/1540-9295(2004)0020035:LEATFI.2.0.CO;2
20. Tsubo, M., Walker, S. and Ogindo, H. (2005) A Simulation Model of Cereal-Legume Intercropping Systems for SemiArid Regions: I. Model development. Field Crops Research, 93, 10-22. http://dx.doi.org/10.1016/j.fcr.2004.09.002
21. Robinson, J. (2005) Pests and Integrated Pest Management in Western Equatoria, Southern Sudan. International Journal of Tropical Insect Science, 25, 224-235. http://dx.doi.org/10.1079/IJT200582
22. Fivawo, N. and Msolla, S. (2012) The Diversity of Common Bean Landraces in Tanzania. TaJONAS: Tanzania Journal of Natural and Applied Sciences, 2, 337-351.
23. Upadhyay, R. and Ahmad, S. (2011) Management Strategies for Control of Stored Grain Insect Pests in Farmer Stores and Public Ware Houses. World Journal of Agricultural Science, 7, 527-549.
24. Phillips, T.W. and Throne, J.E. (2009) Biorational Approaches to Managing Stored-Product Insects. Annual Review of Entomology, 55, 375-397. http://dx.doi.org/10.1146/annurev.ento.54.110807.090451
25. Mwanauta, R.W., Mtei, K.A. and Ndakidemi, P.A. (2014) Prospective Bioactive Compounds from Vernonia amygdalina, Lippia javanica, Dysphania ambrosioides and Tithonia diversifolia in Controlling Legume Insect Pests. Agricultural Sciences, 5, 1129-1139. http://dx.doi.org/10.4236/as.2014.512123
26. Acreman, T. and Dixon, A. (1985) Developmental Patterns in Wheat and Resistance to Cereal Aphids. Crop Protection, 4, 322-328. http://dx.doi.org/10.1016/0261-2194(85)90034-1
27. Way, M. and Banks, C. (1967) Intra-Specific Mechanisms in Relation to the Natural Regulation of Numbers of Aphis fabae Scop. Annals of Applied Biology, 59, 189-205. http://dx.doi.org/10.1111/j.1744-7348.1967.tb04428.x
28. Nderitu, J., Kayumbo, H. and Mueke, J. (1990) Effect of Date of Sowing on Beanfly Infestation of the Bean Crop. Insect Science and Application, 11, 97-101. http://dx.doi.org/10.1017/s1742758400019901
29. Dangl, J.L. and Jones, J.D. (2001) Plant Pathogens and Integrated Defence Responses to Infection. Nature, 411, 826- 833. http://dx.doi.org/10.1038/35081161
30. Cammell, M. and Knight, J. (1992) Effects of Climatic Change on the Population Dynamics of Crop Pests. Advances in Ecological Research, 22, 117-162. http://dx.doi.org/10.1016/S0065-2504(08)60135-X
31. Stoddard, F., Nicholas, A., Rubiales, D., Thomas, J. and Villegas-Fernández, A. (2010) Integrated Pest Management in Faba Bean. Field Crops Research, 115, 308-318. http://dx.doi.org/10.1016/j.fcr.2009.07.002
32. Flint, M.L., Dreistadt, S.H. and Clark, J.K. (1998) Natural Enemies Handbook: The Illustrated Guide to Biological Pest Control. University of California Press, Oakland.
33. Chapman, R., Bernays, E. and Simpson, S. (1981) Attraction and Repulsion of the Aphid, Cavariella

aegopodii, by Plant Odors. Journal of Chemical Ecology, 7, 881-888. http://dx.doi.org/10.1007/BF00992385

34. Dadd, R. (1963) Feeding Behaviour and Nutrition in Grasshoppers and Locusts. Advances in Insect Physiology, 1, 47- 109. http://dx.doi.org/10.1016/S0065-2806(08)60174-5

35. anauta, R.W., Mtei, K.M. and Ndakidemi, P.A. (2015) Potential of Controlling Common Bean Insect Pests (Bean Stem Maggot (Ophiomyia phaseoli), Ootheca (Ootheca bennigseni) and Aphids (Aphis fabae) Using Agronomic, Biological and Botanical Practices in Field. Agricultural Sciences, 6, 489-497. http://dx.doi.org/10.4236/as.2015.65048

36. Copping, L.G. and Menn, J.J. (2000) Biopesticides: A Review of Their Action, Applications and Efficacy. Pest Management Science, 56, 651-676. http://dx.doi.org/10.1002/1526-4998(200008)56:83.0.CO;2-U

37. Alavanja, M.C. (2009) Introduction: Pesticides Use and Exposure, Extensive Worldwide. Reviews on Environmental Health, 24, 303-310. http://dx.doi.org/10.1515/REVEH.2009.24.4.303

38. Williamson, S., Ball, A. and Pretty, J. (2008) Trends in Pesticide Use and Drivers for Safer Pest Management in Four African Countries. Crop Protection, 27, 1327-1334. http://dx.doi.org/10.1016/j.cropro.2008.04.006

39. Nyirenda, S.P., Sileshi, G.W., Belmain, S.R., Kamanula, J.F., Mvumi, B.M., Sola, P., Nyirenda, G.K. and Stevenson, P.C. (2011) Farmers' Ethno-Ecological Knowledge of Vegetable Pests and Pesticidal Plant Use in Malawi and Zambia. African Journal of Agricultural Research, 6, 1525-1537.

40. Kapeya, E., Chirwa, R. and Mviha, P. (2005) Development of an Integrated Pest and Resource Management Package for the Control of Bean Stem Maggot (Ophiomyia spp.) in Malawi. Paper Presented at the PABRA Millennium Workshop, 182.

41. Scaife, A. and Turner, M. (1983) Diagnosis of Mineral Disorders in Plants: Volume 2, Vegetables. Her Majesty's Stationery Office, London.

42. Elserougy, S., Salama, S., Salama, I., Shaaban, F. and Sami, S. (2012) Effects of Obesity and Some Sociodemographic Determinants on Health-Related Quality of Life in School-Aged Children in Greater Cairo. Medical Research Journal, 11, 12-19. http://dx.doi.org/10.1097/01.MJX.0000414712.54546.57

43. Prakash, A., Rao, J. and Nandagopal, V. (2008) Future of Botanical Pesticides in Rice, Wheat, Pulses and Vegetables Pest Management. Journal of Biopesticides, 1, 154-169.

44. Isman, M.B. (2006) Botanical Insecticides, Deterrents, and Repellents in Modern Agriculture and an Increasingly Regulated World. Annual Review of Entomology, 51, 45-66. http://dx.doi.org/10.1146/annurev.ento.51.110104.151146

45. Offor, C., Okaka, A. and Aja, P. (2014) Comparative Analyses on the Effects of Ethanol Leaf-Extracts of Vernonia amygdalina and Azadirachta indica on Lipid Profile in Albino Rats. World Applied Sciences Journal, 32, 61-65.

46. Arannilewa, S., Ekrakene, T. and Akinneye, J. (2006) Laboratory Evaluation of Four Medicinal Plants as Protectants against the Maize Weevil, Sitophilus zeamais (Mots). African Journal of Biotechnology, 5, 2032-2036.

47. Roy, B., Amin, R., Uddin, M., Islam, A., Islam, M. and Halder, B. (2005) Leaf Extracts of Shiyalmutra (Blumea lacera DC.) as Botanical Insecticides against Lesser Grain Borer and Rice Weevil. Journal of Biological Science, 5, 201-204. http://dx.doi.org/10.3923/jbs.2005.201.204

48. Koona, P. and Dorn, S. (2005) Extracts from Tephrosia vogelii for the Protection of Stored Legume Seeds against Damage by Three Bruchid Species. Annals of applied biology, 147, 43-48. http://dx.doi.org/10.1111/j.1744-7348.2005.00006.x

49. Sharma, O.P., Sharma, S., Pattabhi, V., Mahato, S.B. and Sharma, P.D. (2007) A Review of the Hepatotoxic Plant Lantana camara. Critical Reviews in Toxicology, 37, 313-352. http://dx.doi.org/10.1080/10408440601177863

50. Kareru, P., Maina, E.W. and Rotich, Z.K. (2013) Use of Botanicals and Safer Insecticides Designed in Controlling Insects: The African Case. INTECH Open Access Publisher, Winchester.

51. Bhatt, P., Thodsare, N. and Srivastava, R. (2014) Toxicity of Some Bioactive Medicinal Plant Extracts to Asian Army Worm, Spodoptera litura. Journal of Applied and Natural Science, 6, 139-143.

52. Hinson, D. (2000) Rotenone Characterization and Toxicity in Aquatic Systems. Principles of Environmental Toxicity. University of Idaho, Moscow.

53. Khater, H.F. (2012) Prospects of Botanical Biopesticides in Insect Pest Management. Pharmacologia, 3, 641-656. http://dx.doi.org/10.5567/pharmacologia.2012.641.656

54. Aerts, R.J. and Mordue, A.J. (1997) Feeding Deterrence and Toxicity of Neem Triterpenoids. Journal of Chemical Ecology, 23, 2117-2132. http://dx.doi.org/10.1023/B:JOEC.0000006433.14030.04

55. Khalil, M.S. (2013) Abamectin and Azadirachtin as Eco-Friendly Promising Biorational Tools in Integrated Nematodes Management Programs. Journal of Plant Pathology and Microbiology, 1, 1-7. http://dx.doi.org/10.4172/2157-7471.1000174

56. Ambrósio, S.R., Oki, Y., Heleno, V.C.G., Chaves, J.S., Nascimento, P.G.B.D., Lichston, J.E., Constantino, M.G., Varanda, E.M. and Da Costa, F.B. (2008) Constituents of Glandular Trichomes of Tithonia diversifolia: Relationships to Herbivory and Antifeedant Activity. Phytochemistry, 69, 2052-2060. http://dx.doi.org/10.1016/j.phytochem.2008.03.019

57. Chagas-Paula, D.A., de Oliveira, R.B., da Silva, V.C., Gobbo-Neto, L., Gasparoto, T.H., Campanelli, A.P., Faccioli, L.H. and Da Costa, F.B. (2011) Chlorogenic Acids from Tithonia diversifolia Demonstrate Better Anti-Inflammatory Effect Than Indomethacin and Its Sesquiterpene Lactones. Journal of Ethnopharmacology, 136, 355-362. http://dx.doi.org/10.1016/j.jep.2011.04.067

58. Passoni, F.D., Oliveira, R.B., Chagas-Paula, D.A., Gobbo-Neto, L. and Da Costa, F.B. (2013) Repeated-Dose Toxicological Studies of Tithonia diversifolia

(Hemsl.) A. Gray and Identification of the Toxic Compounds. Journal of Ethnopharmacology, 147, 389-394. http://dx.doi.org/10.1016/j.jep.2013.03.024

59. Obafemi, C., Sulaimon, T., Akinpelu, D. and Olugbade, T. (2009) Antimicrobial Activity of Extracts and a Germacranolidetype Sesquiterpene Lactone from Tithonia diversifolia Leaf Extract. African Journal of Biotechnology, 5, 1254- 1258.

60. Ross, I.A. (1999) Medicinal Plants of the World: Chemical Constituents, Traditional and Modern Medicinal Uses. Humana Press, New York.

61. Barceloux, D.G. (2012) Medical Toxicology of Natural Substances: Foods, Fungi, Medicinal Herbs, Plants, and Venomous Animals. John Wiley & Sons, Hoboken.

62. Yuan, Z. and Hu, X.P. (2012) Repellent, Antifeedant, and Toxic Activities of Lantana camara Leaf Extract against Reticulitermes flavipes (Isoptera: Rhinotermitidae). Journal of Economic Entomology, 105, 2115-2121. http://dx.doi.org/10.1603/EC12026

63. Begum, S., Wahab, A., Siddiqui, B.S. and Qamar, F. (2000) Nematicidal Constituents of the Aerial Parts of Lantana camara. Journal of Natural Products, 63, 765-767. http://dx.doi.org/10.1021/np9903548

64. Rajashekar, Y., Raghavendra, A. and Bakthavatsalam, N. (2014) Acetylcholinesterase Inhibition by Biofumigant (Coumaran) from Leaves of Lantana camara in Stored Grain and Household Insect Pests. BioMed Research International, 2014, 1-6. http://dx.doi.org/10.1155/2014/187019

65. Abdullahi, M., Uba, A., Yaro, A., Maxwell, O., Yusuf, A., Kabir, S., Alhasan, A., Umar, A., Bello, S. and Nasir, I. (2015) Phytochemical Screening, Acute Toxicity Study and Evaluation of Antidiabetic Properties of the Methanolic Leaf Extract of Vernonia glaberrima (Asteraceae). Journal Pharmacology, Chemical and Biological Sciences, 3, 169- 177.

66. Erasto, P., Adebola, P., Grierson, D. and Afolayan, A. (2005) An Ethnobotanical Study of Plants Used for the Treatment of Diabetes in the Eastern Cape Province, South Africa. African Journal of Biotechnology, 4, 1454-1460.

67. Erasto, P., Grierson, D. and Afolayan, A. (2006) Bioactive Sesquiterpene Lactones from the Leaves of Vernonia amygdalina. Journal of Ethnopharmacology, 106, 117-120. http://dx.doi.org/10.1016/j.jep.2005.12.016

68. Favi, F., Cantrell, C.L., Mebrahtu, T. and Kraemer, M.E. (2008) Leaf Peltate Glandular Trichomes of Vernonia galamensis ssp. galamensis var. ethiopica Gilbert: Development, Ultrastructure, and Chemical Composition. International Journal of Plant Sciences, 169, 605-614. http://dx.doi.org/10.1086/533598

69. Asawalam, E. and Osondu, H. (2013) Control of Field Insect Pests of Cowpea (Vigna unguiculata L. Walp) in Umudike Nigeria Using Medicinal Plant Extracts. Journal of Sustainable Agriculture and the Environment, 14, 31-37.

70. Ijeh, I. and Obidoa, O. (2004) Effect of Dietary Incorporation of Vernonia amygdalina. del on AFB1 Induced Hepatotoxocity in Weanling Albino Rats. Jamaican Journal of Science and Technology, 15, 32-36.

71. Belmain, S.R., Amoah, B.A., Nyirenda, S.P., Kamanula, J.F. and Stevenson, P.C. (2012) Highly Variable Insect Control Efficacy of Tephrosia vogelii Chemotypes. Journal of Agricultural and Food Chemistry, 60, 10055-10063. http://dx.doi.org/10.1021/jf3032217

72. Mordue, A. and Blackwell, A. (1993) Azadirachtin: An Update. Journal of Insect Physiology, 39, 903-924. http://dx.doi.org/10.1016/0022-1910(93)90001-8

73. Zoubiri, S. and Baaliouamer, A. (2012) Chemical Composition and Insecticidal Properties of Lantana camara L. Leaf Essential Oils from Algeria. Journal of Essential Oil Research, 24, 377-383. http://dx.doi.org/10.1080/10412905.2012.692910

74. Sharma, O., Dawra, R. and Makkar, H. (1987) Isolation and Partial Purification of Lantana (Lantana camara L.) Toxins. Toxicology Letters, 37, 165-172. http://dx.doi.org/10.1016/0378-4274(87)90153-6

75. Sharma, O.P., Vaid, J., Pattabhi, V. and Bhutani, K.K. (1992) Biological Action of Lantadene C, a New Hepatotoxicant from Lantana camara var. aculeata. Journal of Biochemical Toxicology, 7, 73-79. http://dx.doi.org/10.1002/jbt.2570070204

CHAPTER 10

BOTANICAL PESTICIDES

Manas Sarkar and Rajendra Kshirsagar

INTRODUCTION

In ancient culture and in different parts of world, there exist empirical knowledge of the use of plants for pest control for thousands of years (approx. 1500 BC), for example rotenone in east Asia and south America, neem in India, sabadilla in central and south America, and pyrethrin in Persia (Iran). Subsequently, botanical insecticides were introduced in prominent continents like Europe and United states [1]. Botanicals (phytochemicals) are basically secondary metabolites that serve as means of defense mechanism of the plants to withstand the continuous selection pressure from herbivore predators and other environmental factors. Many groups of phytochemicals such as steroids, alkaloids, terpenoids, phenolics and essential oils from more than 2000 plant species have been reported previously for their insecticidal activities. Applications of phytochemicals in mosquito control were in use since 1920, but the discovery of synthetic insecticides such as DDT in 1939 sidetracked the application of phytochemicals in different segments. Synthetic insecticides spread easily and rapidly because it won space on market as products being effective, cheaper, long lasting and easily available. After facing problems due to injudicious and over application of synthetic insecticides in nature, refocus on phytochemicals that are easily biodegradable and have no ill effects on non-target organisms was appreciated. From 1990s, there is renewed interest for use of botanical pesticides due to concerns raised about the rampant supply of synthetic insecticides and its irreversible impact on health and environment; its presence in food web is quite worrying matter for health professionals. At present, phytochemicals make up to 1% of world's pesticide market [2].

Citation: Sarkar M. and Kshirsagar R. (2014). Botanical pesticides. J Biofertil Biopestici 5:e125. DOI:10.4172/2155-6202.1000e125.

CURRENT RESEARCH AND COMMERCIALIZATION CHALLENGES

In a quest of finding solutions for ubiquitous harmful insects through botanicals, there has been considerable research happening in the world. However, very few research projects translate in to distinct product launches. Everybody wants to have quick results for saving cost, time and energy. Hence, most of the academic research is perhaps incomplete because of its design having short term goals. As a result, the outcome iseffected by poorly conducted study, inadequate data, poor analytical work, invalidated data, superficial experimentation having exclusive objective to have a publication. Therefore, funding authorities should give serious thought as to how the research reaches to the end point through systematic research and team efforts of different stake holders. Plant based research in India is no longer piece of cake for anybody in view of Biodiversity legislation, 2002. Anybody plans to work on plants, one has to get permission from State Biodiversity Board (SBB). SBB will hold meetings with local management committees and experts from the field of plant taxonomy, social forestry, zoology, geology, ornithology, etc. and then permission is granted. Then actual collection, processing and screening samples for proposed bioactivity is initiated. Once bioactivity is discovered in a plant, then the commercialization of that plant based product remains to be great challenge. First and foremost priority should be generating proof of concept data by collecting same species from different districts covering different phytogeographical locations, collection in different seasons, collecting species in different stages, collection of different plant parts of same species, processing of the same up to extraction and lastly systematic biological evaluations on concerned pest of interest. Side by side, one also have to rule out the possible external causes of bioactivity (e.g., whether or not the efficacy is due to residual pesticides that may have been sprayed on crops in agricultural land).

Consumption of herbal/Ayurveda medicine is widespread and increasing. The chief source of raw material is harvesting herbs from the wild which is causing loss of genetic diversity and damage to plant habitat. Therefore, domestic cultivation is a viable alternative that offers opportunity to overcome challenges of inconsistency in plant materials and extracts due to factors like wrong identification, phenotypic and genetic variability, extract variability and unstable nature for longer timespan, toxic components and contaminants. For commercialization of any plant based product, raw material sourcing is a real challenge. There are challenges for cultivation and propagation of any wild species. If propagation is by seeds, then it has go through germination, domestication studies, developing package and practices for cultivation, determining pesticidal and manure requirements, re-establishing the bioactivity profile of the plants produced from ex-situ cultivation, addressing factors like change in climate, soil, rainfall, temperature where cultivation is taking place, mobilization of hundreds of hectares of land, testing soil for deciding manure/minerals etc. requirements, assessment of impact of the cultivation on the people residing around those villages, assessment of impact of large scale cultivation on environment vis-à-vis pollen allergies, species (under cultivation) competing with native species, any adverse effect on soil fertility if species under cultivation is having affinity with water and also crop rotation studies. Cultivation of the bioactive plant species will depends on habit of the plant, habitat, size of the seeds, dormancy pattern, seasonality, annual/perennial, secondary metabolites generation, climate, geology, rainfall, endemism, status of the specie (rare, threatened, endangered, critically endangered), rules and regulations for that species germplasm, seed coat pattern, etc. Alternative route can be through controlled environments which will solve problems of cultivation. This can also manipulate phenotypic variation in bioactive compounds and toxins. Conventional plant-breeding programme may improve both agronomic and medicinal traits. There has been significant progress in biotechnological methods like tissue culture and genetic engineering in altering biosynthesis of target molecules.

REVERSE PHARMACOLOGY: A WAY TO BOTANICAL PESTICIDE DISCOVERY

As per Katiyar et al. [3], total number of higher plants species (comprising angiosperms and gymnosperms) is approximately 250,000, out of them only 6% have been screened for biological activity. Initial listing of the candidate species for screening of biological activity is a major task of specific importance in itself. For the purpose, Fabricant and Farnsworth [4], proposed following approaches.

First is random approach in which the plants collected randomly and screened them irrespective of their contents or reports in literature. National Cancer Institute (NCI) of National Institute of Health, USA, thoroughly studied about 35,000 plant species for anticancerous segment investing over two decades (1960- 1980), discovered two molecules viz., paclitaxel and camptothecin.

Second is traditional system of medicine approach; countries like India and China have a rich heritage of well-documented traditional system of medicine. Discovery of artemisinin from *Artemesia alba* for malaria; guggulsterones from *Commiphoramukul* for hyperlipidemia; boswellic acids from *Boswelliaserrata* for anti-inflammatory properties; bacosides from *Bacopamonnieri* as nootropic and memory enhancement; reserpine from *Rauwolfiaserpentina* was based on the leads from these codified systems of medicine prevailing in China and India.

Third is ethnopharmacology approach; this essentially depends on empirical experiences related to the use of botanical drugs for the discovery of NCEs having target specific activity. Whole process comprises observation, description, and in-vitro or in-vivo experimental investigation of indigenous formulations, and is based on chemistry, botany, biochemistry, pharmacology and other different disciplines like archaeology, anthropology, history and linguistics. This approach based on ethnomedicinal usage history that has seen some successes like *Andrographispaniculata* being used for dysentery in ethnomedicine and the compounds responsible for the activity were isolated as andrographolide; Morphine from *Papaversomniferum*Berberine from *Berberisaristata* and *Picroside*from Picrorrhizakurroa are some examples of this approach. This approach seems to be appropriate because, biological, ecological and cultural diversity of any nation leads to generation of empirical and scientific knowledge, the first has its origin in ancient times with practices of observation of the nature and experimentation through trial and error, and inheritance transmitted from generation to generation through test and stories. Botanists have been interested in reviewing corroborated and enriched ethnoflora stating biological significance and the search for new pesticidal phytocompounds useful in protecting and preserving food. Ethnobotany is considered as source for research of phytopesticides of interest to agricultural as well as protection from fatal diseases caused by mosquitoes. Hence, ethnobotanical studies are now recognized to be most viable methods of identifying new medicinal plants. In 2010, Kshirsagar et al. [5], discovered antioxidant and anti-inflammatory compounds from tubers of*Eulophiaochreata*, namely; 9,10-Dihydro-2,5-Dimethoxyphenanthrene-1,7-diol and 5,7-Dimethoxyphenanthrene- 2,6-diol Since time immoral, the tubers have been used by Pawra tribe of Satpuda mountain ranges of Maharashtra for rejuvenating, aphrodisiac and anti-inflammatory properties

However, in India, ethnopharmacological approach for finding biopesticidal botanicals has been considerably underexplored. In 1991, SK Jain [6] published dictionary of Indian Folk Medicine and Ethnobotany based on compilation of research of preceding 30 years. The dictionary reports just 4 plants as biopesticides out of 2532 species which clearly indicates that this field remains significantly unexplored. In last two decades only few research papers throw some light on this topic. For example, Phani Kumar and AlkaChaturvedi [7] tell about 2 plants for insecticidal property. Bhardwaj et al. [8], deals with various plants are being used as ethnomedicines against insects and worms in Aravalli hill range of India. Senthilkumar et al. [9] deals with insecticidal properties of certain flora based on ethnobotanical records against Teak defoliator. Anjali Mathur and Hema Joshi [10] mention about 8 plants being insecticidal. Considering the gap in ethnobotanical knowledge documentation and its systematic biopesticidal exploration by using modern techniques and sophisticated instruments, the efforts should be made to bridge the gap by initiating aggressive studies, since, India harbors around 400 different tribes amounting to 7.5% population of whole country.

REFERENCES

1. Elisa GPL, Laura P, Graciela BZ, Víctor M, VelázquezH (2012) In:Current Status: Mexican Medicinal Plants with Insecticidal Potential, Bioactive Compounds in Phytomedicine, Prof. IrajRasooli (Ed.)
2. Anupam G, Nandita C, Goutam C (2012) Plant extracts as potential mosquito larvicides. Indian J Med Res 135: 581-598.
3. Chandrakant, Arun G, Kanjilal, Shefali K (2012) Drug discovery from plant resources: An integrated approach. An International Quarterly Journal of Research in Ayurveda 33: 10–19.
4. Fabricant DS, Farnsworth NR (2001) The value of plants used in traditional medicine for drug discovery. Environ Health Perspect 109 (Suppl 1): 69–75.
5. Rajendra K D, YogeshK B, Suresh J D, Shakti U N, Rajesh R, et al.(2010) Phenanthrenes of EulophiaochreataLindl. International Journal of Green Pharmacy: July-September; 147-152.
6. Jain SK (1991) Dictionary of Indian Folk Medicine and Ethnobotany. Deep Publications, New Dehli.
7. PhaniPG, Alka C (2010) Ethnobotanical Observations of Euphorbiaceae Species from Vidarbha region, Maharashtra, India. Ethnobotanical Leaflets 14: 674-680.
8. Bhardwaj M, Bharadwaj L, Trigunayat K, Trigunayat MM (2011) Insecticidal and wormicidal plants from Aravalli hill range of India. J Ethno pharmacol 136:103-110.
9. Senthilkumar N, Murugesan S, Vijayalakshmi KB, Monisha M, Suresh Babu D, et al. (2012) Insecticidal Properties of Certain Flora based on Ethnobotanical Records Against Teak defoliator, H. puera Cramer (Lepidoptera: Hybaeidae). European Journal of Experimental Biology 2:513-519.
10. Mathur A and Joshi H (2013) Ethnobotanical Studies of the Tarai Region of Kumaun, Uttarakhand, India. Ethnobotany Research & Applications 11:175-203.

CHAPTER 11

EXTRACTION AND SEPARATION OF ACTIVE INGREDIENTS IN SCHISANDRA CHINENSIS (TURCZ.) BAILL AND THE STUDY OF THEIR ANTIFUNGAL EFFECTS

Haijing Yi, Yan Chen, Jun Liu, Jie Zhang, Wei Guo, Weilie Xiao, and Yuncong Yao

INTRODUCTION

Bioactive substances derived from plants play important roles in environmental improvement, food production, food security, and biodiversity conservation. Increasing attention is being paid to the development and application of botanical pesticides, such as botanical insecticides, fungicides, antiseptic agents, and herbicides. An important reason for this interest is that botanical pesticides are accepted as being more environmentally friendly than synthetic organic pesticides with high residues, which can cause environmental pollution, threats to human health, and other issues [1–5].

Over the years, research regarding the application of botanical pesticides has shown that such substances are advantageous because of their high selectivity, low toxicity to humans, and biodegradability. Today they are one of the most important potential sources for the creation of new pesticides [6–9]. Several studies have shown that plant extracts contain active antibacterial ingredients, many of which have been extracted, developed, and applied as botanical fungicides [10–13]. However, identifying which species in the voluminous plant kingdom contain effective concentrations of active substances, and which of the ingredients in the 400,000 known secondary metabolites derived from plants have insecticidal activity [14], has been and will continue to be a puzzle to be solved.

In the USA, there are currently more than 245 registered biopesticide-active ingredients used in hundreds of products. In Korea, 33 biopesticides are registered, of which 19 are fungicides, 13 are insecticides, and 1 is an herbicide [15]. Industrial versions of ingredients, such as azadirachtin, matrine, rotenone, nicotine, celangulin, veratramine, ethylicin, physcion, aconitine, and eucalyptol have become the backbone of the botanical pesticide industry [10]. These substances are principally applied for their insecticidal effects. They eliminate pests through mechanisms such as acute poisoning, shutdown of the digestive system, hunger inhibition, sterilization, growth inhibition, and by acting as an attractant or repellent for the pest species. Some of these mechanisms are also active in the antiseptic and antibacterial processes of the botanical pesticides. The study of the effectiveness of botanical fungicides against crop disease remains at an early stage, and hence only a few marketable botanical fungicides have been developed thus far. To develop commercially competitive botanical fungicides, additional basic research into active antimicrobial substances is urgently needed. Thus, the immediate problem to be addressed is how to extract active ingredients from plants, develop new botanical fungicides, and improve plant disease control.

Family Schisandraceae, in the sub-class Magnoliidae, are perennial deciduous vines[16]. Plants of this family have been commonly used as medicinal and ornamental landscape plants. The family, comprising approximately 60 species, includes two genera, *Schisandra*Michx and *Kadsura* Kaempf Ex Juss. They are distributed discontinuously in Southeast Asia and North America [17]. In China, they are the most abundant plant resources and *S.chinensis* (Turcz.) Baill is a common species in northeastern China. The fruits, seeds, shoots, and leaves of *S. chinensis* (Turcz.) Baill have been used in traditional Chinese medicine since ancient times and are valued for their various curative properties [18]. The use of *S. chinensis* as a high-grade herbal medicine was first recorded in the ancient pharmaceutical book, "The Divine Husbandman's

Citation: Yi H, Chen Y, Liu J, Zhang J, Guo W, Xiao W, et al. (2016) Extraction and Separation of Active Ingredients in Schisandra chinensis (Turcz.) Baill and the Study of their Antifungal Effects. PLoS ONE 11(5): e0154731. DOI:10.1371/journal.pone.0154731.

Herbal Foundation Canon (Shén Nóng Běn Cǎo Jīng)" [19]. Recently, *S. chinensis* fruit extract (SE) and its ingredients have gained substantial attention. They may play a potentially important role in the treatment of cardiovascular diseases, such as hypertension and myocardial infarction [20, 21]. A water decoction of *S. chinensis* is believed to act as an astringent for the lungs and kidneys. It is effective in the treatment of diarrhea, arresting excessive sweating, calming the spirit by refreshing the heart and kidneys, generating body fluid, and reducing thirst [22, 23]. Pharmacological studies on animals have shown that *S. chinensis* increases immunity and affords a stress-protective effect against a broad spectrum of harmful factors, including heat shock, sunburn, hypothermia, frostbite, aseptic inflammation, irradiation, and heavy metal intoxication [24, 25]. Previous studies have also proven that *Schisandra* contains numerous special kinds of lignans and organic acids. These include Schizandrin, deoxyschizandrin, γ-Schizandrin, isoschizandrin, Schizandrol A and B, schisantherin A-E, citricacid, malicacid, and tartaricacid, substances that could play a key role in the treatment of fatty liver, hepatitis virus infection, and chemical-induced hepatitis [26, 27]. They also help to relieve stress and anxiety [28–30], and increase insulin sensitivity [31].

However, current research on the *Schisandra* genus has primarily focused on species differences, medicinal ingredients, and pharmacological effects [32–35]. Less research has focused on how *S. chinensis* can be applied in agriculture to inhibit plant pathogens or its potential for the development of botanical fungicides. In this study, a new method of extracting active substances from *S. chinensis* is proposed and the effects of these extracts on *Botryosphaeria dothidea*, *Monilinia fructicola*, and *Fusarium oxysporum* are tested, as these are causal agents of three important diseases (apple ring rot, peach brown rot, and*Fusarium* wilt of strawberry, respectively) [36–38]. The tests included the detection of the effects of the extract on mycelial growth, spore germination, and disease infection. Moreover, active ingredients in the *S. chinensis* extracts that inhibit the three plant pathogens were determined, and the mechanism of inhibition employed by the pathogenic fungi was studied.

Therefore, this study is meaningful because it produced alternatives to chemical pesticides and thereby could reduce the negative impacts of chemical pesticide use. It highlights the potential for development of botanical pesticides from *S. chinensis* compounds.

MATERIALS AND METHODS

Plant and Microbes Materials

The experimental material was annual branches and leaves of *S. chinensis* collected from Hong You *Schisandra chinensis* Farm (124.8°E; 40.8°N) (http://www.hywwz.cn; Kuandian County, Liaoning Province, China). Samples of *Botryosphaeria dothidea*, *Monilinia fructicola*, and *Fusarium oxysporum* were obtained from the Plant Pathology Laboratory of Beijing University of Agriculture. They were stored in a refrigerator at a constant temperature of 4°C. Apple fruits (*Malus domestica* 'Fuji'/*Malus prunifolia* (Willd.) Borkh) were obtained from the Chang Ping District (116.23°E, 40.22°N), Beijing, China. Peach fruits (*Prunus persica*) were obtained from Qingzhou City (118.4° E, 36.4°N) in Shandong, China.

Ethics Statement:The study was carried out on private land called the Hong You *Schisandra chinensis* Farm (http://www.hywwz.cn). We confirm that the owner of the land, Guangming Lin, gave permission for his site to be sampled. No specific permissions were required for locations and field studies, as our study did not involve endangered or protected species.

Experimental Design

Extraction and Separation of Active Substances from S. Chinensis

To completely remove moisture, the annual branches and leaves of *S. chinensis* were dried at 28°C for 10 days. Ten kilograms of air-dried materials were weighed and crushed into powder with a grinder (Huangcheng HC-1500Y2, Yongkang, China;http://shop1396630696864.cn.china.cn/supply/). Next, the powdered materials was mixed with 70% acetone at a powder:solution ratio of 1:2 at room temperature in a 70-L stainless steel container (http://www.jmxwm.com/xwm/en/index.asp; Xinweiming Stainless Steel Products Factory, Jiangmen, China). After 24 h, the solution was transferred to another container and an additional 70% acetone was added. This procedure was repeated three times, after which the soaking solution was placed in a rotary evaporator to distill at 40°C and reduced pressure (Büchi Rotavapor R-220, Essen, Germany).

The remaining water solution contained all the active substances extracted from *S.chinensis*. To filter impurities, the water solution was allowed to stand overnight and then filtered using a circulating water vacuum pump (http://www.gyyuhua.com; Gongyi City Yuhua Instrument Co., Gongyi, China). The filtering process was repeated three times. The filtrate was then extracted three times with pure ethyl acetate with a filtrate:ethyl acetate ratio of 1:2. The organic layer was separated from the aqueous layer and was concentrated by evaporating the organic solvent *in vacuo* at 40°C in a rotary vacuum evaporator (Büchi Rotavapor R-124, Essen, Germany). This process produced the primary crude extract, which was further separated along a gradient according to polarity and extraction rate. The primary crude extract was then accurately calculated. The primary extract totaled 240 g and all samples were stored in air-tight brown bottles at 4°C in a refrigerator for the duration of the subsequent experiments.

A portion of the extract (100 g) was subjected to separation using silica gel (200–300 mesh) column chromatography (CC) and sequentially eluted with a series of chloroform-acetone-methanol mixtures (1:0:0, 10:1:0, 9:1:0, 2:1:0 1:1:0, 0:1:0, and 0:0:1) to obtains seven fractions (from SE-I to SE-VII) (Table 1). For set SE-I, for example, the chromatographic column was eluted with chloroform at 20 mL min^{-1} for 2 h, and all the eluent was concentrated in the vacuum distillation unit (http://www.gyyuhua.com; Gongyi City Yuhua Instrument Co., Gongyi, China).

Table 1. Method of separating active substances extracted from *S. chinensis*.

S. chinensis extract	Organic solvent ratio	Elution time (h)	Velocity of flow(mL min^{-1})
SE-I	Chloroform	2	20
SE-II	Chloroform:Acetone = 10:1	2	20
SE-III	Chloroform:Acetone = 9:1	2.5	20
SE-IV	Chloroform:Acetone = 2:1	1.5	20
SE-V	Chloroform:Acetone = 1:1	1	20
SE-VI	Acetone	0.8	20
SE-VII	Methanol	0.8	20

doi:10.1371/journal.pone.0154731.t001

The elution time and velocity of flow to obtain all the factions are listed in Table 1. The seven fractions were concentrated *in vacuo* at 35°C in the vacuum rotary evaporator (Büchi Rotavapor R-124, Essen, Germany). The yield of the seven fractions from SE-Ito SE-VII was 5.1 g, 6.6 g, 8.2 g, 7.1 g, 10.2 g, 6.9 g, and 13.1 g, respectively.

Preparation of Spore Suspensions of the Three Fungi Strains

The three test strains were inoculated on potato dextrose agar medium (PDA) and cultivated in a biochemical incubator (Fuma SPX-160B-2, Shanghai, China) at 28°C with alternating light and dark every 12 h for 4 d. Following this, sterile water was continually added to the medium to wash off and collect the conidia of the fungi. To remove fragments of mycelium and medium, the solution was filtered through gauze and condensed in a centrifuge (Sigma 3K-15, Osterode am Harz, Germany) at 1,000 r min^{-1} for 1 min. The concentration of spores was then determined under a microscope (Leica DM750, Wetzlar, Germany) and adjusted to 1 × 103 spores mL-1.

Analysis of the Mycelial Growth Rate of the Three Pathogenic Fungi

Fractions from SE-Ito SE-VII (1 g) were dissolved in 1 mL dimethyl sulfoxide (DMSO) and the mother liquor (1g mL^{-1}) for each fraction was obtained. Sterile liquid PDA (3 mL) was placed in each well of the six-well culture plate and different volumes (1, 10, or 20 µL) of mother liquor of extracts were added. The final concentrations of the medium-containing extracts were 0.3, 3, and 6 mg mL^{-1}, respectively. Control wells contained the same volumes of dimethyl sulfoxide (DMSO). All concentrations were run in triplicate. All solutions were allowed to solidify for 4 h and a 10-µL suspension of different pathogen spores (1× 10^3spores mL^{-1}) was added to the center of each SE-containing medium. After 3 h, the water in the spore suspension had completely evaporated and all six wells in the culture plates were placed in an incubator (Fuma SPX-160B-2, Shanghai, China) at 28°C for 48 h. The colony diameter was measured three times by the crossing method, and inhibition rates were calculated for each pathogens and extracts using the following equation1:

$$E_1 = \frac{CC - CT}{CC} * 100\%$$

E_1, The inhibition rate of mycelial growth

CC, Colony diameter of control group

CT, Colony diameter of treatment group

Observing Mycelial Morphology with Scanning Electron Microscopy (SEM)

Scanning electron microscopy (SEM) was performed to determine morphological changes in the mycelia on the treated media. First, the growing mycelia (3mm × 5mm) on medium treated with SE-III were sampled with a double-sided knife, and samples for SEM were fixed in 2.5% glutaraldehyde stationary liquid for 12 h. Next, samples were irrigation three times with 0.1 mol L^{-1} phosphate buffer. The samples were dehydrated with a gradient concentration of alcohol systematically at 30%, 50%, 70%, 80%, 90%, 95%, and 100% with 15 min at each gradient. Anhydrous acetone was then applied three times for dehydration and each bout lasted 15 min. A critical point drying apparatus (EMS 850, Pennsylvania, USA) was used to dry the sample completely. The dried sample was then pasted onto the sample stage to be sputtered with gold (Hitachi ion sputter E-1010, Tokyo, Japan) and observed via SEM (Tescan VEGA TS 5136XM, Brno, Czech) under 2,500× magnification.

Analysis of the Spore Generation Rate of the Pathogenic Fungi

The method for preparation of the mother liquor (1 g mL-1) of every fraction was the same as described above. Spore suspensions (0.5 mL) of the three fungi (*B. dothidea*, *M.fructicola*, and *F. oxysporum*) were mixed with 0.1 mL of 1 g mL^{-1} SE-I, SE-II, and SE-III, respectively, and the mixtures were then cultured on a concave slide to retain moisture at 28°C for 24–48 h. The same volume of DMSO was mixed with spore suspensions as a control. Each treatment was replicated three times. Spore germination was observed under a microscope (40×, Leica-DM750, Wetzlar, Germany) after 24 h and 48 h. Inhibition rate of spore germination was calculated by the following equation2:

$$E_2 = \frac{TC - NT}{TC} * 100\%$$

E_2, The inhibition rate of spore germination

TC, Total number of checked spores

NG, Number of germinated spores

The minimum number of spores examined was 200. Germination was defined as occurring when the germ tube of the conidia was longer than the short radius of the spore.

Analysis of the Infection Status of the Fruit Pericarp by Pathogenic Fungi

The extracts SE-I, SE-II, and SE-III in an amount of 5 µL (0.1 g mL^{-1}) were sprayed evenly on the pericarp of apple and peach fruits in a 3-cm diameter circle. The control group was treated with a corresponding volume of DMSO and all sprayed fruits were placed in a biochemical incubator (Fuma SPX-160B-2, Shanghai, China) at 28°C. Each treatment was replicated a total of 12 times for each of the four processing periods (0 h, 24 h, 48 h, and 72 h). After the processing period elapsed, a 0.1-cm sample of the apple pericarp was taken from the sprayed area and all samples were frozen in-80°C liquid nitrogen for the next analysis of ROS, SOD, CAT, and MDA.

Pathogenic fungi was activated and cultivated for 4 d and 0.5-cm diameter colonies were transplanted by pasting them on the center of the sprayed area. This process was repeated for pathogenic fungi corresponding to extracts of SE-I, SE-II, and SE-III, as described above. The apples and peaches were wrapped with cling film to retain moisture and then placed in a biochemical incubator (Fuma SPX-160B-2, Shanghai, China) at 28°C for 96 h. Observations took place 3 d later. To test the protective effect of the SEs in preventing *M.fructicola* and *B. dothidea* from infecting peach and apple fruits, the average diameters of lesions were measured.

Antioxidant Analysis

Measurement of Intracellular Reactive Oxygen Species (ROS)

A Reactive Oxygen Species Assay Kit (Nanjing Jiancheng Bioengineering Institute, Nanjing, China) was used to measure the level of intracellular reactive oxygen species (ROS) following the manufacturer's instructions. The probe 2',7'-dichlorofluorescin diacetate (DCFH-DA) is oxidized by ROS to 2′, 7′-dichlorofluorescein (DCF), which is highly fluorescent at 530 nm. Briefly, a homogenate of apple pericarp was mixed with 1 mol L^{-1} DCFH-DA (V/V = 19:1). The relative levels of fluorescence were quantified by a multi-detection microplate reader with 485 nm excitation and 535 nm emission (Thermo Scientific, Varioskan Flash).

Thiobarbituric Acid (TBA) Test for Lipid Peroxidation

A Malondialdehyde (MDA) Assay Kit (Nanjing Jiancheng Bioengineering Institute) was used to measure the level of lipid peroxidation following the manufacturer's instructions. MDA, the product of lipid peroxide degradation, can condense with thiobarbituric acid (TBA) to form red substances that are highly fluorescent at 532 nm.

Measurement of Superoxide Dismutase (SOD)

A Superoxide Dismutase (SOD) Assay Kit (Nanjing Jiancheng Bioengineering Institute) was used to measure the level of intracellular ROS following the manufacturer's instructions.

Measurement of Catalase (CAT)

A Measurement of Catalase (CAT) Assay Kit (Nanjing Jiancheng Bioengineering Institute) was applied to measure the level of intracellular ROS following the manufacturer's instructions.

Total RNA Extraction and Real-time PCR Analysis

Total RNA was extracted from the pericarp of apples treated by the SEs for 0 h, 24 h, 48 h, and 72 h using a RN42-EASYspin Plus Plant Total RNA Kit (Aidlab, Beijing, China) according to the manufacturer's instructions and then treated with DNase I (TaKaRa, Tokyo, Japan) to remove the residual DNA. The total RNA was then subjected to a reverse transcription reaction to generate first-strand cDNA using oligo (dT) primers (TaKaRa, Tokyo, Japan).

Expression analysis of genes was conducted by real-time PCR with SYBR Green qPCR Mix (TaKaRa, Tokyo, Japan) and a Bio-Rad CFX96 Real-Time PCR System (BIO-RAD, Hercules, USA). Actin (GenBank accession number: DQ341382) was selected as a ribosomal reference gene. Sequences of genes *MdPR4*, *MdAFS*, *MdPPO*, and *MdPAL* special primers used in this study are listed in S1 Table. The PCR mixture (20 μL total volume) for the reaction contained 10 μL SYBR Green Mix, 2 μL cDNA (diluted 10 times), 1 μL of each primer (10 μM), and 6 μL RNase-free water. Samples were amplified for 40 cycles at 95°C for 15 s and at 59°C for 20 s, and the reactions were all run with three biological replicates and two technical replicates. The real-time PCR data were calculated using the 2^ (-ΔΔCt) analysis method with error bars representing the standard deviation (SD).

HPLC-ESI (±)-MS2 Analysis and Identification of Lignans

High-Performance Liquid Chromatography-ElectroSpray Ionization-Mass Spectrometry (HPLC-ESI-MS) analyses of the lignans were performed using an Agilent-1100 HPLC system equipped with a UV detector coupled to a LC-MSD Trap VL ion-trap mass spectrometer with an ESI source (Agilent Technologies, Palo Alto, CA, USA). The HPLC separation conditions were identical to those used for HPLC-DAD analysis. Negative-ion (PI) mode was used for the MS analysis of lignans.

ESI was performed using the following conditions: capillary voltage, 5.0 kV; nebulization pressure, 241.3 kPa; gas (N^2) temperature, 350°C; and flow rate, 8.0 L min^{-1}. The capillary offset and exit voltage were 68.4 V and 100.6 V, respectively. The skim 1 voltage was 29.5 V, the skim 2 voltage was 6.0 V, and the MS and MS^2 spectra were recorded over an m/z range of 150 to 2000 NI. The MS^2 scan of the target ions used normalized collision energy of 20%–45%. The experimental standards of Schizandrol A, Schizandrol B, Schizandrin B, Schizandrin A, and isoschizandrin were purchased from Sigma-Aldrich (Steinheim, Germany); data obtained for experimental standards and information regarding the identified compounds were combined to identify the lignans.

Data Analysis

Three replicates were performed for each experiment. Data are presented as the mean ± standard deviation (SD). Experimental data processing and analysis were completed using SPSS 20.0 (SPSS Inc., Chicago IL, USA). Analysis of variance of all values was used to assess differences in the means among samples ($P < 0.05$). The virulence regression equation was also calculated using SPSS 20.0. Inhibition ratio was converted to the probability value (y) and concentration of the SEs was converted to logarithm (x). A virulence regression equation is then calculated as y = a + bx, where "a" is the intercept and "b" is slope. According to the equation, the EC_{50} value was calculated. Graphs were prepared in OriginPro 8.0 software (Origin Lab Corporation, Northampton, MA, USA).

RESULTS

The Inhibiting Effect of SE-I to SE-VII on the Mycelial Growth of Three Pathogenic Fungi

In the study, according to gradient elution, seven fractions were obtained from crude *S.chinensis* extract and were named SE-I, SE-II, SE-III, SE-IV, SE-V, SE-VI, and SE-VII (SE series, SEs) according to polarity sequence.

In order to investigate the inhibiting effect of SE-I to SE-VII on the mycelial growth of *M.fructicola*, *F. oxysporum*, and *B. dothidea*, antifungal tests with different concentrations (0.3 mg L^{-1}, 3 mg L^{-1}, and 6 mg L^{-1}) of the SE series using the plate cultivation method were conducted. The results indicated that SE-I to SE-III had inhibitive effects against *M.fructicola*, *F. oxysporum*, and *B. dothidea*, but their activities varied among the different pathogenic fungi, with the inhibitive ability of SE-III being best of the three candidate fractions based on their EC_{50} values. However, SE-IV to SE-VII had no critical impact on the growth of pathogenic fungi (Table 2).

Table 2. Percentage of inhibitive effect for seven SEs on the mycelial growth of three pathogenic fungi.

Schisandra extract	Strains	Concentration			Toxicity regression equation	r	EC_{50}(mg/mL)	EC_{50} ($\bar{x}$ ± SD /(mg/mL)
		0.3mg mL^{-1}	3 mg mL^{-1}	6mg mL^{-1}				
	M. fructicola	23.21%	47.27%	50.94%	y = 4.59 + 0.60x	0.99	4.96	
SE-I	*F. oxysporum*	26.90%	44.44%	51.39	y = 4.65 + 0.48x	1.00	5.54	4.67 ± 1.05 (**P = 0.01**)
	B. dothidea	29.25%	39.63%	63.55%	y = 4.70 + 0.59x	0.87	3.51	
	M. fructicola	25.00%	41.82%	58.49%	y = 4.62 + 0.67x	0.99	3.82	
SE-II	*F. oxysporum*	21.72%	61.81%	66.67%	y = 4.48 + 0.97x	0.99	1.91	2.95 ± 0.97 (**P = 0.04**)
	B. dothidea	23.58%	52.99%	54.19%	y = 4.67 + 0.71x	1.00	3.13	
	M. fructicola	55.36%	70.91%	77.36%	y = 5.36 + 0.46x	1.00	0.18	
SE-III	*F. oxysporum*	47.93%	65.28%	77.78%	y = 5.27 + 0.52x	0.95	0.31	0.34 ± 0.18 (**P = 0.08**)
	B. dothidea	37.26%	65.44%	100%	y = 5.92 + 3.27x	0.78	0.53	
	M. fructicola	—	—	—	y = 0.61 + 1.78x	0.66	305.02	
SE-IV	*F. oxysporum*	—	—	—	y = 3.24 + 0.10x	0.98	—	—
	B. dothidea	—	—	—	—	—	—	
	M. fructicola	—	—	—	y = 3.26 + 0.24x	0.73	—	
SE-V	*F. oxysporum*	—	—	—	—	—	—	—
	B. dothidea	—	—	—	—	—	—	
	M. fructicola	—	—	—	y = 1.48 + 2.78x	1	19.75	
SE-VI	*F. oxysporum*	—	—	—	—	—	—	—
	B. dothidea	—	—	—	—	—	—	
	M. fructicola	—	—	—	—	—	—	
SE-VII	*F. oxysporum*	—	—	—	—	—	—	—
	B. dothidea	—	—	—	—	—	—	

doi:10.1371/journal.pone.0154731.t002

For the SE-I, the EC_{50} values were 4.96, 5.54, and 3.51 mg mL^{-1}, respectively, for *M.fructicola*, *F. oxysporum*, and *B. dothidea* (Table 2), showing that the inhibitive effect of SE-I was best on *B. dothidea*. For SE-III, the EC_{50} values were 0.18, 0.31, and 0.53 mg mL^{-1}, respectively, and the colony diameters of the three pathogenic fungi cultured on the SE-III medium were the smallest (Fig 1: C (P = 0.001), G (P = 0.002), K (P = 0.022), D, H, L).

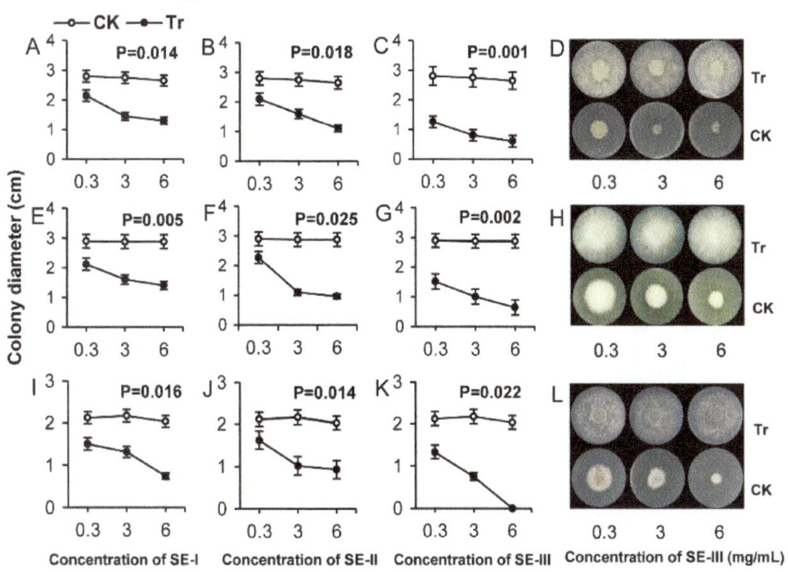

Figure 1. Inhibiting effects of *S. chinensis* extracts SE-I, SE-II, and SE-III on mycelial growth of pathogenic fungi. Inhibiting effects of *S. chinensis* extracts SE-I (A, E, and I), SE-II (B, F, and J), and SE-III (C, G, and K) on colony diameter of *M. fructicola* (A, B, C, and D), *F.*

oxysporum (E, F, G, and H), and *B. dothidea* (I, J, K, and L). The data represent the mean ± SD and P values indicate a significant difference at $P < 0.05$. D, H, and L corresponded to C, G, and K, respectively.

The inhibition rates for SE-I, SE-II, and SE-III at three different treatment concentrations (0.3, 3, and 6 mg mL^{-1}) were significantly different and inhibition rate increased with larger concentration of SEs. Thus, the higher the concentration of SE-I, SE-II, and SE-III in the medium, the higher the inhibition rate impacting the growth of the pathogenic fungi. The inhibition rate of SE-III at the three treatment concentrations was higher than that of SE-I and SE-II, and reached 100% when the treatment concentration was 6 mg mL^{-1} on *B. dothidea*(Fig 1K and 1L).

The Effect of SE-III on the Mycelial Morphology of the Three Tested Pathogenic Fungi

When the treatment concentration of SE-III was 6 mg mL^{-1}, the physiological condition of the three tested pathogenic fungi changed. Under the electron microscope (Fig 2), hyphae on the control medium were smooth and plump, having normal morphology (Fig 2A, 2B and 2C), whereas the hyphae from the SE-III-treated medium was shriveled and dehydrated, indicating that the mycelia of the pathogenic fungi presented abnormal morphological changes, such as deformation, rupture, and dissolution of the hyphae (Fig 2D, 2E and 2F). Based these results, there is direct evidence that the SEs displayed broad inhibition activity against pathogenic fungi infecting plants.

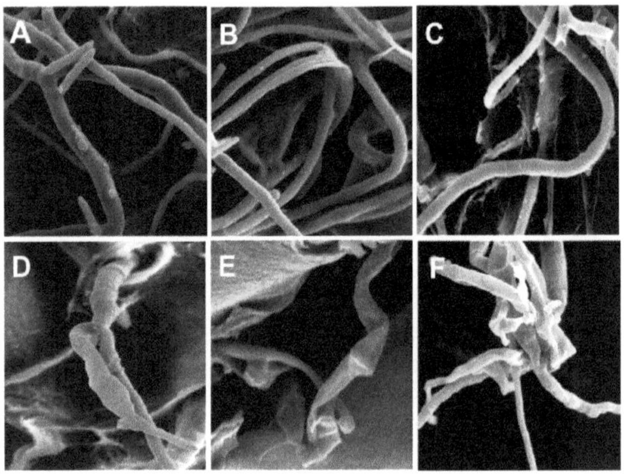

Figure 2. Mycelial morphology of the three tested pathogenic fungi under the electron microscope. Mycelial morphology of the three tested pathogenic fungi from SE-III-treated medium under the electron microscope. A, B, and C represent the normal condition of *M.fructicola*, *F. oxysporum*, and *B. dothidea*, respectively. D, E, and F represent the hypha of *M. fructicola* (D), *F. oxysporum* (E), and *B. dothidea* (F) from the treated medium.

The Inhibiting Effect of SEs on the Spore Germination and Infection of Pathogenic Fungi

To further investigate the inhibitive effect of the SEs on the pathogenic fungi, spore germination rates of the three pathogenic fungi were calculated after the spores were treated with SEs for 24 h and 48 h. The results indicated that the SE-III treatment significantly suppressed the germination rate of the three kinds of spores after inoculation for 24 h and 48 h, and the inhibition rate reached 90% for the germination of all three kinds of spores (Fig 3C). Extract SE-I only inhibited spore germination of *B. dothidea* by more than 90%, whereas its inhibitive effects on spore germination of *M. fructicola* and *F. oxysporum*were not apparent (Fig 3A). The average inhibition rate of SE-II on spore germination of the three kinds of pathogenic fungi was higher than that of SE-I, but lower than that of SE-III during both treatment periods (Fig 3B). This indicated that SEs might prevent fruit from damage caused by pathogenic fungi through inhibiting spore generation of the fungi.

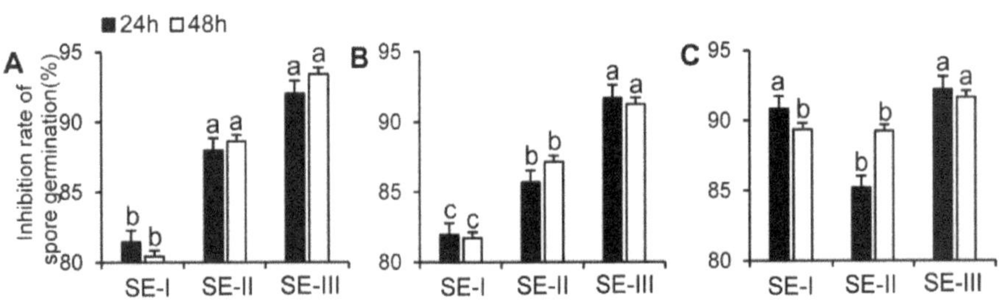

Figure 3. Inhibiting effect of *S. chinensis* extracts SE-I, SE-II, and SE-III on the spore germination of pathogenic fungi. Inhibiting effect of *S. chinensis* extracts SE-I, SE-II, and SE-III on the spore germination of *M. fructicola* (A), *F. oxysporum* (B) and *B. dothidea* (C). The black bar represents spores cultivated in the SE-containing medium for 24 h and the white for 48 h. The data represent the mean ± SD and letters indicate a significant difference ($P < 0.05$).

The Preventive Effect of SEs on Fruit Diseases

The SEs were sprayed on the pericarp of apple fruits and the fruits treated with SEs tended to be more resistant to *B. dothidea* infection. The results indicated that the diameter of the disease spots on apples treated with SEs was smaller than those on the controls (Fig 4A and 4C). Disease spots on apples without treatment with the SEs expanded to more than 2 cm. However, the disease spots on the SE-treated apples were significantly restricted with withered fungal hyphae.

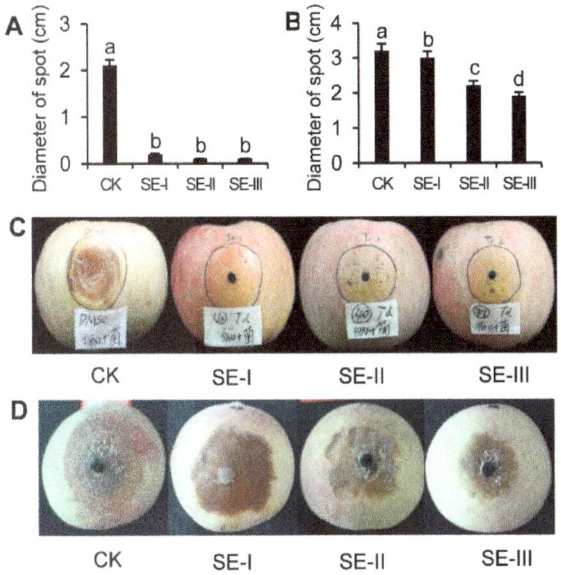

Figure 4. The preventative effect of *S. chinensis* extracts SE-I, SE-II, and SE-III on apple ring rot and peach brown rot. The preventive effect of *S. chinensis* extracts SE-I, SE-II, and SE-III on apple ring rot (C) and peach brown rot (D). A and B, respectively, correspond to C and D indicating the scab diameter with treatment with *S. chinensis* extracts. The data represent the mean ± SD and letters indicate a significant difference ($P < 0.05$).

The SEs were sprayed on the pericarp of peach fruits, and the fruits treated with SEs tended to be more resistant to *M. fructicola* infection. Compared with the control disease spot diameter (3.2 cm), peaches treated with SEs displayed smaller disease spot diameters, which were 3 cm, 2.2 cm, and 1.9 cm for SE-I, SE-II, and SE-III, respectively, to pathogen infection (Fig 4B and 4D). Results showed that among the three treatments, SE-III performed best, with the size of disease spots on the peach fruits treated with SE-III only 50% of that on the controls. These results indicated that SEs have substantial potential for direct application to prevent fruits from infection by pathogenic fungi.

Analysis of Compounds Derived from Fractions of *S. Chinensis* Extract by LC-UV-ESI-MS^2

In the study, SE-I, SE-II, and SE-III exhibited outstanding performance among the seven fractions in the preliminary toxicity experiment. Furthermore, the HPLC-ESI (±)-MS^2 was employed to identify the compounds included in these three fractions (SE-I, SE-II, and SE-III) by standards.

From the results shown in Table 3, Schisanhenol B, Schizandrin A, Schizarin D, Schizandrin B, and isomeride Schizandrin B were found in the SE-I fraction. Schisanhenol B, Gomisin L, Schizandrin A, and Schizandrin B were in the SE-II fraction, and Schizarin D, Schizandrol A, Schizandrol B, Isoschizandrin, Schisanlignone A, Kadsulignan M, and isomeride Schizandrol B were in SE-III.

Table 3. Main compound types identified by LC-UV-ESI-MS^2 in stems and leaves of *Schisandra chinensis*.

NO.	MW	$[M+H]^{-a}$	$MS^2[M-H]-(m/z)^a$	Putative identifier	SE-I	SE-II	SE-III
1	386	387	[387]	Schisanhenol B	√	√	ND
2	386	387	[387]:409	Gomisin L	ND	√	ND
3	416	417	[417]:439	Schizandrin A	√	√	ND
4	484	485	[387]:507, 467, 436	Schizarin D	√	ND	√
5	400	401	[401]:439, 386	Schizandrin B	√	ND	ND
6	400	401	[401]:423, 386, 371	Schizandrin B	√	√	ND
7	432	433	[433]:415, 400, 384	Schizandrol A	ND	ND	√
8	416	417	[417]:399, 369	Schizandrol B	ND	ND	√
9	432	433	[433]:455, 415	Isoschizandrin	ND	ND	√
10	430	431	[431]:399, 387	Schisanlignone A	ND	ND	√
11	430	431	[431]:453, 387, 413	Kadsulignan M	ND	ND	√
12	416	417	[417]:369, 399, 384	Schizandrol B	ND	ND	√

ND, none detected.

√, detected.

[a] obtained by ion trap mass spectrometry.

doi:10.1371/journal.pone.0154731.t003

Differences in Antioxidant Content and Profile of the Related Genes Expression between SEs Treated Fruits and Control Fruits

The results showed that in the apples treated with SEs, ROS production rose in the first stage and then decreased in both the control and treatment groups, except for SE-III. However, production in treated pericarps was significantly lower than that in the control, especially for SE-III. In the SE-III treated group, ROS production reached a minimum level at 72 h. ROS production in the control group was 3.65 times that of the SE-III treated group (Fig 5).

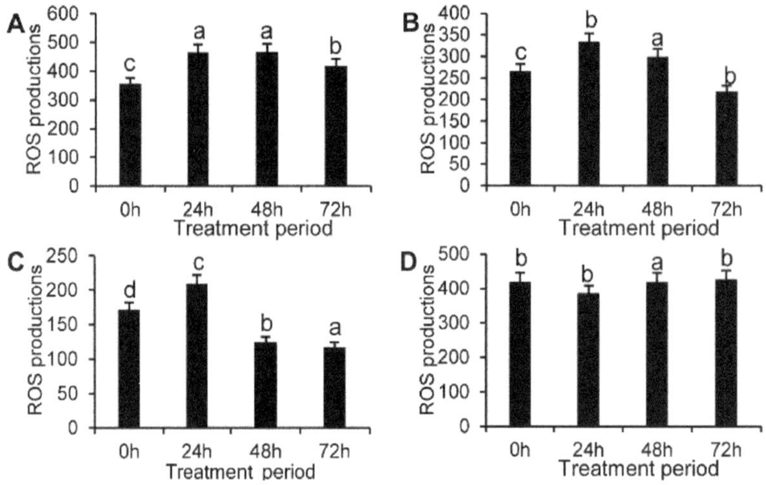

Figure 5. Measurement of ROS production in apple pericarps treated by the SEs. Measurement of ROS production in apple pericarps treated by SE-I(A), SE-II (B), SE-III(C), and control (D) for four different treatment periods. The data represent the mean ± SE and letters indicate a significant difference ($P < 0.05$).

MDA (a product of lipid peroxidation in organisms) content in the apple pericarp first decreased, and then increased in the SE-treated groups during the treatment period, although the content was significantly lower in the treated apple pericarp at all test stages compared with that of the control (Fig 6). MDA content was lowest at 24 h in the SE-III treated group and was 6.7% percent of that of the control (Fig 6).

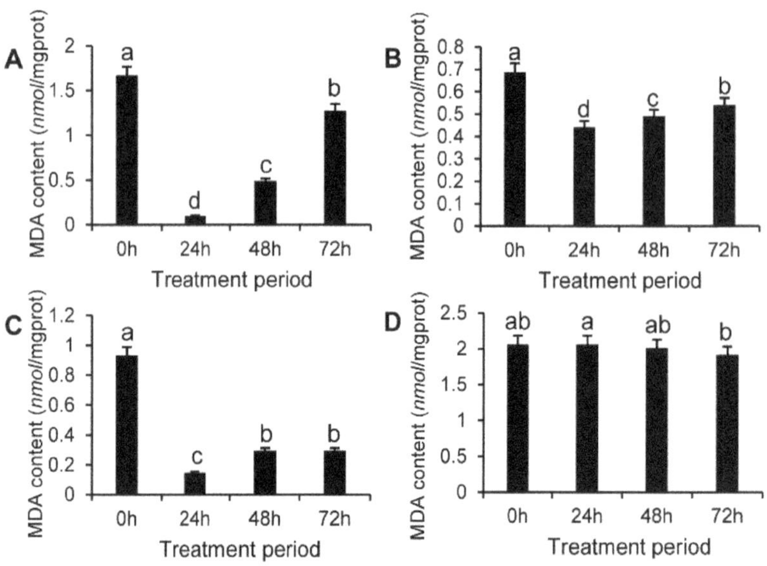

Figure 6. Measurement of MDA content in apple pericarps treated by SEs. Measurement of MDA content in apple pericarps treated by SE-I (A), SE-II (B), SE-III (C), and control (D) for four different treatment periods. The data represent the mean ± SE.

CAT activity was markedly upgraded by the SEs and initially exhibited an increase followed by a decreasing tendency with processing time for all the treated groups. CAT activity in pericarps treated with SE-III was 2.37, 9.51, and 1.68 times that of the control during the first three sampling times, respectively, whereas there was no apparent difference between treatment and control groups at 72 h (Fig 7).

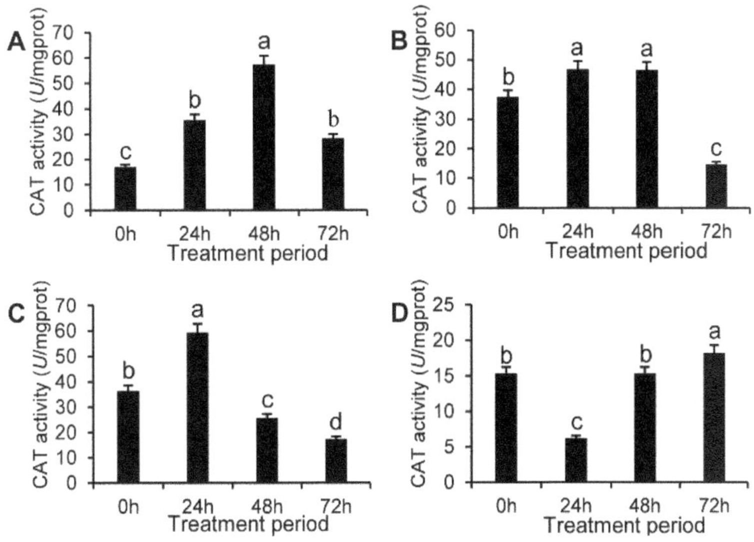

Figure 7. CAT activity in apple pericarps treated by SEs. CAT activity in the apple pericarp treated by SE-I (A), SE-II(B), SE-III(C), and control (D) for four different treatment periods. The data represent the mean ± SD and letters indicate a significant difference ($P < 0.05$).

SOD activity of apple pericarps treated with SEs was improved and positively correlated with time except for SE-I. In the SE-I treated group, SOD activity reached a peak and was 2.13 times that of the control group at 24 h (Fig 8). In the SE-II and SE-III treated groups, SOD activity peaked at 72 h and was 152% and 49% higher than that of the control, respectively.

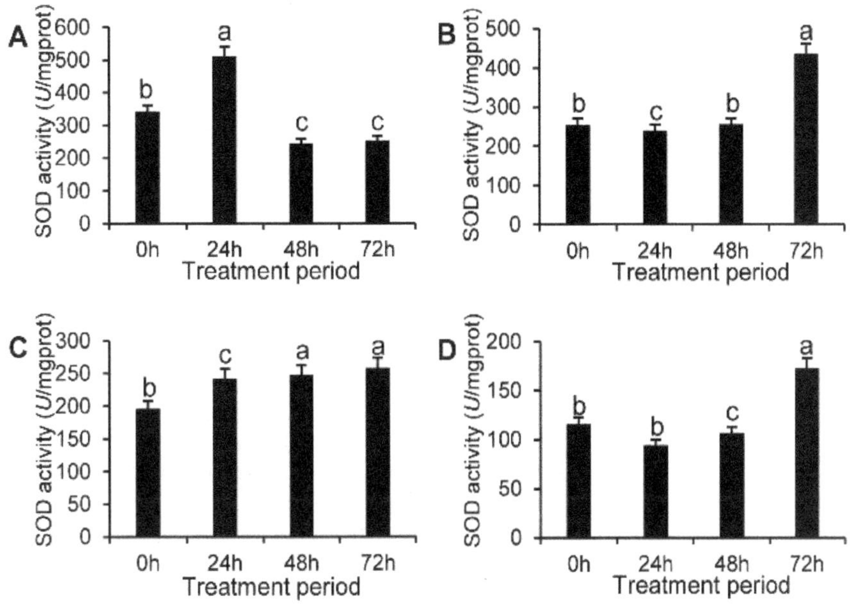

Figure 8. SOD activity in apple pericarps treated by SEs. SOD activity in the apple pericarps treated by SE-I (A), SE-II(B), and SE-III(C), and control group (D) for four different treatment periods. The data represent the mean ± SD and letters indicate a significant difference ($P < 0.05$).

The results indicated that peroxidation caused by ROS in the infected tissues rose gradually and led to MDA accumulation caused by lipid peroxidation. Furthermore, as the pathogenic fungi infected the fruit, SOD and CAT activity in treated apples rose and then, changing trend, dropped, with the activity of the two enzymes in the treated fruit displaying high performance levels compared to those of control fruits at all test stages. This indicated that SE-III possibly induced plant tissues to promote activities of preventative enzymes, inhibiting the accumulation of ROS, and then downgraded lipid peroxidation in apple pericarp tissue.

In this study, the genes related to pathogenic fungi infection and plant infection prevention were selected, and we investigated their profile in SE-III treated apple fruits. Relative gene expressions for both treated and control groups were determined by qRT-PCR, in an attempt to identify differential expression. Four genes were found to be differentially expressed between the parents, including *MdPR4*(A), *MdAFS* (B), *MdPAL* (C), and *MdPPO* (D). All were more highly expressed in the SEs treated group than in the control group and expression was also positively correlated with treatment period.

Compared with that in control groups, *MdPR4* expression in apples was significantly up-regulated by SE-I, SE-II, and SE-III during the treatment period and nearly all expression increased with time, especially for SE-II. In the SE-II treated group, *MdPR4* expression was 30.5, 5.45 and 44.67 times that of the control group at 24 h, 48 h, and 72 h, respectively (Fig

9A), although *MdPR4* expression decreased when treated by SE-III after 48 h.

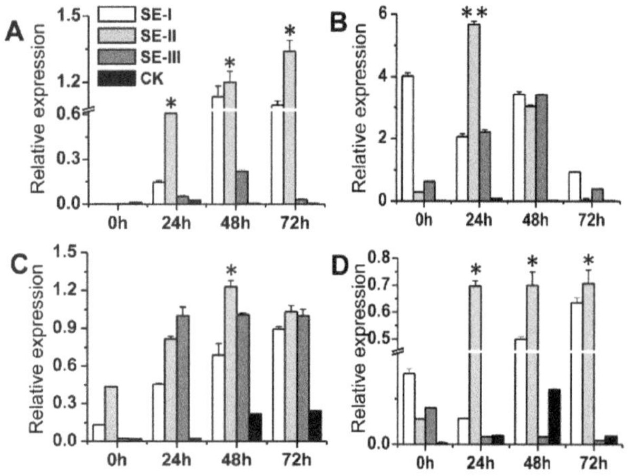

Figure 9. The relative expression of the four genes under treatment of three SEs. The relative expression of the four genes (A: *MdPR4*; B: *MdAFS*; C: *MdPAL*; D:*MdPPO*) under the treatment of three SEs for 4 different treatment period (0 h, 24 h, 48 h, and 72 h). The data represent the mean ± SD and asterisks indicate a significant difference ($P < 0.05$).

Compared with that of control groups, expression of *MdAFS* was evidently up-regulated by the SEs. For example, in the SE-I treated group, *MdAFS* expression was 402, 29.43, 171, and 93 times of the control, respectively with time (Fig 9B). In SE-II and SE-III treated group,*MdAFS* expression showed a downward trend, after an initial first rise, with processing time.*MdPAL* expression increased with processing time in the SE-I, SE-II, and SE-III treated groups(Fig 9C). *MdPPO* expression was evidently up-regulated by SE-I and SE-II, whereas expression in the SE-I treated group initially increased before decreasing slightly with time. The expression of*MdPPO* in the SE-II treated group increased with processing time, and SE-III did not play a key role (Fig 9D).

DISCUSSION

The clinical effects of*S. chinensis* on pathogenic bacteria, fungi, and viruses has been widely demonstrated. However, active substances extracted from the stems and leaves of *S.chinensis* have been rarely reported. Whether the active substances have the potential to effectively inhibit the growth of pathogenic fungi and the potential to be a new broad-spectrum fungicide has also not yet been determined. Fruits of *S. chinensis* are typically chosen as experimental materials [39–42]. However, Yu (1998) reported there was no significant difference in the lignin content of the fruits, stems, and leaves [43]. Therefore, the relatively low cost and rarely studied stems and leaves of *S. chinensis* were chosen as our experimental material.

This study proposed a new and modern method to make full use of the previously worthless stems and leaves of *S. chinensis*. They not only greatly reduced extraction costs and saved resources, but also provided a new method for the future development of botanical fungicides. To obtain active substances with different polarities, different ratios of the polar organic solvent were applied to separate the primary crude extract. According to the antifungal experiment, SE-I, SE-II, and SE-III were effective, although SE-IV to SE-VII had almost no antifungal activity.

When investigating the inhibitory effects of SE-I, SE-II, and SE-III on the mycelial growth of*M. fructicola*, *F. oxysporum* and *B. dothidea*, the SEs show inhibitive effects against the fungi, and the effect of SE-III was the highest according to the EC_{50} values. At the same time, the inhibition rate is positively correlated to the concentration of SE applications. The inhibition rate of SE-III reached 100% against *B. dothidea* growth. These results were also verified by the fact that SE-III remarkably changed the growth status of test fungal hypha cultivated on the medium, causing the hypha to shrivel and dry up.

To varying degrees, the SE fractions largely suppressed the germination rates of the three kinds of spores after inoculation for 24 h and 48 h. Consistent with previous results, SE-III had the greatest effect in the experiments on spore germination. As the development and use of biological fungicide, *S. chinensis* extract should be further applied to postharvest disease control on fruit. Thus, this work further explored the preventive effect of SEs against pathogen invasion. As we expected, SEs can indeed prevent the invasion of pathogenic fungi to a certain degree. After spraying the SEs on the pericarp of apples and peaches, the fungal scab area of apple ring rot and peach brown rot visibly decreased. All of these factors provided evidence that the SEs can inhibit pathogenic fungi in both the *in vitro* and*in vivo* experiments. However, there was a slight injury on the surface of peach fruit. Application for pathogen defense needs to be further improved. After comprehensive analysis of its effect in fungistasis and disease prevention, our results indicated that the optimal active substance was obtained by chloroform and acetone at the ratio of 9:1.

Differences in the principle composition of the substances could be the key reason for their different efficacy. In the present study, to determine the differences in active ingredient composition among SE-I, SE-II, and SE-III, analyses of the lignans were conducted with an evaluation of HPLC-MS. According to our results, there were significant differences in lignan composition among the extracts and SE-I, SE-II, and SE-III were found to contain 5, 4, and 7 kinds of lignans, respectively. The SE-III fraction had the highest lignan abundance, and this may explain its stronger biological activity. In recent experiments on the preventative effects of lignans on fungal infection, results indicated that lignans from *S. chinensis* exhibited obvious inhibition and regulation of fungal infection. For example, some lignans, such as gomisin A, gomisin G, Schizandrin, and schisanhenol were recently reported to possess valuable anti-tumor promoting bioactivities by inhibiting Epstein-Barr virus early antigen (EBV-EA) activation[44], reversing P-glycoprotein-mediated multidrug resistance (Pgp-MDR) in cancer cells [45], and enhancing doxorubicin-induced apoptosis in human hepatic cancer cells, as well as through the inhibition of platelet aggregation and anti-HIV (human immunodeficiency virus) effects [46, 47]. In our study, a variety of lignans were also detected; therefore, we speculated that lignans may play a key role in suppressing pathogens. In the next experiment, we further explored physiological and molecular antifungal mechanisms.

Previous studies indicated that lignan compounds can effectively improve an organism's antioxidant levels. Research concerning SB (Schisandrin B), a major lignan isolated from S. chinensis, showed it has a high antioxidant potential in both in vitro and in vivo experiments [48–51]. Schisandrin B has been shown to prevent ischemia-reperfusion injuries in isolated perfused rat hearts by enhancing the glutathione antioxidant response and mitochondrial ATP generation capacity, or by increasing the expression levels of cardiac protective proteins, such as heat shock proteins 25 (HSP25) and HSP70. Moreover, recent studies have shown that gomisin N has anti-apoptotic activity in H9c2 cells through suppressing mitochondrial permeability transition [52, 53]. In a study on rats, You (2006) showed that SCE (S. chinensis extract) attenuates doxorubicin-induced cardiomyopathy by elevating myocardial glutathione peroxidase and superoxide dismutase (SOD) activities and reducing lipid peroxidation [54]. Extracted solutions of S. chinensis can significantly scavenge free radicals and suppress the growth of different kinds of pathogenic bacteria, such as Staphylococcus aureus, Pneumobacillus, and Enteritis bavoil [55]. In addition, SCE suppressed doxorubicin-induced toxicity by reducing ROS production and lipid peroxidation and also regulated the expression levels of glutathione metabolism and detoxification-related genes [56]. Our results were consistent with those detailed above. The study revealed that the antioxidant system of fruits was pre-induced and activated by active substances of Schisandra. It may be the most likely mechanism for postharvest fruit to effectively defend against disease pathogen invasion. Experimental results suggested that with pathogenic fungi (B. dothidea) infection, ROS production was lower in the control, especially in the SE-III treated group. MDA content was significantly lower in the SE-treated apple pericarp when compared with that of the control. SOD and CAT were widely present in the plant cells. Within a cell, SOD constitutes the first line of defense against ROS. The typical catalase reaction is the dismutation of two molecules of H2O2 to water and O2 [57]. SOD and CAT enzymes directly remove ROS to regulate its level and convert it into less cytotoxic substances. SODs are seen as a sensor of intracellular ROS [58–60]. Furthermore, SOD and CAT activity in induced apple pericarp presented higher levels compared to those recorded in the control. This indicated that the SEs might induce plant tissues to promote activities of preventative enzymes, inhibiting the accumulation of ROS and then downgrading lipid peroxidation in the infected apple pericarp.

The study found that tested genes (*MdPR4, MdAFS, MdPPO, and MdPAL*) related to the defense reaction were significantly up-regulated by the SEs. Previous studies have shown that the pathogenesis-related protein gene PR, as the terminal gene in the plant defense pathway, produces a variety of ROS and toxic secondary metabolites that cooperatively kill pathogens, and some PR gene over expression can significantly improve plant disease resistance [61]. There are other functions of PR genes, such as those involved in senescence [62], osmotic stress and mechanical injury [63]. *MdPPO* (a polyphenol oxidase gene) is considered to be the plant defense protein [64–67] and plays an important role in plant defense responses. *MdPAL* (phenylalanine ammonia lyase) plays a catalytic role in flavonoid biosynthesis, and flavonoids have antibacterial, antiviral, anti-allergic, anti-inflammatory, vasodilatation inhibiting, and other biological functions[68]. Moreover, flavonoids can also inhibit lipid peroxidation and the enzyme systems [69], as α-farnesene synthase (AFS) is the ultimate limiting enzyme in the synthesis of α-farnesene plant mevalonate pathway [70].

The present study indicated that spraying with SEs up-regulated the expression of the four genes to different degrees during the treatment period. We speculated that the active ingredients in the fractions from *S. chinensis* plants not only induced the up-regulation of the four genes, but also activated the antioxidant system in fruits to prevent infection, although the response of the plant itself occurs in case of infection.

Fruit quality characteristics in the process of maturity directly affect the economic value of the commercial crop. However, during this period several physiological disorders may occur. Fruit need to be induced to be resistant to infection through promoting preventative enzyme activities and regulating other scavengers or protectors, decreasing ROS production and reducing lipid peroxidation. This mechanism may extend storage time of fresh apples. According to this study, fractions from *S. chinensis* plants could improve antioxidant levels and promote the capacity for self-regulation while directly inhibiting disease and infection. This indicates that compounds of *S. chinensis* can be used as preservative or protective agents for future commercial development.

Herbal ingredients and their chemically modified derivatives have been invaluable resources for the development of therapeutic substances [71]. Technological advances in lead-generation strategies have drawn notable attention to the chemical complexity and diversity of herbal ingredients [71]. Accordingly, bioactive lignans may be a promising source of new-lead compounds for future drug development. On the other hand, lignans could be exploited as chemical tools for the starting point of new scientific discoveries—assisting in the dissection of complex phenomena into their constituent parts and the identification of the hidden molecular targets and mechanisms underlying pathophysiological phenotypes. Extracts of *S. chinensis* as promising resources for the development of safe, effective and multi-targeted agents against pathogenic fungi, will offer us a novel insight into future challenges and perspectives on *S. chinensis* research and future clinical investigations and strategies.

CONCLUSIONS

The results described in this work showed that active ingredients were extracted with 70% acetone in *S. chinensis* and were separated with different organic solvents. The fraction SE-III-separated by a ratio of chloroform to acetone of 9: 1 was the most potent and antifungal active substance extracted. *S. chinensis* extraction was verified to successfully suppress the growth of three tested pathogens, inhibit the germination of fungal spores, and protect apple fruit from infection by pathogenic fungi. A possible mechanism to explain the results may be that SEs induced and activated the antioxidant system and also up-regulated defense-related genes. Thus, active ingredients in plants could play a major role in the development and exploitation of botanical fungicides.

SUPPORTING INFORMATION

S1 Table. Genetic sequences of primers involved in this study.

Primer	Base sequence (5' to 3')	The number of bases
actin-F1	5'-CTGAACCCAAAGGCTAATCG-3'	26
actin-R1	5'- ACTGGCGTAGAGGGAAAGAA-3'	27
MdPR4-F	5' - ATACCACCTCTACAATCCACA-3'	29
MdPR4-R	5'-GTCCAAGTCCAATCCTCC-3'	24
MdPAL1-F	5'-TTGACGCACAAGTTGAAGCA-3'	26
MdPAL1-R	5'-CCACTGAGGTGATGTTCGGA-3'	26
MdPPO-F	5'-CCTACTCACAAAGCCCAAGC-3'	26
MdPPO-R	5'-CCTCCAAGACCAAGAAGCAC-3'	26
MdAFS1-F	5'-GTTCAGGCCCAACTCCAAAAA-3'	27
MdAFS1-R	5'-AGGCTCGAATGCTACTCCCACAG-3'	29

ACKNOWLEDGMENTS

We thank the Fruit Tree Key Laboratory at the Beijing University of Agriculture and the Beijing Nursery Engineering Research Center for Fruit Crops for providing basic experimental facilities.

Financial support was provided by scientific research projects "Extraction and utilization of active ingredients in *Schisandra chinensis* (Turcz.) Baill and the study of their antifungal effect" (KM201510020004) of Beijing municipal education commission in 2015. Financial support was also provided by the project of construction of innovative teams and teacher career development for universities and colleges under Beijing municipality (IDHT20140509; IDHT20150503), and by the NSFC for Outstanding Young Scholar (81422046). Beijing Collaborative innovation center for eco-environmental improvement with forestry and fruit trees' (CEFF-PXM2016-014207-000038) also provide financial support.

AUTHOR CONTRIBUTIONS

Conceived and designed the experiments: YY YC HY. Performed the experiments: HY YC. Analyzed the data: HY WX. Contributed reagents/materials/analysis tools: YC JZ WG JL YY WX. Wrote the paper: HY YC YY.

REFERENCES

1. Isman MB. Botanical insecticides, deterrents, and repellents in modern agriculture and an increasingly regulated world. Annual Review of Entomology. 2006; 51: 45–66. doi: 10. 1146/annurev. ento. 51. 110104. 151146 pmid: 16332203.
2. Isman MB. Perspective botanical insecticides: For richer, for poorer. Pest Management Science. 2008; 64: 8–11. doi: 10. 1002/ps. 1470
3. Kudom AA, Mensah BA, Botchey MA. Aqueous neem extract versus neem powder on Culex quinquefasciatus: Implications for control in anthropogenic habitats. Journal of Insect Science. 2011; 11(142). doi: 10. 1673/031. 011. 14201.
4. Lv C, Zhong B, Zhong G, Weng Q, Chen S, Hu M, et al. Four botanical extracts are toxic to the hispine beetle, Brontispa longissima, in laboratory and semi–field trials. Journal of Insect Science. 2012; 12(58). doi: 10. 1673/031. 012. 5801.
5. Ladhari A, Laarif A, Omezzine F, Haouala R. Effect of the extracts of the spiderflower, Cleome arabica, on feeding and survival of larvae of the cotton leafworm, Spodoptera littoralis. Journal of Insect Science. 2013; 13(61). doi: 10. 1673/031. 013. 6101. pmid: 23906290
6. Copping LG, Duke SO. Natural products that have been used commercially as crop protection agents. Pest Manage Science. 2007; 63: 524–554. pmid: 17487882 doi: 10. 1002/ps. 1378
7. Dayan FE, Cantrell CL, Duke SO. Natural products in crop protection. Bioorganic & Medicinal Chemistry. 2009; 17: 4022–4034. doi: 10. 1016/j. bmc. 2009. 01. 046
8. Duke SO, Cantrell CL, Meepagala KM, Wedge DE. Tabanca N, Schrader KK. Natural toxins for use in pest management. Toxins (Basel). 2010; 2:1943–1962. doi: 10. 3390/toxins2081943. pmid: 22069667
9. Cantrell CL, Dayan FE, Duke SO. Natural products as sources for new pesticides. Journal of Natural Products. 2012; 75: 1231–1242. doi: 10. 1021/np300024u pmid: 22616957.
10. Tomar A, Singh Rathore H, Singh K. Scope and limitations of neem products and other botanicals in plant protection. Handbook of Pesticides: CRC Press: 2009; 67–89.
11. Nabavi SM, Marchese A, Izadi M, Curti V, Daglia M, Nabavi SF. Plants belonging to the genus Thymus as antibacterial agents: from farm to pharmacy. Food Chemistry. 2015; 173: 339–47. doi: 10. 1016/j. foodchem. 2014. 10. 042 pmid: 25466031.
12. Zabka M, Pavela R, Prokinova E. Antifungal activity and chemical composition of twenty essential oils against significant indoor and outdoor toxigenic and aeroallergenic fungi. Chemosphere. 2014; 112: 443–8. doi: 10. 1016/j. chemosphere. 2014. 05. 014 pmid: 25048938.

13. Jing Li, Lei Zhentian, Li Ligai, Xie Rangjin, Xi Wanpeng, Guan Yu, et al. Antifungal Activity of Citrus Essential Oils. Journal of Agricultural and Food Chemistry. 2014; 62 (14): 3011–33. doi: 10. 1021/jf5006148 pmid: 24628448.
14. Swain T. Secondary compouds as protective agents. Annual Review of Plant Physiology. 1977; 28: 479–501. doi: 10. 1146/annurev. pp. 28. 060177. 002403
15. Yoon MY, Cha B, Kim JC. Recent trends in studies on botanical fungicides in agriculture. Plant Pathology Journal. 2013 29(1):1–9. doi: 10. 5423/PPJ. RW. 05. 2012. 0072. pmid: 25288923
16. The Angiosperm Phylogeny Group. An update of the Angiosperm Phylogeny Group classification for the orders and families of flowering plants: APG III. Botanical Journal of the Linnean Society. 2009; 161(2): 105–121. doi: 10. 1111/j. 1095-8339. 2009. 00996. x.
17. Hancke JL, Burgos RA, Ahumada F. *Schisandra chinensis* (Turcz.) Baill. Fitoterapia. 1999; 70: 451–71. doi: 10. 1016/s0367-326x(99)00102-1
18. Chun JN, Cho M, So I, Jeon JH. The protective effects of *Schisandra chinensis* fruit extract and its lignans against cardiovascular disease: A review of the molecular mechanisms. Fitoterapia. 2014; 97: 224–33. doi: 10. 1016/j. fitote. 2014. 06. 014. pmid: 24976588
19. Lee KH, Morris-Natschke S, Qian K, Dong Y, Yang X, Zhou T, et al. Recent progress of research on herbal products used in traditional Chinese medicine: the herbs belonging to The Divine Husbandman's Herbal Foundation Canon (Shén Nóng Běn Cǎo Jīng). Journal of Traditional Complementary and Medicine. 2012; 2(1): 6–26. pmid: 24716110.
20. Alexander JS, Wang Y. Therapeutic potential of *Schisandra chinensis* extracts for treatment of hypertension. Introduction to: 'antihypertensive effect of gomisin A from*Schisandra chinensis* on angiotensin II induced hypertension via preservation of nitric oxide bioavailability' by Park et al. Hypertension Research. 2012; 35: 892–3. doi: 10. 1038/hr. 2012. 101 pmid: 22763484.
21. Park EJ, Chun JN, Kim SH, Kim CY, Lee HJ, Kim HK, et al. Schisandrin B suppresses TGFbeta1 signaling by inhibiting Smad 2/3 and MAPK pathways. Biochemical Pharmacology 2012; 83(3): 378–84. doi: 10. 1016/j. bcp. 2011. 11. 002 pmid: 22100726.
22. Panossian A, Wikman G. Pharmacology of *Schisandra chinensis* Bail: an overview of Russian research and uses in medicine. Journal of Ethnopharmacology. 2008; 118(2): 183–212. doi: 10. 1016/j. jep. 2008. 04. 020 pmid: 18515024.
23. Jiang Y, Fan X, Wang Y, Tan H, Chen P, Zeng H, et al. Hepato-protective effects of six Schisandra lignans on acetaminophen-induced liver injury are partially associated with the inhibition of CYP-mediated bioactivation. Chemico-Biological Interactions. 2015; 231: 83–9. doi: 10. 1016/j. cbi. 2015. 02. 022 pmid: 25753323.
24. Zhong S, Nie YC, Gan ZY, Liu XD, Fang ZF, Zhong BN, et al. Effects of *Schisandra chinensis* extracts on cough and pulmonary inflammation in a cough hypersensitivity guinea pig model induced by cigarette smoke exposure. Journal of Ethnopharmacology. 2015; 165:73–82. doi: 10. 1016/j. jep. 2015. 02. 009 pmid: 25681545.
25. Kim EY, Baek IH, Rhyu MR. Cardioprotective effects of aqueous *Schizandra chinensis* fruit extract on ovariectomized and balloon-induced carotid artery injury rat models: effects on serum lipid profiles and blood pressure. Journal of Ethnopharmacology. 2011; 134(3):668–75. doi: 10. 1016/j. jep. 2011. 01. 019. pmid: 21256204
26. Cyong JC, Ki SM, Iijima K, Kobayashi T, Furuya M. Clinical and pharmacological studies on liver disease treated with Kampo herbal medicine. American Journal of Chinese Medicine. 2000; 28: 351–60. doi: 10. 1142/ S0192415X00000416 pmid: 11154048.
27. Pan SY, Yu Q, Zhang Y, Wang XY, Sun N, Yu ZL, et al. Dietary Fructus Schisandrae extracts and fenofibrate regulate the serum /hepatic lipid-profile in normal and hypercholesterolemic mice, with attention to hepatotoxicity. Lipids in Health and Disease. 2012; 11: 1186–94. doi: 10. 1186/1476-511X-11-120 pmid: 22989092.
28. Lee S, Kim DH, Jung JW, Oh JH, Park HJ, Park C, et al. *Schizandra chinensis* and*Scutellaria baicalensis* counter stress behaviors in mice. Phytotherapy Research. 2007; 21(12): 1187–1192. doi: 10. 1002/ptr. 2233 pmid: 17639560.
29. Panossian A, Hambardzumyan M, Hovhanissyan A, Wikman G. The adaptogens rhodiola and Schizandra modify the response to immobilization stress in rabbits by suppressing the increase of phosphorylated stress-activated protein kinase, nitric oxide and cortisol. Drug Target Insights. 2007; 2: 39–54. Available: http://www. la-press. com/copyright. htm. pmid: 21901061.
30. Chen WW, He RR, Li YF, Li SB, Tsoi B, Kurihara H. Pharmacological studies on the anxiolytic effect of standardized Schisandra lignans extract on restraint-stressed mice. Phytomedicine. 2011; 18(13): 1144–47. doi: 10. 1016/j. phymed. 2011. 06. 004 pmid: 21757327.
31. Kwon DY, Kim DS, Yang HJ, Park S. The lignan-rich fractions of Fructus Schisandrae improve insulin sensitivity via the PPAR-γ pathways in in vitro and in vivo studies. Journal of Ethnopharmacology, 2011; 135(2): 455–62. doi: 10. 1016/j. jep. 2011. 03. 037 pmid: 21440615.
32. Opletal L, Sovová H, Bártlová M. Dibenzo[a,c] cyclooctadiene lignans of the genus Schisandra: importance, isolation and determination. Journal of Chromatography B. 2004; 812(1–2): 357–71. doi: 10. 1016/j. jchromb. 2004. 07. 040 pmid: 15556508.
33. Huang TH, Shen PN, Shen YG. Preparative separation and purification of deoxyschisandrin and γ-schisandrin from *Schisandra chinensis* (Turcz.) Baill by high-speed counter-current chromatography. Journal of Chromatography A. 2005; 1066(1–2): 239–42. pmid: 15794577. doi: 10. 1016/j. chroma. 2005. 01. 025
34. Tian K, Qi SD, Cheng YQ, Chen XG, Hu Z. Separation and determination of lignans from seeds of

Schisandra species by micellar electrokinetic capillary chromatography using ionic liquid as modifier. Journal of Chromatography A. 2005; 1078(1–2): 181–7. pmid: 16007996. doi: 10. 1016/j. chroma. 2005. 05. 018

35. He F, Li XY, Yang GY, Li XN, Luo X, Zou J, et al. Nortriterpene constituents from*Schisandra sphenanthera*. Tetrahedron. 2012; 68(2): 440–6. doi: 10. 1016/j. tet. 2011. 11. 026

36. Tang W, Ding Z, Zhou ZQ, Wang YZ, Guo LY. Phylogenetic and pathogenic analyses show that the causal agent of apple ring rot in China is *Botryosphaeria dothidea*. Plant Disease. 2012; 96:486–96. doi: 10. 1094/ pdis-08-11-0635

37. Hu M-J, Cox KD, Schnabel G, Luo C-X. Monilinia species causing brown rot of peach in China. PLoS ONE. 2011; 6(9): e24990. doi: 10. 1371/journal. pone. 0024990 pmid: 21980371.

38. Koike ST, Kirkpatrick SK, Gordon TR. Fusarium wilt of strawberry caused by*Fusarium oxysporum* in California. Plant Diseae, 2009; 93(10): 1077. doi: 10. 1094/PDIS-93-10-1077A.

39. Cheng Z, Song H, Yang Y, Zhou H, Liu Y, Liu Z. Smashing Tissue Extraction of Five Lignans From the Fruit of *Schisandra chinensis*. Journal of Chromatographic Science. 2016; 54(2):246–56. doi: 10. 1093/chromsci/ bmv116. pmid: 26254462

40. Lee HJ, Jo S, Ryu J, Jeong HS, Lee G, Ryu MH, et al. Effects of *Schisandra chinensis*Turcz. fruit on contact dermatitis induced by dinitrofluorobenzene in mice. Molecular Medicine Reports. 2015; 12(2): 2135–9. doi: 10. 3892/mmr. 2015. 3618. pmid: 25892286

41. Xue Y, Li X, Du X, Li X, Wang W, Yang J, et al. Isolation and anti-hepatitis B virus activity of dibenzocyclooctadiene lignans from the fruits of *Schisandra chinensis*. Phytochemistry. 2015; 116: 253–61. doi: 10. 1016/j. phytochem. 2015. 03. 009. Epub 2015 Apr 13. pmid: 25882501

42. Cheng Z, Song H, Yang Y, Liu Y, Liu Z, Hu H, et al. Optimization of microwave-assisted enzymatic extraction of polysaccharides from the fruit of*Schisandra chinensis* Baill. International Journal of Biological Macromolecules. 2015 May;76:161–8. doi: 10. 1016/j. ijbiomac. 2015. 01. 048. pmid: 25661875

43. Yu JL, Li SH, Chen J. A comparative study of lignans among Schisandra root, stems, leaves, and fruit. Ginseng Research. 1998; (1):11–12.

44. Chen DF, Zhang SX, Kozuka M, Sun QZ, Feng J, Wang Q, et al. Interiotherins C and D, two new lignans from *Kadsura interior* and antitumor-promoting effects of related neolignans on Epstein-Barr virus activation. Journal of Natural Products. 2002; 65(9): 1242–5. pmid: 12350139. doi: 10. 1021/np0105127

45. Yoo HH, Lee M, Lee MW, Lim SY, Shin J, Kim DH. Effects of Schisandra lignans on P-glycoprotein-mediated drug efflux in human intestinal Caco-2. Planta Medica. 2007; 73: 444–50. pmid: 17566147 doi: 10. 1055/s-2007-967178

46. Chen M, Kilgore N, Lee KH, Chen DF. Rubrisandrins A and B, Lignans and related anti-HIV compounds from *Schisandra rubriflora*. Journal of Natural Products. 2006; 69(12): 1697–701. doi: 10. 1021/np060239e pmid: 17190445.

47. Chen DF, Zhang SX, Xie L, Xie JX, Chen K, Kashiwada Y, et al. Anti-AIDS agents—XXVI. Structure-activity correlations of gomisin-G-related anti-HIV lignans from*Kadsura interior* and of related synthetic analogues. Bioorganic & Medicinal Chemistry. 1997; 5(8): 1715–23. pmid: 9313872. doi: 10. 1016/s0968-0896(97)00118-1

48. Hou W, Gao W, Wang D, Liu Q, Zheng S, Wang Y. The Protecting Effect of Deoxyschisandrin and Schisandrin B on HaCaT Cells against UVB-Induced Damage. PLoS One. 2015; 10(5):e0127177. doi: 10. 1371/journal. pone. 0127177. pmid: 25978330

49. Chang CY, Chen YL, Yang SC, Huang GC, Tsi D, Huang CC, et al. Effect of Schisandrin B and sesamin mixture on CCl(4)-induced hepatic oxidative stress in rats. Phytotherapy Research. 2009; 23(2): 251–6. doi: 10. 1002/ptr. 2602. pmid: 18979496

50. Giridharan VV, Thandavarayan RA, Bhilwade HN, Ko KM, Watanabe K, Konishi T. Schisandrin B, attenuates cisplatin-induced oxidative stress, genotoxicity and neurotoxicity through modulating NF-κB pathway in mice. Free Radical Research. 2012; 46(1): 50–60. doi: 10. 3109/10715762. 2011. 638291 pmid: 22059853.

51. Giridharan VV, Thandavarayan RA, Sato S, Ko KM, Konishi T. Prevention of scopolamine-induced memory deficits by schisandrin B, an antioxidant lignan from*Schisandra chinensis* in mice. Free Radical Research. 2011; 45(8): 950–8. doi: 10. 3109/10715762. 2011. 571682 pmid: 21615274.

52. Chiu PY, Luk KF, Leung HY, Ng KM, Ko KM. Schisandrin B stereoisomers protect against hypoxia/ reoxygenation-induced apoptosis and inhibit associated changes in Ca^{2+}-induced mitochondrial permeability transition and mitochondrial membrane potential in H9c2cardiomyocytes. Life Sciences. 2008; 82(21–22): 1092–101. doi: 10. 1016/j. lfs. 2008. 03. 006 pmid: 18452952.

53. Chiu PY, Leung HY, Poon MK, Mak DH, Ko KM. (–) Schisandrin B is more potent than its enantiomer in enhancing cellular glutathione and heat shock protein production as well as protecting against oxidant injury in H9c2cardiomyocytes. Molecular and Cellular Biochemistry. 2006; 289(1–2):185–91. doi: 10. 1007/ s11010-006-9163-1 pmid: 16718369.

54. You JS, Pan TL, Hou YC. *Schisandra chinensis* protects against adriamycin induced cardiotoxicity in rats. Chang Gung Medical Journal. 2006; 29(1): 63–70. pmid: 16642728.

55. Li B, Meng XJ, Xue X, Wu Q, Li YS, Wang YQ. Schisandrin B from Schisandra scavenging free radical and bacteriostasis in vitro experiment. Food Science, 2011; 32 (5): 79–82.

56. Choi EH, Lee N, Kim HJ, Kim MK, Chi SG, Kwon DY, et al. *Schisandra fructus* extract ameliorates doxorubicin-induce cytotoxicity in cardiomyocytes: altered gene expression for detoxification enzymes. Genes and Nutrition. 2008; 2(4): 337–45. doi: 10. 1007/

s12263-007-0073-y pmid: 18850228.

57. Zamocky M, Furtmüller PG, Obinger C. Evolution of catalases from bacteria to humans. Antioxidants and Redox Signaling. 2008; 10(9): 1527–47. doi: 10. 1089/ars. 2008. 2046 pmid: 18498226.

58. Alscher RG, Erturk N, Heath LS. Role of superoxide dismutases (SODs) in controlling oxidative stress in plants. Journal of Experimental Botany. 2002; 53(372): 1331–41. doi: 10. 1093/jexbot/53. 372. 1331 pmid: 11997379.

59. Del Río LA. Peroxisomes as a cellular source of reactive nitrogen species signal molecules. Archives of Biochemistry and Biophysics. 2011; 506(1): 1–11. doi: 10. 1016/j. abb. 2010. 10. 022. pmid: 21055384

60. Mhamdi A, Queval G, Chaouch S, Vanderauwera S, Breusegem FV, Noctor G. Catalase function in plants: a focus on Arabidopsis mutants as stress-mimic models. Journal of Experimental Botany. 2010; 61(15): 4197–220. doi: 10. 1093/jxb/erq282 pmid: 20876333.

61. Sarowar S, Kim YJ, Kim EN, Kim KD, Hwang BK, Islam R, et al. Over expression of a pepper basic pathogenesis-related protein 1 gene in tobacco plants enhances resistance to heavy metal and pathogen stresses. Plant Cell Reports. 2005; 24(4): 216–24. doi: 10. 1007/s00299-005-0928-x pmid: 15719238.

62. Buchanan-Wollaston V, Earl S, Harrison E, Mathas E, Navabpour S, Page T. The molecular analysis of leaf senescence a genomics approach. Plant Biotechnology Journal. 2003; 1(1):3–22. pmid: 17147676. doi: 10. 1046/j. 1467-7652. 2003. 00004. x

63. Slusarenko AJ, Fraser RSS, Loon LC. Mechanisms of resistance to plant diseases. 2000 (Dordrecht: Kluwer).

64. Constabel CP, Bergey DR, Ryan CA. Systemin activates synthesis of wound-inducible tomato leaf polyphenol oxidase via the octadecanoid defense signaling pathway. Proceedings of the National Academy of Sciences of the United States of America. 1995; 92(2): 407–11. pmid: 7831300. doi: 10. 1073/pnas. 92. 2. 407

65. Pinto MST, Siqueira FP, Oliveira AEA, Fernandes KVS. A wounding-induced PPO from cowpea (Vigna unguiculata) seedlings. Phytochemistry. 2008; 69(12): 2297–302. doi: 10. 1016/j. phytochem. 2008. 06. 003 pmid: 18675435.

66. Stewart RJ, Sawyer BJB, Bucheli CS, Robinson SP. Polyphenol oxidase is induced by chilling and wounding in pineapple. Australian Journal of Plant Physiology. 2001; 28(3): 181–91. doi: 10. 1071/PP00094.

67. Thipyapong P, Hunt MD, Steffens JC. Antisense downregulation of polyphenol oxidase results in enhanced disease susceptibility. Planta. 2004; 220(1): 103–17. doi: 10. 1007/s00425-004-1330-6.

68. Freitas M, Ribeiro D, Sara TM, Artur SM, Fernandes E. Anti-inflammatory and pro-apoptotic activities of chlorinated flavonoids in human neutrophils. Free Radical Biology and Medicine. 2014, 75: S30. doi: 10. 1016/j. freeradbiomed. 2014. 10. 760.

69. Robak J, Korbut R, Shridi F, Swies J, Rzadkowska-Bodalska H. On the mechanism of antiaggregatory effect of myricetin. Polish Journal of Pharmacy and Pharmacology. 1988; 40: 337–40. pmid: 3241773.

70. Rupasinghe HPV, Almquist KC, Paliyath G, Murr DP. Cloning of hmg1 and hmg2 cDNAs encoding 3-hydroxy-3-methylglutaryl coenzyme A reductase and their expression and activity in relation to α-farnesene synthesis in apple. Plant Physiology and Biochemistry. 2001; 39(11): 933–47. doi: 10. 1016/S0981-9428(01)01316-X.

71. Koehn FE, Carter GT. The evolving role of natural products in drug discovery. Nature Reviews Drug Discovery. 2005; 4(3): 206–20. doi: 10. 1038/nrd1657 pmid: 15729362.

CHAPTER 12 VALIDATION OF THE TARGET PROTEIN OF INSECTICIDAL DIHYDROAGAROFURAN SESQUITERPENE POLYESTERS

Lina Lu, Zhijun Qi, Qiuli Li, and Wenjun Wu

Keywords: Dihydroagarofuran sesquiterpene polyesters, target protein, target validation, V-ATPase, subunit H of V-ATPase, *Mythimna separata* Walker

INTRODUCTION

Identification of target proteins is the basis for development of new pesticides. The discovery of novel targets may result in a series of new pesticides. Moreover, in the development of pesticides, natural products are useful probes in providing new targets. A common example is nicotine; the identification and elucidation of the molecular structure of the target protein of nicotine (*i.e.*, insect nAChR) promoted the development of neonicotinoid pesticides [1,2,3].

A series of insecticidal compounds, namely, sesquiterpene polyesters, sharing a dihydro-*b*-agarofuran sesquiterpenoid skeleton were isolated and characterized by Wu *et al.* from the root bark of *Celastrus angulatus* Max (Celastraceae) [4,5,6]. These insecticidal compounds mainly affect the digestive system of pests, presenting a series of symptoms, such as excitement, twitching, emesis, and loss of body fluid after oral administration [7,8]. Transmission electron microscopy (TEM) analysis showed that the midgut epithelial cells of *Mythimna separata* Walker larvae that ingested celangulin V (CV) were damaged, showing visible vacuolization of cytoplasm, serious disruption of microvilli, fragmentation of rough endoplasmic reticulum cisternae, and rupture of plasma membrane. Subsequently, these morphological changes induce leakage of cytoplasm contents into the midgut lumen, resulting in appearance of numerous lysosome-like vacuoles and secretion [8,9].

Citation: Lina Lu, Zhijun Qi, Qiuli Li, and Wenjun Wu. Validation of the target protein of insecticidal dihydroagarofuran sesquiterpene polyesters, Toxins (2016); 8(3), 79. DOI:10.3390/toxins8030079.

However, the mechanisms of action and the target protein of the dihydroagarofuran sesquiterpene polyesters remain unknown. In our previous study, 11 binding proteins were isolated by affinity chromatography using a derivative of CV (one of the insecticidal compounds) as ligand [10]. Considering the functions of these proteins and the symptoms caused by these compounds, we speculated that V-ATPase and aminopeptidase N (APN)-3 are the putative target proteins. In the present study, we measured the insecticidal toxicity and the enzyme-inhibiting activity of the 12 dihydroagarofuran sesquiterpene polyesters against *M. separata* larvae. The target protein was then validated based on correlation analysis. Results showed that the subunit H of V-ATPase is the target protein of the dihydroagarofuran sesquiterpene polyesters. As with Tulipaline A, as one compound of lactones and aromatic aldehydes that could be exploited as novel nematicdes through inhibiting the activity of V-ATPase [11], this finding also provides ideas for the development of novel pesticides.

RESULTS

Insecticidal Activity

For the subsequent correlation analysis, the insecticidal activity of 12 dihydroagarofuran sesquiterpene polyesters against*M. separata* larvae was evaluated. Results showed that CV-6-*N*-methylisatoic, CV-6-isobutyric acid ester, CV-6-ketone, NW62, and NW57 did not have insecticidal activity at a dose of 668.45 μg/g, whereas CV, CV-6-α-aminopropanoic acid ester, CV-6-aminoacetic acid ester, wilforine, NW69, NW03, and NW70 were all toxic to the fifth instar larvae. The LD_{50} of the insecticidal compounds was then measured (Table 1). Among the seven insecticidal compounds, the LD_{50} value of wilforine was the lowest. CV-6-α-aminopropanoic acid ester and NW70 also had high toxicity with LD_{50} of 33.605 and 86.271 μg/g, respectively, whereas CV and CV-6-aminoacetic acid ester had relatively lower insecticidal activity.

Table 1. Insecticidal activity of dihydroagarofuran sesquiterpene polyesters against the fifth instar larvae of *M. separate*.

Samples	Regression Equation	Correlation Coefficient	LD_{50} (μg/g)	95% Confidence Limits (μg/g)
Celangulin V (CV)	$Y = 3.6544 + 0.47509x$	0.9813	679.479	171.7784–2687.7183
CV-6-α-aminopropanoic acid ester	$Y = 1.9537 + 1.99575x$	0.9488	33.605	27.1526–41.5919
NW03	$Y = 2.9385 + 0.79935x$	0.9922	379.286	173.1693–830.7353
NW70	$Y = 1.3655 + 1.87692x$	0.9829	86.271	60.8162–122.3788
NW69	$Y = -3.6165 + 3.45942x$	0.9912	309.559	257.2790–372.4620
Wilforine	$Y = 5.3414 + 1.52479x$	0.9972	0.597	0.3752–0.9506
CV-6-aminoacetic acid ester	$Y = 3.2465 + 0.6077x$	0.9805	767.811	189.6674–894.5342

Effects on the Activity of APN and V-ATPase Activity

Effects on the Activity of APN

Our previous study indicated that the symptoms caused by dihydroagarofuran sesquiterpene polyesters are similar to those of Bt toxin; APN is the receptor of Bt toxin [12,13,14]. Here, we chose CV as the representative compound to measure the effect on APN activity. The results showed that APN activity of the group treated with CV has no significant difference with that of the group treated with DMSO. Thus, CV had no effect on the activity of APN (Figure 1).

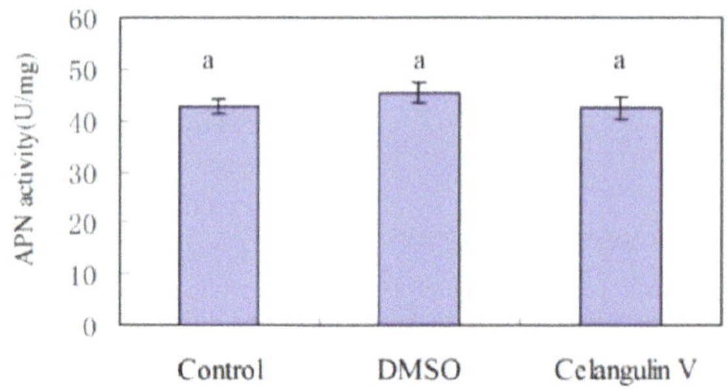

Figure 1. Activity of aminopeptidase N (APN). Results are shown as mean ± SE. The bars show the ranges of standard errors. Letters on the error bars indicate the significant differences by ANOVA analysis ($p < 0.05$).

Effects on the Activity of V-ATPase

The results of the effects of the 12 dihydroagarofuran sesquiterpene polyesters on the V-ATPase activity of *M. separata*larvae are shown in Table 2.

Table 2. Effects of dihydroagarofuran sesquiterpene polyesters on V-ATPase from the midgut of *M. separata* larvae.

Compounds	Concentration (μM)	Specific Activity (nmol Pi/min/mg)	Inhibition Rate (%)
DMSO	-	451.65 ± 19.2740^{a}	-
Bafilomycin A1	3	233.52 ± 14.3258^{h}	48.29
Celangulin V (CV)	100	387.57 ± 12.6793^{cde}	14.19
CV-6-aminoacetic acid ester	100	394.19 ± 29.1940^{bcd}	12.72
CV-6-α-aminopropanoic acid ester	100	362.05 ± 21.6747^{bcd}	19.83

Wilforine	100	204.25 ± 34.6747 h	54.78
NW03	100	298.19 ± 12.3455 g	33.98
NW69	100	394.61 ± 35.3014 bcd	12.63
NW70	100	317.80 ± 17.9918 fg	29.64
NW57	100	443.07 ± 34.4936 abc	1.90
NW62	100	420.49 ± 5.5092 ab	6.90
CV-6-*N*-methylisatoic	100	343.06 ± 11.3159 efg	24.04
CV-6-isobutyric acid ester	100	396.10 ± 11.2234 cd	12.30
CV-6-ketone	100	421.308 ± 27.3712 abc	6.72

Note: different lowercase letters (a, b, c, d, e, f, g, h) indicate the significant differences by ANOVA analysis ($p < 0.05$).

Table 2 shows that the positive control, bafilomycin A1, has an inhibition rate of 48.29% at a dose of 3 μM. Among the dihydroagarofuran sesquiterpene polyesters, wilforine displays high inhibitory effect against V-ATPase, with an inhibition rate of 54.78% at a concentration of 100 μM. CV, CV-6-aminoacetic acid ester, CV-6-α-aminopropanoic acid ester, NW03, NW69, NW70, CV-6-*N*-methylisatoic, and CV-6-isobutyric acid ester showed relatively lower inhibition rate. Compounds NW57, NW62, and CV-6-ketone barely had any effect on V-ATPase.

Comparison of Table 1 and Table 2 shows that, in general, the tested dihydroagarofuran sesquiterpene polyesters that had high insecticidal activity also had high inhibitory effect on V-ATPase. Moreover, correlation analysis demonstrated that the Pearson correlation coefficient between the LD_{50} and the probit value of the inhibition rate was −0.816 ,which was significant at 0.05 significance level (two-tailed) ($p = 0.025$; Figure 2).

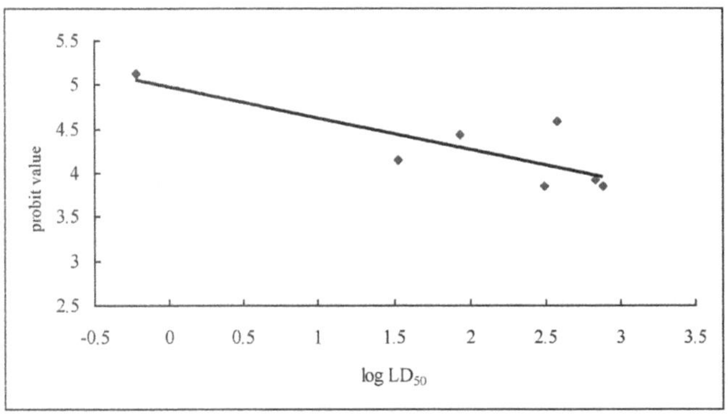

Figure 2. Correlation curve of LD_{50} and inhibition rate of V-ATPase.

Interaction between Subunit H and Dihydroagarofuran Sesquiterpene Polyesters

From the results of V-ATPase assay and the correlation analysis, we could roughly conclude that V-ATPase was the target protein of dihydroagarofuran sesquiterpene polyesters. However, the proteins separated by affinity chromatography include subunit a, B, and H of V-ATPase. Subunit H of V-ATPase is essential for the catalysis but not for the assembly of the enzyme [15,16,17]. Furthermore, it acts as an inhibitor of ATP hydrolysis in the free V_1 complex [16], and is probably the binding site for other proteins which interact with V-ATPase [17,18]. Owing to the important function of subunit H, we firstly cloned, expressed, and purified subunit H to study the interaction between subunit H and the 12 dihydroagarofuran sesquiterpene polyesters.

Expression and Purification of Subunit H of V-ATPase

Subunit H was expressed at 18 °C after inducing by isopropyl-β-d-thiogalactoside (IPTG) for 18 h. After twice purification by using Ni-NTA, subunit H (~55 kDa) was obtained and concentrated using an ultra membrane (Figure 3). Figure 3 shows that the purity of recombinant subunit H can meet the requirements for interaction analysis.

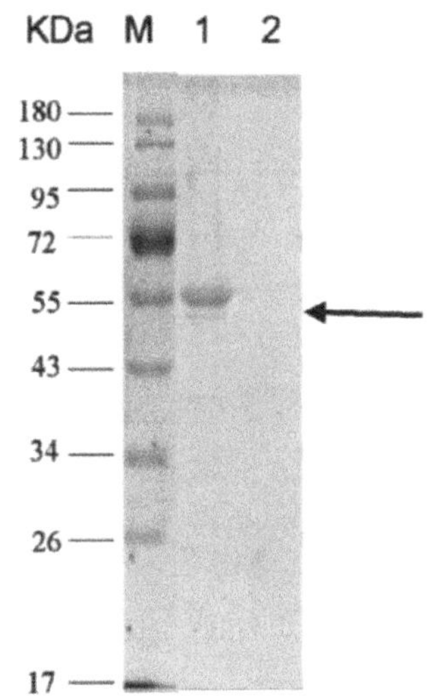

Figure 3. Recombinant subunit H purified by Ni-NTA. M: Protein Marker; 1: concentrated subunit H purified by Ni-NTA; 2: protein flow through the ultramembrane.

Interaction between Subunit H and Small Molecules

After purifying the subunit H, the interaction between subunit H and dihydroagarofuran sesquiterpene polyesters was evaluated by microscale thermophoresis (MST, Monolith NT.115). The results indicated that four compounds cannot bind with the recombinant subunit H, whereas the others can bind with subunit H; K_D values were obtained through the binding curve (Table 3).

Table 3. Nteraction between dihydroagarofuran sesquiterpene polyesters and subunit H of *M. separata*.

Samples	Interaction	K_D Value (μM)
Wilforine	binding	59.0
CV-6-α-aminopropanoic acid ester	binding	144.0
CV-6-aminoacetic acid ester	binding	156.0
NW70	binding	164.0
NW69	binding	186.0
NW03	binding	203.0
CV-6-*N*-methylisatoic	binding	213.0
Celangulin V (CV)	binding	253.0
CV-6-isobutyric acid ester	no-binding	-
CV-6-ketone	no-binding	-
NW57	no-binding	-
NW62	no-binding	-

Based on the data in Table 1, Table 2 and Table 3, the correlation of K_D and insecticidal toxicity, as well as the inhibition rate of V-ATPase activity, were analyzed. First, the logarithm value of LD_{50} and K_D can be fitted in the regression curve $y = 53.174x + 47.634$. The Pearson correlation coefficient was 0.870, and the *p*-value was 0.011 (two-tailed), which is significant at 0.05 significance level (Figure 4a). Second, the corresponding probit values of inhibition rate of V-ATPase and K_D value also were also correlated by fitting in the regression equation of $y = -95.912x + 580.47$. The Pearson correlation coefficient was −0.730, which is significant at a 0.05 significance level ($p = 0.04$) (Figure 4b).

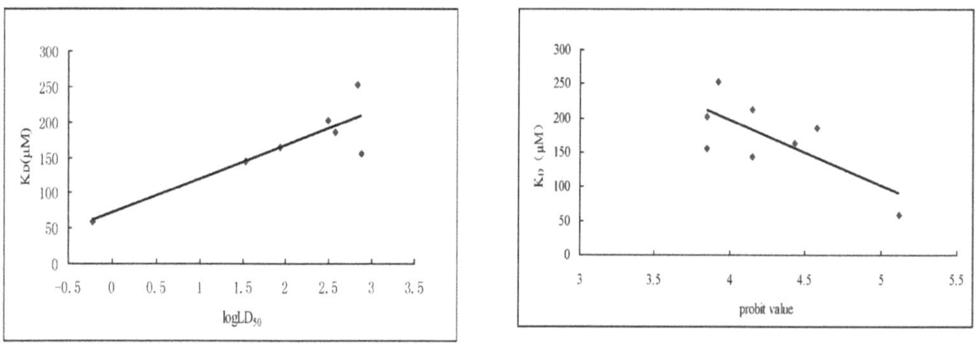

(a) (b)

Figure 4. (a) Correlation curve of K_D and LD_{50}; (b) Correlation curve of K_D and probit value of inhibition rate of V-ATPase.

CV-6-*N*-methylisatoic and CV-6-isobutyric acid ester were exceptions among all these compounds. CV-6-*N*-methylisatoic did not show insecticidal activity, but it can inhibit the activity of V-ATPase, with an inhibition rate of 24.04% and bind with subunit H with a K_D value of 213 μM. Whereas, CV-6-isobutyric acid ester did not show toxicity to *M. separata* larvae and cannot bind with subunit H, but it showed low inhibitory activity of V-ATPase. The possible reason is that the insecticidal activity was conducted *in vivo*, where penetration, storage, and degradation by detoxification enzymes occur, thereby inhibiting the insecticidal activity. By contrast, the experiments of V-ATPase activity and interaction with subunit H were conducted *in vitro*, where the compounds form direct contact with proteins.

Except for wilforine, the other dihydroagarofuran sesquiterpene polyesters had lower inhibition rate against V-ATPase from midgut of *M. separata* larvae. One of the possible reasons is that the insecticidal compounds are polyesters, which have poor solubility in aqueous solution. Owing to the difficulty of increasing the solubility, these compounds showed lower inhibition of V-ATPase. In addition, the low solubility of dihydroagarofuran sesquiterpene polyesters may lead to inaccuracy of measurement of K_D.

DISCUSSION

Correlation analysis is one of the methods used for target validation. The goal of correlation analysis is to analyze the correlation between the different affinities for target proteins after getting a series analog of small molecules of interest and the potency of causing the corresponding phenotype by these molecules. The keys in applying this method is to synthesize analogs of the bioactive small molecule, which has a wide range of activity (IC_{50} or LD_{50} values) spanning at least three orders of magnitude and to make the values more or less equally distributed over those three orders of magnitude of activity instead of being concentrated around one particular value [19]. In general, synthetic bioactive molecules can easily meet the above requirements; by contrast, obtaining sufficient numbers of natural bioactive products for analysis is difficult. In this study, the insecticidal activity (LD_{50} value) of the dihydroagarofuran sesquiterpene polyesters vary in three orders of magnitude, but the number is insufficient.

One important function of midgut epithelial cells of Lepidoptera insect is to transport K^+ from the haemolymph to the gut lumen. The process is performed by V-ATPase, which is located at the goblet cell apical membranes, as well as the V-ATPase's partner, the K^+/nH^+ antiporter [20]. Then, the combined action of the V-ATPase and $K^+/2H^+$ antiporter generates a transepithelial voltage [21]. Herein, the measurement of transmembrane potential difference can indicate the effect of a bioactive molecule on V-ATPase. We evaluated the influence of CV on transmembrane potential of goblet cell apical membranes of the sixth instar larvae of *M. separata* Walker. The results demonstrated that CV can induce depolarization of midgut apical membranes potential, that is, a decrease of the potential difference, which is another indication that CV can inhibit V-ATPase activity [22].

In addition to correlation analysis, genetic approaches can be employed for target validation. Through binding to a specific protein, the bioactive small molecule causes a cellular phenotype and loss of its function. Thus, over-expression of the target protein produce resistance to the small molecule; deletion of the target protein causes the same cellular phenotype with the small molecule; and reduction of the target protein causes hypersensitivity to the small molecule [19]. Subunit H has physiological function only after assembling with other subunits because V-ATPase is a multi-subunits enzyme. The over-expression of subunit H cannot induce the excess assembly of V-ATPase, which then causes resistance to the small molecule, whereas the reduction of subunit H decreases the assembly of V-ATPase, which causes hypersensitivity to small molecules. By using RNAi technology, we injected the dsRNA of subunit H into the third instar larvae of *M. separata* Walker, which significantly decreased the expression of subunit H. Moreover, silencing of the genes of subunit H induced the death of larvae, which showed the same symptoms as those induced by CV and other dihydroagarofuran sesquiterpene polyesters [23].

The target validation showed that subunit H of V-ATPase is one of the target proteins of the insecticidal dihydroagarofuran sesquiterpene polyesters. Based on the above results, we speculate that the mechanism of action of the insecticidal dihydroagarofuran sesquiterpene polyesters. Small molecular compounds are taken by larvae and then transported from the peritrophic membrane to the midgut cells. Then, the small molecules bind with subunit H of V-ATPase on the plasma membrane, which impacts the assembly of subunit H and other subunits so as to inhibit the activity; or the subunit H with small bioactive molecules become a part of V-ATPase, which lead to the malfunction of the whole enzyme. According to the function of V-ATPase, inhibition of its activity may lead to three kinds of results: (1) Amino acid cannot be transported to the gut cells, thereby influencing the protein synthesis and the function of the whole cell; (2) The high alkaline environment in the midgut cannot be maintained, thereby hindering numerous enzymes that function in high-pH environment; and (3) K^+ cannot enter the gut lumen from cells, thereby inducing accumulation of K^+ in cells. As a result, the osmotic pressure of cells will be imbalanced and excessive amount of water will enter the cells, resulting in swollen and cracked cells. Subsequently, the haemolymph will enter the gut lumen, which might be the reason for the loss of body fluid.

CONCLUSIONS

The correlation analysis of target protein in the current study, along with the RNAi and electrophysiologic assays in a previous study, showed that subunit H of V-ATPase is the target protein of insecticidal dihydroagarofuran sesquiterpene polyesters.

MATERIALS AND METHODS

Insects

Laboratory-adapted *M. separata* (Walker) was obtained from the Institute of Pesticide Science, Northwest A & F University (NWAFU, Yangling, China). The strain was reared on wheat and corn leaves under laboratory conditions for about 20 years, and was never treated with insecticides.

Chemicals

CV and its derivatives and analogs (purity >98% according to HPLC analysis) were provided by the Institute of Pesticide Science, NWAFU. Wilforine was purchased from PureOne Biotechnology (Shanghai, China). The structures of all the 12 compounds are presented below (Figure 5). Tris-ATP, orthovanadate, HEPES, $NaHCO_3$, KCl, $CaCl_2$, and Mes were purchased from Sigma (Shanghai, China). Protease inhibitor cocktail (EDTA-free) was purchased from Roche (Shanghai, China). Bafilomycin A1 was purchased from Santa Cruz Biotechnology (Shanghai, China). Ether was purchased from Xilong Chemical (Shantou, China). The substrate of APN, L-leucine-*p*-nitroanilide (Leu-pNA), was purchased from J & K Chemical Ltd. (Shanghai, China). Tris, NaCl, HCl, EDTA, $MgSO_4$, and all other chemicals were purchased from Guanghua Sci-Tech (Guangzhou, China).

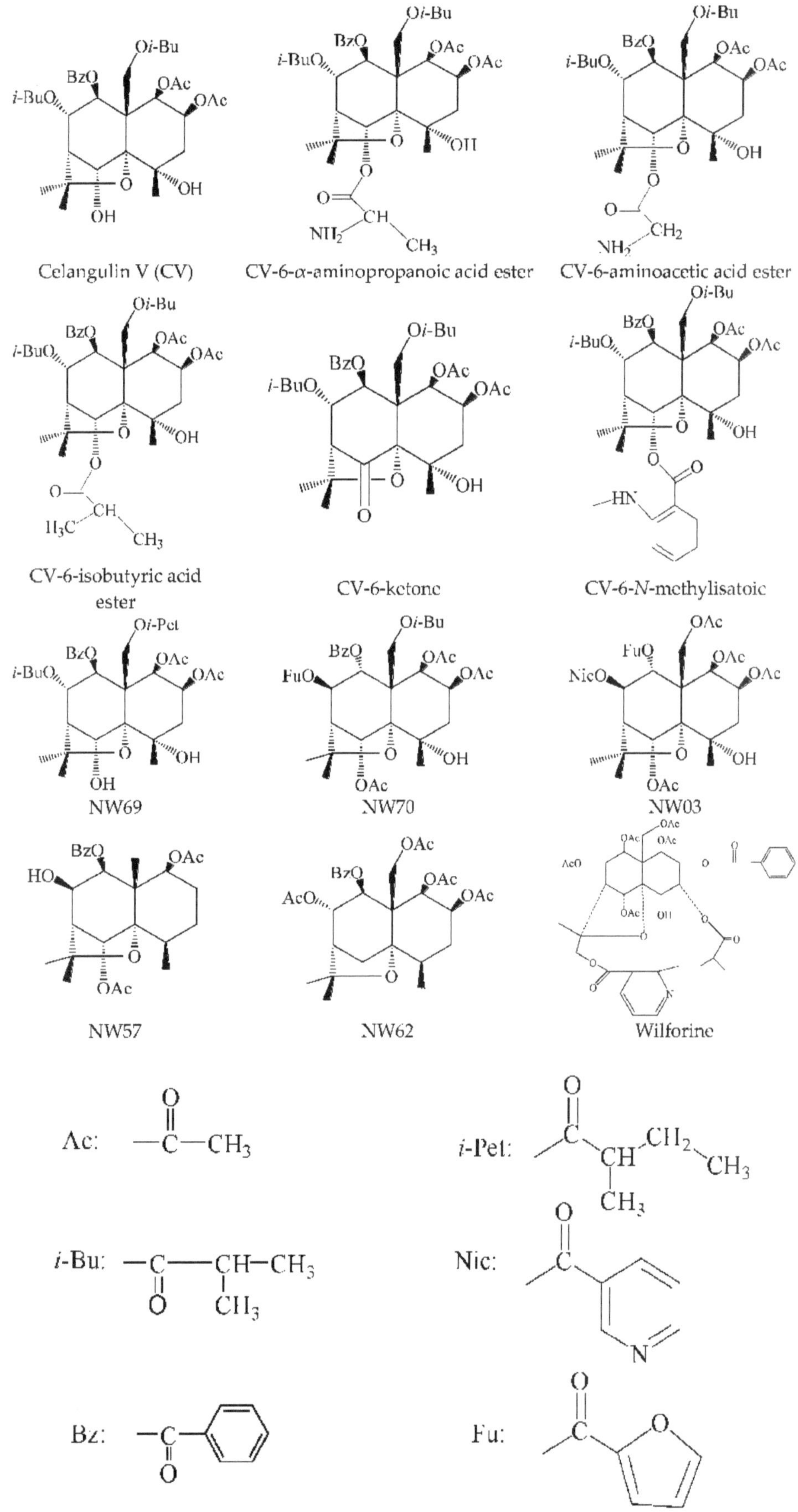

Figure 5. Structures of dihydroagarofuran sesquiterpene polyesters.

Bioassay of Insecticidal Activity

Force-feeding bioassay was conducted to test the insecticidal activity of the 12 dihydroagarofuran sesquiterpene polyesters against the fifth instar larvae of *M. separata* [24]. The 12 tested compounds were dissolved in DMSO and diluted to 5 concentrations by using serial dilution method. Second-day fifth instar larvae were selected and narcotized by cotton ball dipped with ether in Petri dish. Then, 0.5 μL of tested compounds were pipetted into the mouthparts of each larva. The larvae were then transferred into a 24-well plate after swallowing the compounds. For each concentration of each compound, 12 larvae were tested with three replicates. The symptoms presented by the larvae were observed after forced feeding, and the mortality was calculated 24, 48, and 72 h later. At the same time, 20 larvae were randomly selected and weighed to calculate the average body weight. Finally, the LD_{50} (μg/g) of each compound was calculated according to the LC_{50} value, volume of compounds, and average weight of larvae.

APN and V-ATPase Activity Assays

APN Activity Assays

The midgut BBMV of *M. separata* was isolated according to the $MgCl_2$ precipitation method [25], as modified by Ferre *et al.*[26]. The final pellet was dissolved in buffer C (150 mM NaCl, 5 mM EGTA, 1 mM PMSF, 20 mM Tris-HCl, and 1% CHAPS) [27]. The protein concentration of BBMV was measured by Bradford Assay.

APN activity was measured according to the method published by Liang *et al.* [28], Watanabe *et al.* [29], Takesue *et al.* [30], and Silva-Filha *et al.* [31]. The enzyme assay system includes 1 mL of buffer (0.25 mol/L Tris-HCl, 26 M NaCl, pH 7.8), 10 μL of BBMV extract, and 1.6 μL of CV (30.2 mM, dissolved in DMSO). The final concentration of CV in the reaction solution is 47 μM according to the amount added into the buffer, as some CV has dissolved out from the buffer. After incubation of the mixture for 30 min at 37 °C, 16 μL of substrate (15.96 mg of Leu-pNA was dissolved in 1 mL of methanol) was added. The absorbance at 405 nm was measured after incubation for 60 min at 37°C. The assay was repeated for three times, and DMSO was used as control.

V-ATPase Activity Assays

The V-ATPase activity was measured according to the method published by Tiburcy *et al.* [32]. The tested dihydroagarofuran sesquiterpene polyesters were dissolved in DMSO, and the final concentration of the polyesters in the reaction system was 100 μM. Bafilomycin A1 was used as the positive control with the final concentration of 3 μM.

Midguts of sixth instar larvae of *M. separata* were removed from the larvae in the Ringer solution; the peritrophic membrane, gut contents, and malphigian tubules were discarded. The rear parts of the midguts were placed in ice-cold low-adhesion Eppendorf tubes and frozen in liquid nitrogen. Then, the midguts were homogenized and centrifuged twice according to the method published by Tiburcy *et al.* [32]. The V-ATPase assays were performed with three replications. A 160 μL reaction solution consisted of 50 μL of PO buffer (160 mM Tris-Mes, pH 6.9), 20 μL of MV buffer (30 mM $MgCl_2$ and 0.8 mM sodium orthovanadate), 20 μL of KN buffer (160 mM KCl and 4 mM NaN_3), and 10 μL of DMSO/Bafilomycin A1/tested compounds. After 10 min of pre-incubation at 30 °C, 20 μL of Tris-ATP (8 mM) was added to start the reaction. The reaction was stopped after 60 min of incubation at 30 °C by freezing the samples in liquid nitrogen. To measure the V-ATPase activity, the produced inorganic phosphate was determined as described by Wieczorek *et al.* [33].

Binding Affinity Measurements

Expression and Purification of Subunit H of V-ATPase

The gene of V-ATPase subunit H of *M. separata* was cloned through RACE technology [34]. To obtain enough recombinant protein for the binding affinity measurements, the CDS of V-ATPase subunit H was cloned to the pET-15b vector by adding a His tag at the *N*-terminal and then heterologously expressed in *E. coli* BL21 (DE3). The construction of expression vector was described by Li *et al.* [35]. When the OD_{600} value of media reaches ~0.6, IPTG was added into the LB broth at a final concentration of 0.4 mmol/L to induce the expression of recombinant protein at 18 °C for 18 h.

Afterward, the bacterial cells were collected by centrifugation and resuspended with lysis buffer (20 mmol/L Tris-HCl, 300 mmol/L NaCl, and 5 mmol/L imidazole). Lysozyme and nuclease were added into the solution, followed by ice incubation for 30 min. Then, the cells were lysed by ultrasonic. After centrifugation at 12,000 rpm for 30 min (4 °C), the supernatant was used to purify the recombinant protein by Ni-NTA.

The supernatant was loaded onto the Ni-NTA column, which was then washed using buffer (20 mmol/L Tris-HCl, 300 mmol/L NaCl, pH 8.0), with 20 mmol/L and 40 mmol/L imidazole in sequence. An elution buffer (20 mmol/L Tris-HCl, 300 mmol/L NaCl, 500 mmol/L imidazole, pH 8.0) was then used to elute all the binding proteins. SDS-PAGE electrophoresis analysis showed that the eluate proteins washed by the elution buffer contained a number of unspecific proteins. Therefore, a second Ni-NTA purification was necessary. After dialysis at 4 °C for 24 h, the eluate was loaded to the Ni-NTA column. Then, each fraction washed by buffer (20 mmol/L Tris-HCl, 300 mmol/L NaCl, pH 8.0) containing 40, 80, 100, and 200 mM imidazole were collected and detected by SDS-PAGE electrophoresis (Liuyi Biotechnology, Beijing, China).

Interaction between Subunit H and Small Molecules

The purified recombinant protein was concentrated using an ultra membrane (Merck Millipore Amicon, Shanghai, China), and then diluted to 16 μM. After that, the protein was labeled by RED-NHS, according to the user's manual of Monolith NT.115 (Nano Temper, Munich, Germany). The labeled protein sample was diluted by buffer (20 mmol/L Tris-HCl, 300 mmol/L NaCl, pH 8.0) to ensure the florescence value was between 200 and 1500. Furthermore, Tween-20 (0.1% final concentration) was used to enhance the sample quality. Each dihydroagarofuran sesquiterpene polyester was resolved in DMSO, and 16 different concentrations were prepared by serial dilution. After mixing the 16 samples with labeled protein at a ratio of 1:1, MST capillaries were filled and placed on a tray for the binding affinity measurements. The K_D values of each compound with the subunit H were calculated according to the binding curve [36].

ACKNOWLEDGMENTS

This research was supported with grants from the National Natural Science Foundation of China (31371958) and the National Basic Research Program (973 Program, 2010CB126105) from Science and Technology, Ministry of China.

AUTHOR CONTRIBUTIONS

W.W. conceived and designed the experiments; L.L., Z.Q. and Q.L. performed the experiments; W.W., L.L. and Z.Q. analyzed the data; L.L. and W.W. wrote the paper.

REFERENCES

1. Nauen, R.; Denholm, I. Resistance of insect pests to neonicotinoid insecticides: Current status and future prospects.*Arch. Insect Biochem. Physiol.* 2005, *58*, 200–215.
2. Matsuda, K.; Buckingham, S.D.; Kleier, D.; Rauh, J.J.; Grauso, M.; Sattelle, D.B. Neonicotinoids-insecticides acting on insect nicotinic acetylcholine receptors. *Trends Pharmacol. Sci.* 2001, *22*, 573–580.
3. Copping, L.G.; Duke, S.O. Natural products that have been used commercially as crop protection agents. *Pest Manag. Sci.* 2007, *63*, 524–554.
4. Gao, J.M.; Wu, W.J.; Zhang, J.W.; Konishi, Y. The dihydro-β-agarofuran sesquiterpenoids. *Nat. Prod. Rep.* 2007, *5*, 1153–1189.
5. Ji, Z.; Zhang, Q.; Shi, B.; Wei, S.; Wang, M.; Wu, W. Three new insecticidal sesquiterpene pyridine alkaloids from*Celastrus angulatus*. *Nat. Prod. Res.* 2009, *23*, 470–478.
6. Ji, Z.; Wu, W.; Yang, H.; Shi, B.; Wang, M. Four novel insecticidal sesquiterpene esters from *Celastrus angulatus*. *Nat. Prod. Res.* 2007, *21*, 334–342.
7. Yang, R.Y.; Liu, H.X.; Wu, W.J.; Wang, J.L. Study on the functioning mechanism of celangulin V. *J. Northwest Sci. Technol. Univ. Agric. For.* 2001, *29*, 77–79.
8. Qi, Z.; Shi, B.; Hu, Z.; Zhang, Y.; Wu, W. Ultrastructural effects of Celangulin V on midgut cells of the oriental armyworm, *Mythimna separata* walker (Lepidoptera: Noctuidae). *Ecotoxicol. Environ. Saf.* 2011, *74*, 439–444.
9. Wu, W.J.; Ji, Z.Q.; Hu, Z.N. Natural products and digestive poisons. *Pesticides* 1997, *36*, 6–9.
10. Lu, L.; Qi, Z.; Zhang, J.; Wu, W. Separation of binding protein of Celangulin V from the Midgut of *Mythimna separata*Walker by affinity chromatography. *Toxins* 2015, *7*, 1738–1748.
11. Caboni, P.; Tronci, L.; Liori, B.; Tocco, G.; Sasanelli, N.; Diana, A. Tulipaline A: Structure-activity aspects as a nematicide and V-ATPase inhibitor. *Pestic. Biochem. Physiol.* 2014, *112*, 33–39.
12. Malik, K.; Mahmood, T.; Riazuddin, S. The receptor for *Bacillus thuringiensis* Cry 1Ac Delta-endotoxin in the brush border membrane of the Lepidopteran *Helicoverpa armigera* is aminopeptidase N. *J. Biol. Sci.* 2001, *1*, 782–784.
13. Knight, P.J.K.; Crickmore, N.; Ellar, D.J. The receptor for *Bacillus thuringiensis* CrylA(c) delta-endotoxin in the brush border membrane of the lepidopteran *Manduca sexta* is aminopeptidase N. *Mol. Microbiol.* 1994, *11*, 429–436.
14. McNall, R.J.; Adang, M.J. Identification of novel Bacillus thuringiensis Cry1Ac binding proteins in *Manduca sexta*midgut through proteomic analysis. *Insect Biochem. Mol. Biol.* 2003, *33*, 999–1010.
15. Liu, M. Structural and functional separation of the *N*- and *C*-terminal domains of the yeast V-ATPase subunit H. *J. Biol. Chem.* 2005, *280*, 36978–36985.
16. Geyer, M.; Fackler, O.T.; Peterlin, B.M. Subunit H of the V-ATPase involved in endocytosis shows homology to β-adaptins. *Mol. Biol. Cell* 2002, *13*, 2045–2056.
17. Margaret, N.H.; Stevens, T.H.; Matthews, B.W.; Ohya, Y.; Takatsukill, A.; Stevens, T.H.; Anraku, Y. *VlMA13* Encodes a 54-kDa Vacuolar H^+-ATPase subunit required for activity but not assembly of the enzyme complex in *Saccharomyces cerevisiae*. *J. Biol. Chem.* 1993, *268*, 18286–18292.
18. Jefferies, K.C.; Forgac, M. Subunit H of the Vacuolar (H^+) ATPase inhibits ATP hydrolysis by the free V1 domain by interaction with the rotary subunit F. *J. Biol. Chem.* 2007, *283*, 4512–4519.
19. Titov, D.V.; Liu, J.O. Identification and validation of protein targets of bioactive small molecules. *Bioorg. Med. Chem.*2012, *20*, 1902–1909.
20. Wieczorek, H.; Beyenbach, K.W.; Huss, M.; Vitavska, O. Vacuolar-type proton pumps in insect epithelia. *J. Exp. Biol.*2009, *212*, 1611–1619.
21. Wieczorek, H.; Grüber, G.; Harvey, W.R.; Huss, M.; Merzendorfer, H.; Zeiske, W. Structure and regulation of insect plasma membrane H^+-V-ATPase. *J. Exp. Biol.* 2000, *203*, 127–135.
22. Wang, Y. A Comparative Study of Digestive Poisons Periplocoside P and Celangulin V on Midgut Cell Transmembrane Potential of *Mythimna Separata* Larvae. Master's Thesis, Northwest A & F University, Yangling, China, 14 May 2015.
23. Zhang, S.; Wang, G.; Liu, S.; Wu, W.; Qi, Z. Silencing of V-ATPase subunit H of *Mythimna separata* through RNA interference. *Acta Agric. Boreali Occident. Sin.* 2015, *24*, 170–174.
24. Abdullah, M.A.F.; Moussa, S.; Taylor, M.D.; Adang, M.J. *Manduca sexta* (Lepidoptera: Sphingidae) cadherin fragments function as synergists for Cry1A and Cry1C *Bacillus thuringiensistoxins* against noctuid moths *Helicoverpa zea*, *Agrotis ipsilon* and *Spodoptera exigua*. *Pest Manag. Sci.* 2009, *65*, 1097–1103.
25. Wolfersberger, M.G.; Luthy, P.; Maurer, A.; Parenti, P.; Sacchi, V.F.; Giordana, B.; Hanozet, G.M. Preparation and partial characterization of amino acid transporting

brush border membrane vesicles from the larval midgut of the cabbage butterfly *(Pieris brassicae)*. *Comp. Biochem. Physiol. A Physiol.* 1987, *86*, 301–308.

26. Ferré, J.; Real, M.D.; Rie, J.V.; Jansens, S.; Peferoen, M. Resistance to the *Bacillus thuringiensis* bioinsecticide in a field population of *Plutella xylostella* is due to a change in a midgut membrane receptor. *Proc. Natl. Acad. Sci. USA* 1991, *88*, 5119–5123.
27. Chen, L.Z.; Liang, G.M.; Zhang, J.; Wu, K.M.; Guo, Y.Y.; Rector, B.G. Proteomic analysis of novel Cry1Ac binding proteins in *Helicoverpa armigera* (Hübner). *Arch. Insect Biochem. Physiol.* 2010, *73*, 61–73.
28. Liang, G.; Xu, G.; Wang, G.; Wu, K.; Guo, Y. Isolation and purification of aminopeptidase N (APNs), the main receptor proteins for *Bacillus thuringiensis* toxin in midgut of *Helicoverpa armigera* larva. *J. Agric. Biotechnol.* 2005, *13*, 102–107.
29. Watanabe, Y.; Shimamori, Y.; Ozaki, M.; Uchida, M.; Fujimoto, Y. Purification of a membrane-bound neutral endopeptidase from rat kidney and its activation by polyamines. *J. Biochem.* 1984, *95*, 1725–1732.
30. Takesue, Y.; Yokota, K.; Nishi, Y.; Taguchi, R.; Ikezawa, H. Solubilization of trehalase from rabbit renal and intestinal brush-border membranes by a phosphatidylinositol-specific phospholipase C. *FEBS Lett.* 1986, *201*, 5–8.
31. Silva-Filha, M.H.; Nielsen-LeRoux, C.; Charles, J.-F. Identification of the receptor for *Bacillus sphaericus* crystal toxin in the brush border membrane of the mosquito *Culex pipiens* (Diptera Culicidae). *Insect Biochem. Mol. Biol.* 1999, *29*, 711–721.
32. Tiburcy, F.; Beyenbach, K.W.; Wieczorek, H. Protein kinase A-dependent and -independent activation of the V-ATPase in Malpighian tubules of *Aedes aegypti*. *J. Exp. Biol.* 2012, *216*, 881–891.
33. Wieczorek, H.; Cioffi, M.; Klein, U.; Harvey, W.R.; Schweikl, H.; Wolfersberger, M.G. Isolation of goblet cell apical membrane from tobacco hornworm midgut and purification of its vacuolar-type ATPase. *Methods Enzymol.* 1990, *192*, 608–616.
34. Lu, L.; Qi, Z.; Wu, W. Cloning, expression and purification of subunit H of vacuolar H^+-ATPase from *Mythimna separata* Walker (Lepidoptera: Noctuidae). *Int. J. Mol. Sci.* 2014, *15*, 15443–15455.
35. Li, Q.; Zhao, J.; Qi, Z.; Wu, W. Prokaryotic expression and purification of V-ATPase subunit H in the midgut of *Mythimna separata*. *Acta Agric. Boreali Occident. Sin.* 2014, *23*, 102–106.
36. Jerabek-Willemsen, M.; Wienken, C.J.; Braun, D.; Baaske, P.; Duhr, S. Molecular interaction studies using microscale thermophoresis. *Assay Drug Dev. Technol.* 2011, *9*, 342–353.

CHAPTER 13

CHITINASES FROM BACTERIA TO HUMAN: PROPERTIES, APPLICATIONS, AND FUTURE PERSPECTIVES

Abhishek Singh Rathore and Rinkoo D. Gupta

INTRODUCTION

Chitin is a polymer of N-acetylglucosamine, linked with β-1,4-glycosidic bonds. It is a major portion of cell walls of fungi, exoskeleton of insects, and crustacean shells. Despite its abundance, chitin does not accumulate in the environment due to presence of chitinolytic enzymes known as "chitinases." Several organisms including bacteria, fungi, insects, plants, and animals produce chitinases. Detailed structural and mechanical differences among the chitinases produced by these organisms are reviewed by Adrangi and Faramarzi [1]. These organisms produce numerous types of chitinolytic enzymes. Many bacteria including Serratia and Bacillus are known to produce four different types of chitinases. However, most of the filamentous fungi are known to produce up to 20 different chitinases [2]. They act together in synergistic manner to catabolize chitin. These enzymes are able to degrade the chitin present in the cell walls of fungi as well as the exoskeletons of insects. They have shown being the potential agents for biological control of the plant diseases caused by various pathogenic fungi and insect pests that can be used as an alternative to chemical pesticides. There has been steady increase in demand of chitin derivatives, obtained by action of chitinases on chitin polymer for various industrial, clinical, and pharmaceutical purposes. However, there are immense challenges for the production of specific and cost-effective enzymes. Possibilities of several potential applications of chitinases make it an interesting target enzyme for protein engineering. Considering the significant contribution of chitinases as biopesticides and in various industrial and pharmaceutical purposes, we aim to articulate properties of chitinases starting from bacteria, which is followed by fungi, insects, plants, and animals. Comprehensive applications of the chitinases are also explored. At the end, we aim to survey the future perspectives of chitinases by applying protein-engineering techniques.

STRUCTURES AND FUNCTIONS OF CHITIN

Chitin is a linear stable polymer of β-1,4-N-acetylglucosamine (GlcNAc), which is 2-acetamido-2-deoxy-D-glucose, and is the second highest occurring biopolymer on the planet after cellulose [3]. It is present in exoskeleton of insect, crabs, shrimp, lobsters, fungi, yeast, diatoms, nematodes, crustaceans, and other invertebrates. Percentage of chitin found varies with different organisms. Highest percentage of chitin is found in crabs and shrimps, which makes 90% of chitinous waste. It is a crystalline polysaccharide that exists in nature in three different forms: α-chitin, β-chitin, and ϒ-chitin. α-Chitin is the most abundant, isomorphic, and more compact form due to the arrangement of chitin chains in antiparallel fashion which favors strong hydrogen bonding. β-Chitin is loosely packed as in this chitin, and chains are arranged in parallel fashion with weaker intermolecular forces which leads to less stable form. Third polymorphic form is ϒ-chitin which is a mixture of both α- and β-chitins [4]. Hydrolysis of glycosidic bonds of chitin is catalyzed by action of chitinases which degrades chitin into disaccharides and longer oligosaccharides. β-chitin is easily hydrolyzed and more soluble as compared to α-chitin. It can be swollen in water as well as dissolved in formic acid [5].

Chitin mainly helps to form exoskeleton as in arthropods and some species of fungi; in addition chitin in several species is associated with many proteins, which determine whether the exoskeleton will be rigid, soft, or flexible. It also associates with non-protein part such as calcium carbonate that forms the structure in shrimp, crab, and lobsters. Chitin and its related materials have extensive applications in drug delivery: wound

Citation: Abhishek Singh Rathore and Rinkoo D. Gupta, "Chitinases from Bacteria to Human: Properties, Applications, and Future Perspectives," Enzyme Research, vol. 2015, Article ID 791907, 8 pages, 2015. DOI:10.1155/2015/791907.

healing, dietary fiber, and wastewater treatment [6]. Also, chitin provides protective benefits by forming exoskeleton in arthropods and fungi.

CLASSIFICATIONS AND CATALYTIC MECHANISM OF CHITINASES

Chitinases (EC 3.2.2.14) are glycosyl hydrolases, characterized for hydrolyzing β-1,4 linkage of N-acetyl glucosamine present in chitin chains which vary between the size ranging from 20 kDa to about 90 kDa [7]. In natural state, chitin is tightly bound with lipid pigments, proteins, and minerals like calcium carbonate; hence preparation of chitin includes deproteinization and demineralization of chitinous waste by strong acid or a base. These processes involve high cost, low yield, and corrosion problems due to which cost of its oligomers is much. Therefore, chitinolytic enzymes are important component in utilization of chitinous waste and solve environmental problems as they are biodegradable and inexpensive. Chitinolytic enzymes have been divided into two main categories: endochitinases (EC 3.2.1.14) and exochitinases. Endochitinases are the class of chitinases that cleaves randomly at internal sites in polymer of chitin generating low molecular mass multimers of glucosamine residues such as chitotriose, chitobiose, and diacetylchitobiose (Figure 1). Exochitinases have been classified into two categories, namely, chitobiosidases (EC 3.2.1.29) which catalyze progressive release of diacetylchitobiose from terminal nonreducing end and N-acetylglucosaminidases (EC 3.2.1.30) which cleaves oligomeric products obtained by endochitinases into monomers of N-acetyl glucosamine (GlcNAc) [8].

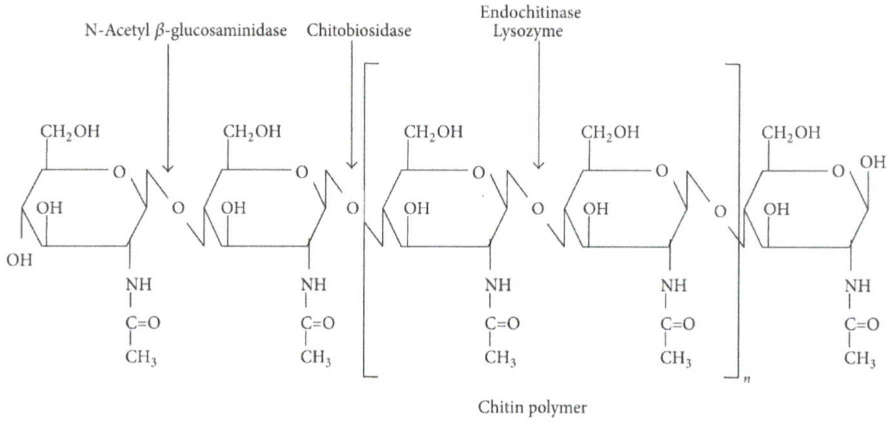

Figure 1: Specificity of different chitinases towards chitin polymer. N-Acetyl β-glucosaminidase cleaves monomeric unit of GlcNAc from nonreducing terminal. Chitobiosidase cleaves dimeric unit of GlcNAc from nonreducing terminal. Endochitinase cleaves glycosidic bond randomly at internal sites in chitin polymer.

Based on amino acid similarity of chitinases from various organisms, they have been divided into five classes and have been categorized into families 18, 19, and 20 of glycosyl hydrolases [11]. Family 18 comprises chitinases from viruses, bacteria, fungi, animals, and certain plant chitinases. It consists of a number of conserved repeats of amino acids and enzyme core, which has 8 strands of parallel β sheets, creating a barrel positioned down α helices, in turn forming a ring towards the outside (Figure 2) [12]. Family 19 comprises some plant chitinases and Streptomyces chitinases [13]. Families 18 and 19 do not share amino acid sequence similarity, and they have completely different 3D structure and therefore have been said to evolve from different ancestors. Family 20 involves N-acetylglucosaminidases from bacteria, certain fungi, and humans.

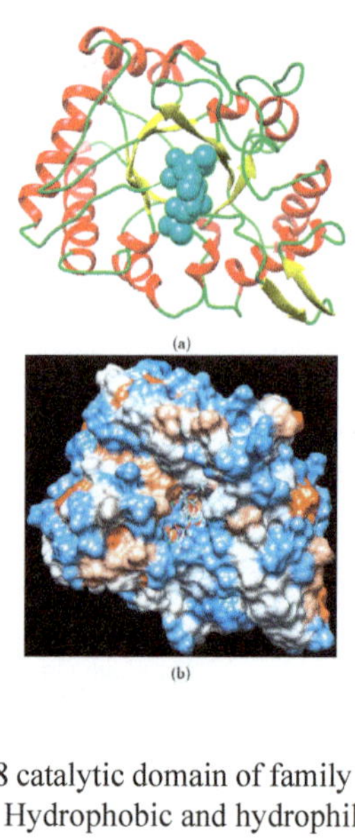

Figure 2: Interaction of GlcNAc dimer TIM-barrel (β/α)8 catalytic domain of family 18 bacterial endochitinase of S. marcescens: (a) Bacterial endochitinase interaction with GlcNAc dimer (blue); (b) Hydrophobic and hydrophilic regions around active site of chitinase C.

ROLE OF CHITINASES IN VARIOUS ORGANISMS

Mainly, organisms require these chitinolytic enzymes for three different purposes. (a) organisms possess a tough layer of chitin and chitinases are expressed during developmental phases to help in the remodeling of their exoskeleton so as to maintain and support body size and shape. (b) Organisms that consume other chitin containing organisms as a source of nutrient express chitinases to digest the insoluble chitin polymer into absorbable metabolites, which gives energy. (c) Organisms that are prone to infection by chitin-coated microorganisms express chitinases to degrade the protective shield of the infecting pathogens, thereby providing immunity. Different forms and specific functions of chitinases in bacteria, fungi, insects, plants, and vertebrates are described below for comprehensive appraisal.

Bacteria

Bacteria mainly produce chitinases in order to supply nitrogen and carbon as a source of nutrients or precursors and parasitism [14, 15]. They are used for degradation of chitin and its utilization as an energy source [16]. Chitinases play an important role in bacterial pathogenesis wherever host contains chitin [17].Serratia marcescens, one of the best studied chitinolytic bacteria, has been reported producing mainly four types of chitinases ChiA, ChiB, ChiC, and CBP21 (chitin binding protein). All three chitinases belong to family 18 of glycosyl hydrolases with (β/α) 8 TIM-barrel catalytic domain with approximately six sugar subsites [18]. ChiA and ChiB have multimodular organization, that is, have an N-terminal chitin binding module with a fibronectin-like fold in ChiA or a C-terminal CBM5 module. CBM modules found in chitinases are distantly related, and they are characterized by presence of conserved exposed tryptophan residues that interact with the substrate [19]. Presence of this domain increases the substrate binding affinity as well as efficiency of chitin hydrolysis, particularly for more crystalline forms of chitin [20].

Marine bacteria such as Vibrio are well studied because they solely live on chitin, which is highly abundant in marine ecosystems. Therefore they serve as an ideal candidate for bioconversion of chitin biomaterials for various purposes. The chitin degradation machinery of Vibrio is proved to be highly efficient. Structure and functions of chitinases and chitin uptake system in Vibrio are well studied. Songsiritthigul has described that the interacting sugars undergo conformational changes prior to hydrolysis by the wild-type enzyme with the help of crystallographic data obtained by four X-ray structures of Vibrio harveyi chitinase A and its catalytically inactive mutant (E315M) in the presence and absence of substrates [21]. Recently, chitoporin (VhChiP), a sugar-specific channel responsible for the transport of chitooligosaccharides through the outer membrane of the marine bacterium Vibrio harveyi, has been reported by Suginta and group [22]. VhChiP is found to be the most potent sugar-specific channel reported to date, with its high efficiency. This reflects apparently the need for the bacterium to take up chitin containing nutrients in aquatic conditions as its sole source of energy. Previous reviews, intensive on bacterial chitinases only, have signified the structural and functional aspects in great detail [1, 7, 14].

Fungi

Fungal chitinases play important role in nutrition, morphogenesis, and developmental process and are known to be produced at various stages during fungus growth [23]. Fungal chitinases belong to GH18 family of glycoside hydrolases which showed little amino acid sequence similarity with class 3 plant chitinases [8]. The fungal chitinases are divided into three groups, being chitinases A, B, and C on the basis of sequence and structural similarities [24]. Fungal chitinase A is processive chitinase with a singular catalytic domain having deep substrate binding site and no CBMs. Type B is nonprocessive chitinases and they have a CBM or a serine/threonine-rich domain on C-terminal of their catalytic domain. Type C fungal chitinases are also processive in nature due to their deep substrate binding site. They have a CBM on the N-terminal of catalytic domain. A special feature of fungal chitinase C is that it comprises several lysine motifs (LysM) also known as CBM 50 [2]. LysM containing receptors in plants play an important role in formation of root nodules in leguminous plants by binding to Nod factor [25]. Fungal chitinase C is found to expressed in many mycoparasitic fungi. A study on mycoparasitic Trichoderma species showed that these enzymes are involved in degradation of both self and non-self-cell walls [26]. Protection of self-cell wall in fungi is achieved by restricting the access to chitin by cell surface proteins.

Entomopathogenic fungi can play a major role as a biocontrol agent to control insects and pests. Overexpression and engineering of chitinases in these fungi increased their efficiency as a biocontrol agent [27]. Nematode parasitic fungi producing chitinase are used as biocontrol agent by targeting eggs and larva of plant and animal infecting nematodes [28]. Chitinase and chitin binding proteins can also be used for detection of invasive fungal infection in immunosuppressed patients [29].

The chitinases, which are associated with fungal cell wall, have role in filamentous fungal sporulation. This has been shown by using chitinase inhibitors demethylallosamidin and allosamidin, causing inhibition of fragmentation of hyphae into arthroconidia. Trichoderma have been considered as biocontrol agents in case of soil borne fungal pathogens amongst various chitinolytic fungi. Chitinases and β-1,3-glucanases fromTalaromyces flavus and Trichoderma sp. have been purified and characterized. Hartl and group have given detailed insight into the diversity, mechanistic properties, and biotechnological potential of several other types of fungal chitinases [2].

Insects

All insect chitinases belong to GH18 family. On the basis of sequence similarity insect chitinases are divided into eight groups, which vary in their molecular weight ranging between 40 kDa and 85 kDa and their pH optima [30]. The overall structure consists of three domains: the catalytic region, a region enriched in the amino acids proline, glutamate, serine and threonine, and a cysteine-rich region. However, the last two domains are not necessary for chitinase activity. Cysteine-rich C-terminal chitin binding domain of insect chitinases anchored the substrate [2, 30].

Insect chitinases are involved in degradation of cuticle layer to smaller oligosaccharides by hydrolyzing and randomly breaking cuticle that is used for synthesis of new cuticle. In this way, these chitinases are required for partial degradation of old cuticle and synthesis of the new one. Expression of these enzymes is required to be under stringent control to avoid any premature exposure, which may lead to growth inhibition [31]. To achieve this goal insect produces specific chitinases that are differentially expressed over the course of time in different stages of metamorphosis [32]. Chitinases are also reported to be present in venom and salivary gland of some insects. In these cases, role of chitinases may be degradation of host cuticle.

Plants

Based on their amino acid sequence plant chitinases have been divided into 7 classes each possessing different structure [33]. Classes III and V of plant chitinase belong to GH18 family while the rest of the five classes belong to GH19 family [34]. The plant chitinases are generally endochitinases and smaller in molecular weight as compared to the insect chitinases. Many plants continuously express chitinases in tuber, stem, and root. In plants they act as pathogenesis related proteins, that is, synthesized in response to self-defense against phytopathogenic attack also at time these chitinases have been reported to take part in vital physiological processes of plants like embryogenesis and ethylene formation [12]. Chitinase is known to play a major role in defense against chitin containing pathogens like fungi and insects. Intensive research is being conducted to develop transgenic plants overexpressing various combinations of chitinases as well as other pathogenesis related proteins [35].

Chitinases are also known to be expressed during high environmental stress conditions such as cold and drought [8]. Some of the plant chitinases are found to show ice structuring tendency, which means they bind to ice crystals reducing their growth thus helping cold stress condition [36]. Some of class III chitinases are reported to play an important role in calcium storage without having more significant effect on their catalytic activity. The structures and catalytic functions of plant chitinases and their role in plant physiology have remarkably been reviewed by Sahai and Manocha [8] and Grover [37].

Vertebrates

In vertebrates, chitinases have been known to be the part of digestive tracts; for example, tGCase functions as a chitinolytic enzyme in the toad stomach [38] and acidic mammalian chitinases (AMCase) in the mouse gastrointestinal tract [39]. Chitotriosidase in human macrophages has been found in higher levels in lysosomal storage diseases as well as in patients with Plasmodium falciparum malarial infection [40] and chitin binding protein b04 (CBPb04) in bovine serum [41]. Since chitin is a main constituent of fungus cell wall and also present in mammalian diet, so role of mammalian chitinases is to provide innate immunity and also to digest the food. Human also produces AMCase; a family 18 chitinase is induced during TH2 inflammation through an interleukin dependent mechanism. Cytokines which appear to be of particular importance in asthma are lymphokines secreted by T-lymphocytes, IL-3 necessary for survival of mast cells and IL-5 important for growth, differentiation, and survival of eosinophils. Therefore, mammalian chitinases can be used as a biomarkers for asthma and hence inhibition of AMCase with chitinase inhibitors such as allosamidin reduces inflammation and can be used as potential target for asthma therapy [42]. Recently published research articles and reviews on human chitinases have implied its role in pathogenesis and as a biomarker [1, 43, 44].

APPLICATIONS OF CHITINASES

Chitinases have several field applications. Chitinases are attaining prominence in the field of biotechnology applied in waste management, pest control in agriculture, and human health care which have been recapitulated in Figure 3 and discussed below in detail.

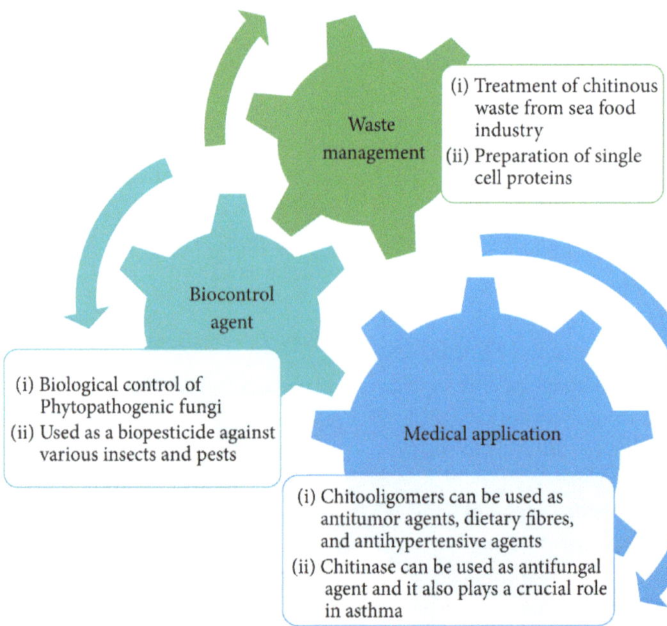

Figure 3: Industrial applications of chitinolytic enzymes.

Waste Management

Recombinant chitinases can be used to convert chitinous biomass, that is, chitinous waste of marine organisms into simpler useful depolymerized components, hence reducing water pollution. Chitooligomers obtained by action of chitinases have a wide range of biotechnological applications in biochemical, food, and various chemical industries. Chitinase can also be used in conversion of chitinous waste into biofertilizers [45].

Another approach to utilize the chitinous waste effectively is production of single cell protein (SCP) [46]. In this approach chitinase degraded chitinous waste is used as carbon or nutritional source for production of biomass. Chitinase producing bacteria and yeast (e.g., S. marcescens and Pichia kudriavzevii) can be used in aqua cultures for SCP production.

Biocontrol Agents

Chitinases are present in plants alongside various pathogenesis related proteins as a part of plant defense mechanism. So overexpression of a combination of various chitinases in transgenic plant may aid against fungal pathogens [35]. Chitinase can also be directly used as biopesticides against various fungi and insects that can be an alternative to chemical pesticides [47]. Other than being used directly as a biocontrol agent, chitinase can act as a target for biopesticide as chitin has a major role in insect metamorphosis as well as in gut of insects. A pseudo-trisaccharide allosamidin is an inhibitor of chitinase enzyme and it can be potentially used as biopesticide [48].

Medical Application

Chitinase is used as antifungal agent in combination with antifungal drugs in therapy for various fungal infections [49]. Human AMCase is found to increase in Th2 inflammation and is considered to play a role in asthma and allergic reactions. It is also involved in effector pathway of IL-13 [50]. Chitinase is also being suggested to be used for detection of invasive fungal infection in humans [51]. Chitooligosaccharides also have an enormous pharmaceutical potential to be used in human medicines because of its antitumor activity (shown by chitohexaose and chitoheptaose), wound healing property, and antihypertensive activity [52]. N-Acetyl glucosamine, which is a monomeric unit of chitin polymer, is also reported to be anti-inflammatory agent [53].

Plasmodium Falciparum is also reported to produce chitinase enzyme in the sporogonic cycle. Chitinase is produced by pathogen in the midgut of anopheline vector to disrupt the peritrophic membrane and let the parasite get to the salivary glands [54]. Inhibition of chitinase can stop the sporogonic cycle, so it can be considered a good target. Plasmodium falciparum forms a membrane sac around the ingested blood meal, that is, chitin containing peritrophic matrix, and thus addition of exogenous chitinases to blood meal prevents the formation of peritrophic matrix [55].

Miscellaneous Applications

Chitinases have been exploited to isolate fungal protoplasts that are used as experimental means to study the synthesis of cell wall, synthesis and secretion of enzymes, and strain improvement for biotechnological applications [56].

The level of chitinases can also be used for the indirect determination of fungal biomass present in the soil. An enzyme of food industry purpose tannase is produced by Aspergillus niger but tannase binds to its cell wall reducing the yield. Chitinase is used for fungal cell wall degradation that releases tannase from cell wall and increases the yield [57].

FUTURE PROSPECTIVE

Chitinase being the hydrolyzing enzyme for the second most abundant naturally occurring polymer holds many applications in the field of environmental, therapeutic, and industrial biotechnology. Economical and practical utilization of chitinases can be achieved either by high yield or better catalytic activity. Main focus in chitinase research is on improving its catalytic activity. Protein engineering of enzyme is providing a way to achieve this goal. Two major approaches for engineering protein are directed evolution and site directed mutagenesis. In directed evolution approach, chitinase can be randomly mutated by error prone PCR and thereby generation of a mutant library. The chitinase gene library can then be screened by using fluorescent activated cell sorter and/or in 96-well plates. Mutants showing higher activity as compared to wild type can be selected and combination of these positive mutations can also be attempted to get the best catalytic activity (Figure 4) [9, 10].

In case of site directed mutagenesis, specific amino acid residues residing in the chitin binding and catalytic sites are mutated. For instance, glutamate present in the active site of chitinase enzyme plays a major role in catalytic activity by acting as a proton donor for glycosidic bond cleavage [58]. So, any mutation near glutamate that enhances its proton donating efficiency can enhance enzymatic activity. To reduce steric hindrance, bulky group carrying amino acids at the edge of active site could be replaced by small noninteractive amino acid, for example, alanine [59]. Some residues far from the active site also play role in binding of chitinase to the enzyme. Processivity of chitinase can be enhanced by placing hydrophobic amino acid at the edge of active site to make a narrow groove above active site. Due to this narrow groove, chitinase will not release the polymeric chain after every cleavage but will continue on cleaving glycosidic bonds in a processive fashion. Domain swapping is another approach for enhancing chitinase activity. Many chitinases showed low catalytic activity due to lack of chitin binding domain. The chitin binding domain from an active chitinase can be swapped by applying protein engineering methodology. This could increase the catalytic activity by improving substrate binding of the engineered chitinase. The above-mentioned approaches of protein engineering hold huge potential for the improvement of chitinolytic enzymes for future applications.

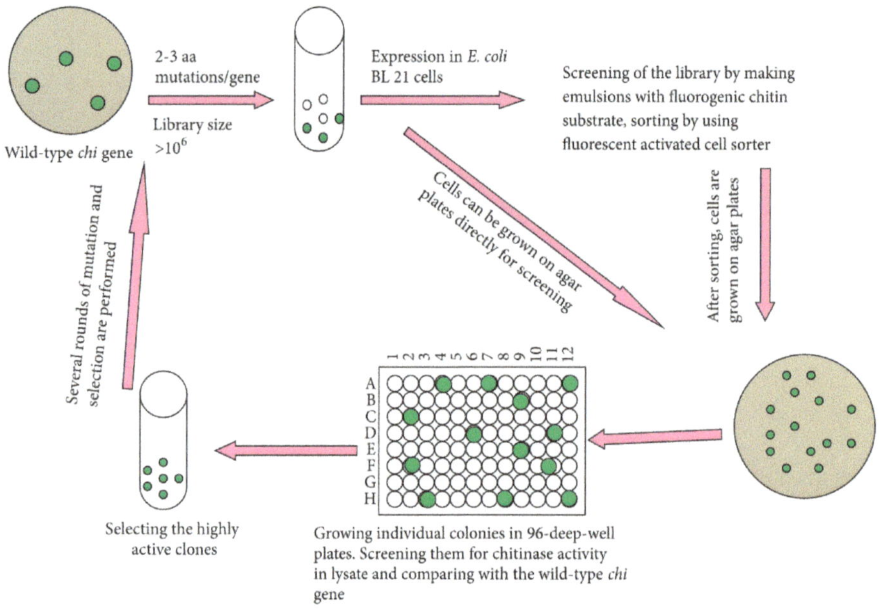

Figure 4: A schematic diagram showing methodology of directed laboratory evolution technique for the creation of new enzymes with enhanced catalytic activity. Chitinase gene library is created by random mutagenic PCR, which randomly incorporates desired number of amino acid mutations. Then the PCR product is cloned, expressed, and screened for higher catalytic activity. For detailed methodology, kindly refer to [9, 10].

ACKNOWLEDGMENTS

The authors are thankful to Department of Science and Technology and South Asian University, New Delhi, India, for financial assistance.

REFERENCES

1. S. Adrangi and M. A. Faramarzi, "From bacteria to human: a journey into the world of chitinases,"Biotechnology Advances, vol. 31, no. 8, pp. 1786–1785, 2013.
2. L. Hartl, S. Zach, and V. Seidl-Seiboth, "Fungal chitinases: diversity, mechanistic properties and biotechnological potential," Applied Microbiology and Biotechnology, vol. 93, no. 2, pp. 533–543, 2012.
3. F. Shahidi and R. Abuzaytoun, "Chitin, chitosan, and co-products: chemistry, production, applications, and health effects," Advances in Food and Nutrition Research, vol. 49, pp. 93–135, 2005.
4. L. Ramírez-Coutiño, M. D. C. Marín-Cervantes, S. Huerta, S. Revah, and K. Shirai, "Enzymatic hydrolysis of chitin in the production of oligosaccharides using Lecanicillium fungicola chitinases,"Process Biochemistry, vol. 41, no. 5, pp. 1106–1110, 2006.
5. K. Kurita, "Controlled functionalization of the polysaccharide chitin," Progress in Polymer Science, vol. 26, no. 9, pp. 1921–1971, 2001.
6. R. A. Muzzarelli, "Clinical and biochemical evaluation of chitosan for hypercholesterolemia and overweight control," EXS, vol. 87, pp. 293–304, 1999.
7. D. Bhattacharya, A. Nagpure, and R. K. Gupta, "Bacterial chitinases: properties and potential," Critical Reviews in Biotechnology, vol. 27, no. 1, pp. 21–28, 2007.
8. A. S. Sahai and M. S. Manocha, "Chitinases of fungi and plants: their involvement in morphogenesis and host-parasite interaction," FEMS Microbiology Reviews, vol. 11, no. 4, pp. 317–338, 1993.
9. R. D. Gupta, M. Goldsmith, Y. Ashani et al., "Directed evolution of hydrolases for prevention of G-type nerve agent intoxication," Nature Chemical Biology, vol. 7, no. 2, pp. 120–125, 2011.
10. R. D. Gupta and D. S. Tawfik, "Directed enzyme evolution via small and effective neutral drift libraries,"Nature Methods, vol. 5, no. 11, pp. 939–942, 2008.
11. B. Henrissat and A. Bairoch, "New families in the classification of glycosyl hydrolases based on amino acid sequence similarities," The Biochemical Journal, vol. 293, no. 3, pp. 781–788, 1993.
12. G. W. Gooday, "Aggressive and defensive roles for chitinases," EXS, vol. 87, pp. 157–169, 1999.
13. P. J. Hart, H. D. Pfluger, A. F. Monzingo, T. Hollis, and J. D. Robertus, "The refined crystal structure of an endochitinase from Hordeum vulgare L. seeds at 1.8 Å resolution," Journal of Molecular Biology, vol. 248, no. 2, pp. 402–413, 1995.
14. N. Dahiya, R. Tewari, and G. S. Hoondal, "Biotechnological aspects of chitinolytic enzymes: a review,"Applied Microbiology and Biotechnology, vol. 71, no. 6, pp. 773–782, 2006.
15. S. Adrangi, M. A. Faramarzi, A. R. Shahverdi, and Z. Sepehrizadeh, "Purification and characterization of two extracellular endochitinases from Massilia timonae," Carbohydrate Research, vol. 345, no. 3, pp. 402–407, 2010.
16. R. Cohen-Kupiec and I. Chet, "The molecular biology of chitin digestion," Current Opinion in Biotechnology, vol. 9, no. 3, pp. 270–277, 1998.
17. J. N. Busby, M. J. Landsberg, R. M. Simpson et al., "Structural analysis of Chi1 chitinase from Yen-Tc: the multisubunit insecticidal ABC toxin complex of Yersinia

entomophaga," Journal of Molecular Biology, vol. 415, no. 2, pp. 359–371, 2012.

18. D. M. F. Van Aalten, B. Synstad, M. B. Brurberg et al., "Structure of a two-domain chitotriosidase from Serratia marcescens at 1.9-Å resolution," Proceedings of the National Academy of Sciences of the United States of America, vol. 97, no. 11, pp. 5842–5847, 2000.
19. G. Vaaje-Kolstad, S. J. Horn, M. Sørlie, and V. G. H. Eijsink, "The chitinolytic machinery of Serratia marcescens—a model system for enzymatic degradation of recalcitrant polysaccharides," The FEBS Journal, vol. 280, no. 13, pp. 3028–3049, 2013.
20. J.-T. Chern and Y.-P. Chao, "Chitin-binding domain based immobilization of d-hydantoinase," Journal of Biotechnology, vol. 117, no. 3, pp. 267–275, 2005.
21. C. Songsiririttthigul, S. Pantoom, A. H. Aguda, R. C. Robinson, and W. Suginta, "Crystal structures ofVibrio harveyi chitinase A complexed with chitooligosaccharides: implications for the catalytic mechanism," Journal of Structural Biology, vol. 162, no. 3, pp. 491–499, 2008.
22. W. Suginta, W. Chumjan, K. R. Mahendran, A. Schulte, and M. Winterhalter, "Chitoporin from Vibrio harveyi, a channel with exceptional sugar specificity," The Journal of Biological Chemistry, vol. 288, no. 16, pp. 11038–11046, 2013.
23. G. W. Gooday, Wei-Yun Zhu, and R. W. O›Donnell, "What are the roles of chitinases in the growing fungus?" FEMS Microbiology Letters, vol. 100, no. 1-3, pp. 387–391, 1992.
24. V. Seidl, "Chitinases of filamentous fungi: a large group of diverse proteins with multiple physiological functions," Fungal Biology Reviews, vol. 22, no. 1, pp. 36–42, 2008.
25. W. Knogge and D. Scheel, "LysM receptors recognize friend and foe," Proceedings of the National Academy of Sciences of the United States of America, vol. 103, no. 29, pp. 10829–10830, 2006.
26. S. G. Gruber and V. Seidl-Seiboth, "Self versus non-self: fungal cell wall degradation in Trichoderma,"Microbiology, vol. 158, no. 1, pp. 26–34, 2012.
27. W. Fang, P. Azimzadeh, and R. J. St Leger, "Strain improvement of fungal insecticides for controlling insect pests and vector-borne diseases," Current Opinion in Microbiology, vol. 15, no. 3, pp. 232–238, 2012.
28. M. A. Baloyi, M. D. Laing, and K. S. Yobo, "Use of mixed cultures of biocontrol agents to control sheep nematodes," Veterinary Parasitology, vol. 184, no. 2–4, pp. 367–370, 2012.
29. A. Lupetti, M. G. J. de Boer, P. Erba, M. Campa, and P. H. Nibbering, "Radiotracers for fungal infection imaging," Medical Mycology, vol. 49, no. 1, pp. S62–S69, 2011.
30. J. Zhang, X. Zhang, Y. Arakane et al., "Comparative genomic analysis of chitinase and chitinase-like genes in the African malaria mosquito (Anopheles gambiae)," PLoS ONE, vol. 6, no. 5, Article ID e19899, 2011.
31. H. Zhang, M. Liu, Y. Tian, and X. Hu, "Comparative characterization of chitinases from silkworm (Bombyx mori) and bollworm (Helicoverpa armigera)," Cell Biochemistry and Biophysics, vol. 61, no. 2, pp. 267–275, 2011.
32. Q. Zhu, Y. Arakane, R. W. Beeman, K. J. Kramer, and S. Muthukrishnan, "Functional specialization among insect chitinase family genes revealed by RNA interference," Proceedings of the National Academy of Sciences of the United States of America, vol. 105, no. 18, pp. 6650–6655, 2008.
33. A. Kasprzewska, "Plant chitinases—regulation and function," Cellular and Molecular Biology Letters, vol. 8, no. 3, pp. 809–824, 2003.
34. T. Ohnuma, T. Numata, T. Osawa et al., "A class V chitinase from Arabidopsis thaliana: gene responses, enzymatic properties, and crystallographic analysis," Planta, vol. 234, no. 1, pp. 123–137, 2011.
35. K. J. Al Ahmadi, M. T. Yazdi, M. F. Najafi et al., "Optimization of medium and cultivation conditions for chitinase production by the newly isolated: Aeromonas sp," Biotechnology, vol. 7, no. 2, pp. 266–272, 2008.
36. M. Hassas-Roudsari and H. D. Goff, "Ice structuring proteins from plants: mechanism of action and food application," Food Research International, vol. 46, no. 1, pp. 425–436, 2012.
37. A. Grover, "Plant chitinases: genetic diversity and physiological roles," Critical Reviews in Plant Sciences, vol. 31, no. 1, pp. 57–73, 2012.
38. H. Oshima, R. Miyazaki, Y. Ohe, H. Hayashi, K. Kawamura, and S. Kikuyama, "Molecular cloning of a putative gastric chitinase in the toad Bufo japonicus," Zoological Science, vol. 19, no. 3, pp. 293–297, 2002.
39. C.-R. Shen, C.-L. Liu, H.-P. Lee, and J.-K. Chen, "The identification and characterization of chitotriosidase activity in pancreatin from porcine pancreas," Molecules, vol. 18, no. 3, pp. 2978–2987, 2013.
40. G. H. Renkema, R. G. Boot, A. O. Muijsers, W. E. Donker-Koopman, and J. M. F. G. Aerts, "Purification and characterization of human chitotriosidase, a novel member of the chitinase family of proteins,"Journal of Biological Chemistry, vol. 270, no. 5, pp. 2198–2202, 1995.
41. M. Suzuki, M. Morimatsu, T. Yamashita, T. Iwanaga, and B. Syuto, "A novel serum chitinase that is expressed in bovine liver," FEBS Letters, vol. 506, no. 2, pp. 127–130, 2001.
42. L. Shuhui, Y.-K. Mok, and W. S. F. Wong, "Role of mammalian chitinases in asthma," International Archives of Allergy and Immunology, vol. 149, no. 4, pp. 369–377, 2009.
43. A. Nagpure, B. Choudhary, and R. K. Gupta, "Chitinases: in agriculture and human healthcare," Critical Reviews in Biotechnology, vol. 34, no. 3, pp. 215–232, 2014.
44. S. Duru, M. Yüceege, and S. Ardiç, "Chitinases and lung diseases," Tuberkuloz ve Toraks, vol. 61, no. 1, pp. 71–75, 2013.

45. K. Sakai, A. Yokota, H. Kurokawa, M. Wakayama, and M. Moriguchi, "Purification and characterization of three thermostable endochitinases of a noble Bacillus strain, MH-1, isolated from chitin-containing compost," Applied and Environmental Microbiology, vol. 64, no. 9, pp. 3397–3402, 1998.

46. S. Revah-Moiseev and P. A. Carroad, "Conversion of the enzymatic hydrolysate of shellfish waste chitin to single-cell protein," Biotechnology and Bioengineering, vol. 23, no. 5, pp. 1067–1078, 1981.

47. L. S. Melchers and M. H. Stuiver, "Novel genes for disease-resistance breeding," Current Opinion in Plant Biology, vol. 3, no. 2, pp. 147–152, 2000.

48. S. Sakuda, "Studies on the chitinase inhibitors, allosamidins," in Chitin Enzymology, R. A. A. Muzzarelli, Ed., vol. 2, pp. 203–212, 1996.

49. N. A. Oranusi and A. P. J. Trinci, "Growth of bacteria on chitin, fungal cell walls and fungal biomass, and the effect of extracellular enzymes produced by these cultures on the antifungal activity of amphotericin B," Microbios, vol. 43, no. 172, pp. 17–30, 1985.

50. Z. Zhu, T. Zheng, R. J. Homer et al., "Acidic mammalian chitinase in asthmatic Th2 inflammation and IL-13 pathway activation," Science, vol. 304, no. 5677, pp. 1678–1682, 2004.

51. K. Vega and M. Kalkum, "Chitin, chitinase responses, and invasive fungal infections," International Journal of Microbiology, vol. 2012, Article ID 920459, 10 pages, 2012.

52. S. H. Park, J.-H. Lee, and H. K. Lee, "Purification and characterization of chitinase from a marine," The Journal of Microbiology, vol. 38, no. 4, pp. 224–229, 2000.

53. P. A. Aloise, M. Lumme, and C. A. Haynes, "N-Acetyl-D-glucosamine production from chitin-waste using chitinases from Serratia marcescens," in Chitin Enzymology, R. A. A. Muzzarelli, Ed., pp. 581–594, European Chitin Society, Grottammare, Italy, 1996.

54. A. Giansanti, M. Bocchieri, V. Rosato, and S. Musumeci, "A fine functional homology between chitinases from host and parasite is relevant for malaria transmissibility," Parasitology Research, vol. 101, no. 3, pp. 639–645, 2007.

55. F. Li, K. P. Patra, and J. M. Vinetz, "An anti-Chitinase malaria transmission-blocking single-chain antibody as an effector molecule for creating a Plasmodium falciparum-refractory mosquito," The Journal of Infectious Diseases, vol. 192, no. 5, pp. 878–887, 2005.

56. N. Dahiya, R. Tewari, R. P. Tiwari, and G. S. Hoondal, "Chitinase production in solid-state fermentation by Enterobacter sp. NRG4 using statistical experimental design," Current Microbiology, vol. 51, no. 4, pp. 222–228, 2005.

57. C. Barthomeuf, F. Regerat, and H. Pourrat, "Improvement in tannase recovery using enzymatic disruption of mycelium in combination with reverse micellar enzyme extraction," Biotechnology Techniques, vol. 8, no. 2, pp. 137–142, 1994.

58. D. M. F. van Aalten, D. Komander, B. Synstad, S. Gåseidnes, M. G. Peter, and V. G. H. Eijsink, "Structural insights into the catalytic mechanism of a family 18 exo-chitinase," Proceedings of the National Academy of Sciences of the United States of America, vol. 98, no. 16, pp. 8979–8984, 2001.

59. H. Ni, S. Zeng, X. Qin et al., "Molecular docking and site-directed mutagenesis of a Bacillus thuringiensischitinase to improve chitinolytic, synergistic lepidopteran-larvicidal and nematicidal activities,"International Journal of Biological Sciences, vol. 11, no. 3, pp. 304–315, 2015.

CHAPTER 14

CONTACT TOXICITY AND REPELLENCY OF THE ESSENTIAL OIL OF LIRIOPE MUSCARI (DECN.) BAILEY AGAINST THREE INSECT TOBACCO STORAGE PESTS

Yan Wu, Wen-Juan Zhang, Ping-Juan Wang, Kai Yang, Dong-Ye Huang, Jian-Yu Wei, Zhao-Fu Tian, Jia-Feng Bai, and Shu-Shan Du

Keywords: Liriope muscari, contact activity, repellency, essential oil composition, stored product insects

INTRODUCTION

With the rapid development of the tobacco industry in China, the size and number of tobacco storage facilities is becoming larger and larger. However, it is very easy for tobacco leaves to be effected by storage insects [1]. Coleopterans have global importance in grain storage and industrial food production and can cause damage to the appearance of products as well as weight loss and reduced nutrient levels, rendering products unfit for human consumption [2]. The red flour beetle (*Tribolium castaneum* Herbst) (Coleoptera: Tenebrionidae) is distributed world-wide and among the most economically important stored product pests [3]. The cigarette beetle (*Lasioderma serricorne* Frbricius) (Coleoptera: Anobiidae) is a widespread and destructive primary insect pests of stored cereals, tobacco, oilseeds, dried fruits and traditional Chinese medicinal materials [4]. The booklouse (*Liposcelis bostrychophila* Badonnel) (Psocoptera) is frequently found in stored product grains and traditional medicines, or other amylaceous products [5]. At present, phosphine is the most widely used fumigant, but threatens human health [6]. In recent years, research and development on botanical pesticides have attracted great attention in the whole world. Volatile plant ingredients which are highly selective, easy to decompose, leave no residue, and cause no pollution or threats to human and environmental harmony, such as plant essential oils, are found in botanical pesticides [7,8]. Botanical pesticides have the advantage of providing novel modes of action against insects that can reduce the risk of cross-resistance as well as offering new leads for the design of target-specific molecules. In previous studies, the bioactivities of a large number of essential oils have been also evaluated against a number of stored product insects [9,10,11,12].

The species *Liriope muscari* (Decne.) Bailey is a perennial herb, belonging to the Liriope of the Liliaceae family, and native to East Asia, including China, Japanese and North Korea. The dried tuberous roots have been used as the medicinal part of this plant to treat cough, insomnia, astriction and other diseases [13]. Genuine medicinal materials containing this material are mainly distributed in Quanzhou and Putian, Fujian Province [14]. A literature survey has shown that *L. muscari* has anti-tumor metastatic activity [15], immune functions [16,17], anti-inflammatory [18], anti-thrombotic activity [19], antioxidant activity [20] and nematocidal activity [14]. The chemical constituents of the root tuber of *L. muscari* have been reported [14,20,21,22,23,24]. However, there is no report on the volatile constituents and bioactivity of the essential oil of *L. muscari*aerial parts. Here, the chemical constituents and bioactivities of the essential oil of *L. muscari* aerial parts were investigated against three stored product insects for the first time.

Citation: Yan Wu, Wen-Juan Zhang, Ping-Juan Wang, Kai Yang, Dong-Ye Huang, Jian-Yu Wei, Zhao-Fu Tian, Jia-Feng Bai, and Shu-Shan Du. Contact Toxicity and Repellency of the Essential Oil of Liriope muscari (DECN.) Bailey against Three Insect Tobacco Storage Pests, Molecules (2015), 20(1), 1676-1685. DOI:10.3390/molecules20011676.

RESULTS AND DISCUSSION

Chemical Composition of the Essential Oil

The chemical composition of the essential oil of *L. muscari* aerial parts is reported here for the first time. The essential oil yield of *L. muscari* was 0.05% (v/w) and the density of the concentrated essential oil was determined as 1.18 g/cm^3. The GC/MS and GC/FID analysis results for the oil are presented in Table 1. The analyses revealed 14 components, representing 96.12% of the oil. The main components were methyl eugenol (42.15%), safrole (17.15%), myristicin (14.18%), 3,5-dimethoxytoluene (10.60%) and 3,4,5-trimethoxytoluene (9.05%).

Table 1. Chemical composition of *Liriope muscari* essential oil.

Peak No.	Compound	RI *	Relative Content (%)
1	Linalool	1075	0.18
2	Camphor	1143	0.05
3	Eucarvone	1146	0.38
4	Borneol	1156	0.33
5	α-Terpineol	1181	0.31
6	Estragole	1195	0.28
7	3,5-Dimethoxytoluene	1264	10.60
8	Safrole	1285	17.15
9	Linalool propionate	1340	0.05
10	Methyleugenol	1403	42.15
11	3,4,5-Trimethoxytoluene	1408	9.05
12	Germacrene D	1476	0.11
13	Myristicin	1523	14.18
14	Elemicin	1554	1.30
	Phenylpropanoids		75.06
	Monoterpenoids		1.30
	Sesquiterpenoids		0.11
	Aromatic compounds		19.65
	Total		96.12

* RI, retention index as determined on a HP-5MS column using the homologous series of *n*-alkanes.

Contact Toxicity

The crude essential oil of *L. muscari* aerial parts showed pronounced contact toxicity against *T. castaneum*, *L. serricorne* and*L. bostrychophila* with LD_{50} values of 13.36, 11.28 μg/adult and 21.37 μg/cm^2, respectively (Table 2).

Table 2. Contact toxicity of *Liriope muscari* essential oil against *Tribolium castaneum*(TC), *Lasioderma serricorne* (LS) and *Liposcelis bostrychophila* (LB) adults.

Insect	Compounds	LD_{50} *	95% FL *	Slope ± SE	Chi square (χ^2)
LS	Essential oil	11.28	10.22–12.43	4.59 ± 0.48	16.10
	Pyrethrins **	0.24	0.16–0.35	1.31 ± 0.20	17.36
TC	Essential oil	13.36	10.94–15.70	1.67 ± 0.26	11.96
	Pyrethrins	0.26	0.22–0.30	3.34 ± 0.32	13.11
LB	Essential oil	21.37	20.17–22.45	8.60 ± 1.01	10.35
	Pyrethrins **	18.72	17.60–19.92	2.98 ± 0.40	10.56

* Concentration (μg/adult)/(μg/cm^2); ** data from Yang *et al.* [25].

According to the data listed in Table 2, the essential oil of *L. muscari* aerial parts exhibited stronger contact toxicity against*L. bostrychophila* than *T. castaneum* and *L. serricorne*. When compared with the positive control, pyrethrins, the essential oil demonstrated 51, 47 and 1.1 times less toxicity against *T. castaneum*, *L. serricorne* adults and *L. bostrychophila*, respectively. This showed that *L. bostrychophila* were most susceptible to the contact toxicity of the essential oil of *L. muscari* and *T. castaneum* adults were more tolerant than *L. serricorne* adults. However, compared with other essential oils mentioned in the literature, the essential oil of *L. muscari* aerial parts possessed stronger contact toxicity against *T. castaneum* adults, e.g., essential oils of *Dracocephalum moldavica* (LD_{50} = 18.28 μg/adult) [26], *Murraya exotica* (LD_{50} = 20.94 μg/adult) [20], *Evodia lepta* (LD_{50} = 166.94 μg/adult) [22]. The essential oil of *L. muscari* aerial parts also exhibited stronger contact toxicity against*L. bostrychophila* than other reported oils, for example essential oils of *Lonicera japonica* (LD_{50} = 64.04 μg/cm^2) [5], *Foeniculum vulgare* (LD_{50} = 90.36 μg/cm^2) [27], *Acorus calamus* (LD_{50} = 100.21 μg/cm^2) [28], *Curcuma wenyujin* (LD_{50} = 208.85 μg/cm^2) [29].

In another reference [30], six compounds, including, eugenol, methyl eugenol, methyl isoeugenol, elemicin, myristicin and safrole isolated from *Myristica fragrans* (LD_{50} = 19.3 μg/adult) showed potent contact toxicity against *L. serricorne* adults with LD_{50} values of 13.2, 12.8, 21.3, 9.8, 20.5 and 14.6 μg/adult, respectively. Methyl eugenol, safrole and myristicin are also the main compounds of the essential oil of *L. muscari*, hence, the revealed toxic properties of the essential oil of *L. muscari* on those three tobacco storage insects might be attributed to the synergistic effects of its diverse major and minor components. Also, the different volatility of oils and different mechanisms of action on different insects might lead to different contact toxicity. However, there is insufficient data at present to conform these hypotheses. In addition, further investigations need to be conducted in the future to determine whether the oil is safe to humans or not.

Repellency

In addition to contact activities, the repellent effect of the essential oil of *L. muscari* aerial parts against the three stored product insects was also investigated. The results are presented in Table 3 and Table 4.

Table 3. Pecentage repellency (PR) after two exposure times for the essential oil against *Tribolium castaneum* (TC) and *Lasioderma serricorne* (LS) adults [a].

Insect	Treatment	2 h/4 h				
		78.63 *	**15.73 ***	**3.15 ***	**0.63 ***	**0.13 ***
TC	Crude oil	84 ± 3a; 82 ± 5a	92 ± 5a; 90 ± 4a	84 ± 7a; 82 ± 10a	74 ± 10a; 68 ± 5a	72 ± 8a; 66 ± 7a
	DEET	100 ± 0a; 94 ± 6a	96 ± 3a; 80 ± 10a	74 ± 16a; 46 ± 17b	66 ± 9a; −2 ± 20b	18 ± 16b; −8 ± 22b
LS	Crude oil	86 ± 6a; 82 ± 3a	68 ± 9a; 52 ± 8a	18 ± 8a; 30 ± 10a	6 ± 17a; 8 ± 12a	8 ± 21a; −12 ± 20a
	DEET	88 ± 7a; 98 ± 4a	76 ± 14a; 78 ± 9b	28 ± 7a; 58 ± 16b	20 ± 14b; 56 ± 14b	16 ± 7a; 46 ± 7b

* Concentration (nL/cm^2); [a] Means in the same column followed by the same letters do not differ significantly ($p < 0.05$) in ANOVA and Tukey's tests. PR was subjected to an arcsine square-root transformation before ANOVA and Tukey's tests.

The essential oil strongly repelled all tested species of stored product insects, but the repellent effect was more marked on *T. castaneum* than on *L. serricorne*. Data showed that at the tested concentration of 0.13 nL/cm^2, the crude oil still showed strong repellency (class IV) against the red flour beetle, *T. castaneum* adults at 2 h, 4 h after exposure (Table 3) while no repellency of the positive control, DEET, was seen, but some insect attractant properties were observed. Moreover, at the assayed concentrations of 78.63, 15.73 and 3.15 nL/cm^2, the essential oil exhibited strong repellency (class V) at 2 h, 4 h after exposure (Table 3).

Table 4. Percentage repellency (PR) after two exposure times for the essential oil against *Liposcelis bostrychophila* (LB) [a].

Insect	Treatment	2 h/4 h				
		31.58 *	**6.32 ***	**1.26 ***	**0.25 ***	**0.05 ***
LB	Crude oil	80 ± 10a; 78 ± 5a	100 ± 0a; 96 ± 3a	92 ± 3a; 86 ± 3a	50 ± 13a; 42 ± 19a	24 ± 7a; 16 ± 6a
	DEET	100 ± 0a; 96 ± 7b	92 ± 7a; 90 ± 10a	92 ± 7a; 90 ± 6a	74 ± 9b; 72 ± 19b	30 ± 10a; 32 ± 7b

* Concentration (nL/cm^2); [a] Means in the same column followed by the same letters do not differ significantly ($p < 0.05$) in ANOVA and Tukey's tests. PR was subjected to an arcsine square-root transformation before ANOVA and Tukey's tests.

The repellency against *L. serricorne* is weaker than against *T. castaneum*. Compared with the positive control, DEET, only at the highest tested concentration of 78.63 nL/cm^2, the essential oil of *L. muscari* possessed the same level and strong repellency (class V) against *L. serricorne* adults (Table 3). As the concentrations were reduced, the repellency of the crude oil against *L. serricorne* also tended to drop off. Moreover, at the lowest concentration, no repellency of the crude oil, but some insect attractant properties were observed at 4 h after exposure (Table 3).

At the same time, the essential oil of *L. muscari* also showed strong repellency against the booklouse, *L. bostrychophila*. At the tested concentration of 0.25 nL/cm^2, the crude oil still exhibited strong repellency (class III) at 2 h, 4 h after exposure, while at the assayed concentrations of 6.32, 1.26 nL/cm^2, the oil showed strongly repellent against booklice at 2 h, 4 h after exposure (Table 4). However, at the lowest concentration, the repellency of the oil against booklice was weaker compared with the positive control, DEET (Table 4). Many essential oils have been evaluated for repellency against insects [31].

In China, essential oils derived from spices and Chinese medicinal herbs were also evaluated for insecticidal activity and repellency against insects [28,29,32,33,34]. In this article, we report the contact and repellent activities of the essential oil of*L. muscari* aerial parts for the first time. Above all, the essential oil of *L. muscari* possessed strong repellency against the three stored product insects. These findings, considered together, suggest that the essential oil of *L. muscari* shows potential for development as a natural repellent for stored products.

EXPERIMENTAL SECTION

Plant Material and Extractions

Fresh aerial parts (2 kg) of *Liriope muscari* were harvested in 2013 from Quanzhou (Fujian Province, China, N latitude: 24°30'–25°56'; E longitude: 117°25'–119°05'). The plant species was identified and a voucher specimen (BNU-dushushan-2013-05-24-06) was deposited at the Herbarium of the College of Resources Science and Technology, Beijing Normal University (Beijing, China). The aerial parts of *Liriope muscari* were air dried for 1 week and ground to a powder. The ground powder of *L. muscari* was subjected to hydrodistillation using a modified Clevenger-type apparatus for 8 h and extracted with *n*-hexane. Anhydrous sodium sulphate was used to remove water after extraction. The essential oil was stored in an airtight container in a refrigerator at 4 °C.

Insects

Red flour beetles (*T. castaneum*), cigarette beetles (*L. serricorne*) and booklice (*L. bostrychophila*) were obtained from laboratory cultures maintained in the dark in incubators at 29–30 °C and 70%–80% relative humidity (RH). The red flour beetles and cigarette beetles were reared on wheat flour mixed with yeast (10:1, w/w) at 12%–13% moisture content while the booklice were reared on a 1:1:1 mixture, by mass, of milk powder, active yeast and flour. The unsexed adult beetles/booklice used in all the experiments were about 1–2 weeks old. All containers housing insects and the Petri dishes used in experiments were made escape proof with a coating of polytetrafluoroethylene (Fluon).

Contact Toxicity

The contact toxicity of the essential oil of *L. muscari* aerial parts against *T. castaneum* and *L. serricorne* adults was measured by using topical application as described by Liu and Ho [35]. Serial dilutions of essential oil (10%–1.97% (v/v) for *T. castaneum* and *L. serricorne*; five concentrations) were prepared in *n*-hexane. Preliminary experiments were conducted to determine appropriate ranges of test concentrations. Aliquots (0.5 μL) of the dilutions were applied topically to the dorsal thorax of the insects. Controls were established using *n*-hexane. Both treated and control insects were then transferred to glass vials (ten insects per vial) with culture media and kept in incubators at 29–30 °C and 70%–80% RH. Mortality of insects was observed daily after treatment. The experiments were repeated three times. While the contact toxicity of the essential oil against *L. bostrychophila* was also tested as described [5]. A 5.5 cm diameter filter paper was treated with 300 μL of the solution of the essential oil. The filter paper after treated with solid glue was placed in a 5.5 cm diameter Petri dish and 10 booklice were put on the filter paper. A cover was put and all the Petri dishes were kept in incubators. *n*-Hexane was used as a negative control. Five concentrations (in *n*-hexane) and five replicates of each concentration were used. As a positive control, pyrethrins (pyrethrin 1: 24%; pyrethrin 2: 13%; cinnerin 1: 2%; cinnerin 2: 2%; jasmolin 1: 1%; jasmolin 2: 1%) was used under the conditions as the oil. Mortality of insects was observed after 24 h. The LD_{50} values were calculated by using Probit analysis [36].

Repellency

The repellent effects of the essential oil against *T. castaneum*, *L. serricorne* and *L. bostrychophila* were assessed by using assays on Petri dishes [37]. Petri dishes 9 cm in diameter were used to confine beetles during the experiment for *T. castaneum*and *L. serricorne*. The essential oil of *L. muscari* aerial parts was prepared in *n*-hexane (78.63, 15.73, 3.15, 0.63 and 0.13 nL/cm^2), and absolute *n*-hexane was used as the control. Filter paper 9 cm in diameter was cut in half and 500 μL of each concentration was applied separately to half of the filter paper as uniformly as possible with a micropipette. The other half (control) was treated with 500 μL of absolute *n*-hexane. Both the treated half and the control half were then air-dried to evaporate the solvent completely (30 s). A full disk was carefully remade by attaching the tested half to the negative control half with tape. Each reassembled filter paper after treatment with solid glue was placed in a Petri dish with the seam oriented in one of four randomly selected different directions to avoid any insecticidal stimuli affecting the distribution of insects. Twenty insects were released in the center of each filter paper disk, and a cover was placed over the Petri dish. Five replicates were used, and the experiment was repeated three times. Counts of the insects present on each strip were made after 2 and 4 h. The percent repellency (PR) of each volatile oil was then calculated using the formula

$$PR\ (\%) = [(Nc - Nt)/(Nc + Nt)] \times 100$$

where *Nc* is the number of insects present in the negative control half and *Nt* is the number of insects present in the treated half. The averages were then assigned to different classes (0 to V) using the following scale (percentage repellency) [32]. Class, % repellency: 0, >0.01 to <0.1; I, 0.1–20.0; II, 20.1–40.0; III, 40.1–60.0; IV, 60.1–80.0 and V, 80.1–100.

As for the booklice, Petri dishes and filter papers were changed to 6 cm in diameter and the concentration of the oil used in the experiments were 31.58, 6.32, 1.26, 0.25, 0.05 nL/cm^2. The half filter paper was treated with 150 μL of the solution. As a positive control, a commercial repellent DEET (*N,N*-diethyl-3-methylbenzamide), was used under the conditions as the oil. Analysis of variance (ANOVA) and Tukey's test were conducted by using SPSS Statistics ver. 20 for Windows 2007. Percentage was subjected to an arcsine square-root transformation before ANOVA and Tukey's tests.

GC/MS and GC/FID Analysis

Components of the essential oil of *L. muscari* were separated and identified using a Bruker-320 gas chromatography/mass spectrometry (GC/MS, Bruker Daltonics Inc., Billerica, MA,

USA) system equipped with a flame ionization detector (GC-FID, Bruker Daltonics Inc.) and a HP-5MS (30 m × 0.25 mm × 0.25 μm, Bruker Daltonics Inc.) capillary column. The GC settings were as follows: the initial oven temperature was held at 50 °C for 4 min and ramped at 10 °C·min^{-1} to 290 °C for 17 min. The injector temperature was maintained at 250 °C. The samples (1 μL, diluted to 1% with hexane) were injected, with a split ratio of 1:60. The carrier gas was helium at flow rate of 1 mL/min. Spectra were scanned from 45 to 650 *m/z* at 2 scans/s. Most constituents were identified by gas chromatography by comparison of their retention indices with those of the literature or with those of authentic compounds available in our laboratories. The retention indices were determined in relation to a homologous series of *n*-alkanes (C_5–C_{35}) under the same operating conditions. Further identification was made by comparison of their mass spectra with those stored in NIST 05 (Standard Reference Data, Gaithersburg, MD, USA) and Wiley 275 libraries (Wiley, New York, NY, USA) or with mass spectra from literature [38]. Component relative percentages were calculated based on GC peak areas, without using correction factors.

CONCLUSIONS

The composition of the essential oil of *L. muscari* aerial parts and the contact toxicity and repellency of the essential oil against three stored product insects for the first time. The study indicates that the essential oil of *L. muscari* aerial parts has potential for development into a natural insecticide/repellent for the control of insects in stored products.

ACKNOWLEDGMENTS

This project was supported by National Natural Science Foundation of China (No. 81374069) and Beijing Municipal Natural Science Foundation (No. 7142093). The authors thank Dr. Liu Q.R. from College of Life Sciences, Beijing Normal University, Beijing 100875, for the identification of the investigated medicinal herb.

AUTHOR CONTRIBUTIONS

Yan Wu, Wen-Juan Zhang, Kai Yang and Shu-Shan Du designed reseach; Ping-Juan Wang, Dong-Ye Huang, Jian-Yu Wei, Zhao-Fu Tian and Jia-Feng Bai performed reseach and analyzed the data; Yan Wu, Wen-Juan Zhang, Kai Yang and Shu-Shan Du wrote the paper. All authors read and approved the final manuscript.

REFERENCES

1. Yan, W.H.; Huang, S.L.; Zhu, G.N. Taxonomy and identification of microorganism causing mold damage of stored tobacco leaf in Guangxi. *Tobacco Sci. Technol.* 2008, *247*, 50–56.
2. Nansen, C.; Subramanyam, B.; Roesli, R. Characterizing spatial distribution of trap captures of beetles in retail pet stores using SADIE® software. *J. Stored Prod. Res.* 2004, *40*, 471–483.
3. Garcìa, M.; Donael, O.J.; Ardanaz, C.E.; Tonn, C.E.; Sosa, M.E. Toxic and repellent effects of *Baccharis salicifolia*essential oil on *Tribolium castaneum*. *Pest Manag. Sci.* 2005, *61*, 612–618.
4. Ebadollahi, A.; Safaralizadeh, M.H.; Pourmirza, A.A.; Gheibi, S.A. Toxicity of essential oil of *Agastache foeniculum*(Pursh) kuntze to *Oryzaephilus surinamensis* L. and *Lasioderma serricorne* F. *J. Plant Prot. Res.* 2010, *50*, 215–219.
5. Zhou, H.Y.; Zhao, N.N.; Du, S.S.; Yang, K.; Wang, C.F.; Liu, Z.L.; Qiao, Y.J. Insecticidal activity of the essential oil of*Lonicera japonica* flower buds and its main constituent compounds against two grain storage insects. *J. Med. Plants Res.*2012, *6*, 912–917.
6. Kim, S.I.; Park, C.; Ohh, M.H.; Cho, H.C.; Ahn, Y.J. Contact and fumigant activities of aromatic plant extracts and essential oils against *Lasioderma serricorne* (Coleoptera: Anobiidae). *J. Stored Prod. Res.* 2003, *39*, 11–19.
7. Dong, Z.J.; Cheng, D.Q.; Dong, S.D.; Zhou, Y.H. Botanical pesticide research and application in tobacco pest control.*Acta Tabacaria Sinica* 2004, *10*, 42–47.
8. Kestenholz, C.; Stevenson, P.C.; Belmain, S.R. Comparative study of field and laboratory evaluations of the ethnobotanical *Cassia sophera* L. (Leguminosae) for bioactivity against the storage pests *Callosobruchus maculatus* (F.) (Coleoptera: Bruchidae) and *Sitophilus oryzae* (L.) (Coleoptera: Curculionidae). *J. Stored Prod. Res.* 2007, *43*, 79–86.
9. Hori, M. Repellency of essential oils against the cigarette beetle, *Lasioderma serricorne* (Frbricius) (Coleoptera: Anobiidae). *Appl. Entomol. Zool.* 2003, *38*, 467–473.
10. Hori, M. Evaluation of practicability of hinokitiol as a repellent against the cigarette beetle, *Lasioderma serricorne*(Frbricius) (Coleoptera: Anobiidae). *Appl. Entomol. Zool.* 2004, *39*, 699–704.
11. Kim, D.H.; Ahn, Y.J. Contact and fumigant activities of constituents of *Foeniculum vulgare* fruit against three coleopteran stored-product insects. *Pest Manag. Sci.* 2001, *57*, 301–306.
12. Mojtaba, G.J.; Ali, A.P.; Mohammad, H.S. Repellent effect of sirinol (garlic emulsion) against *Lasioderma serricorne*(Coleoptera: Anobiidae) and *Tribolium castaneum* (Coleoptera: Tenebrionidae) by three laboratory methods. *Afr. J. Biotechnol.* 2012, *11*, 280–288.
13. China Pharmacopoeia Committee. *Chinese Pharmacopoeia*, 9th ed.; China Medical Science and Technology Press: Beijing, China, 2010; p. 25.
14. Zhang, H.M.; Wang, G.L.; Bai, C.Q.; Liu, Z.M.; Liu, Q.Z.; Wang, Y.Y.; Liu, Z.L.; Du, S.S.; Deng, Z.W. A new eudesmane sesquiterpene glucoside from *Liriope muscari* fibrous roots. *Molecules* 2011, *16*, 9017–9024.
15. Zhang, Y.Y.; Liu, J.H.; Kou, J.P.; Yu, J.; Yu, B.Y. DT-13, a steroidal saponin from *Liriope muscari* L. H. Bailey, suppresses A549 cells adhesion and invasion by inhibiting MMP-2/9. *Chin. J. Nat. Med.* 2012, *10*, 436–440.
16. Liu, J.L.; Chen, T.; Yu, B.Y.; Xu, Q. Ruscogenin glycoside (Lm-3) isolated from *Liriope muscari* inhibits lymphocyte adhesion to extracellar matrix. *J. Pharm. Pharmacol.* 2002, *54*, 959–966.

17. Yu, B.Y.; Yin, X.; Rong, Z.Y.; Yang, T.M.; Zhang, C.H.; Xu, G.J. Biological activities of ruscogenin 1-*O*-[β-d-glucopyranosyl(1→2)][β-d-xylopyranosyl(1→3)]-β-d-fucopyranoside from Tuberous roots of *Liriope muscari* (Decne.) Bailey. *J. China Pharm. Univ.* 1994, *25*, 286–288.
18. Tian, Y.Q.; Kou, J.P.; Li, L.Z.; Yu, B.Y. Anti-inflammatory effects of aqueous extract from radix *Liriope muscari* and its major active fraction and component. *Chin. J. Nat. Med.* 2011, *9*, 222–226.
19. Tian, Y.Q.; Ma, S.T.; Lin, B.Q.; Kou, J.P.; Yu, B.Y. Anti-thrombotic activity of DT-13, a saponin isolated from the root tuber of *Liriope muscari*. *Indian J. Pharmacol.* 2013, *45*, 283–285.
20. Li, W.J.; Cheng, X.L.; Liu, J.; Lin, R.C.; Wang, G.L.; Du, S.S.; Liu, Z.L. Phenolic compounds and antioxidant activities of*Liriope muscari*. *Molecules* 2012, *17*, 1797–1808.
21. Cheng, Z.H.; Wu, T.; Guo, Y.L.; Yu, B.Y.; Xu, L.S. Two new steroidal glycosides from *Liriope muscari*. *Chin. Chem. Lett.*2006, *17*, 31–34.
22. Jiang, C.H.; Liu, Q.Z.; Du, S.S.; Deng, Z.W.; Liu, Z.L. Essential oil composition and insecticidal activity of *Evodia lepta*(Spreng) Merr. Root barks from China against two grain storage insects. *J. Med. Plants Res.* 2012, *6*, 3464–3469.
23. Liu, M.Z.; Yu, B.; Wu, X.Y.; Hui, Y.Z.; Fung, K.P. Synthesis of (25*R*)-ruscogenin-1-yl β-d-xylopyranosyl-(1→3)-[β-d-glucopyranosyl-(1→2)]-β-d-fucopyranoside. *Carbohydr. Res.* 2000, *329*, 745–754.
24. Yu, B.Y.; Hirai, Y.; Shoji, J.; Xu, G.J. Comparative studies on the constituents of ophiopogonis tuber and its congeners. VI. Studies on the constituents of the subterranean part of *Liriope spicata* var. prolifera and *L. muscari*. (1). *Chem. Pharm. Bull.* 1990, *38*, 1931–1936.
25. Yang, K.; Wang, C.F.; You, C.X.; Geng, Z.F.; Sun, R.Q.; Guo, S.S.; Du, S.S.; Liu, Z.L.; Deng, Z.W. Bioactivity of essential oil of *Litsea cubeba* from China and its main compounds against two stored product insects. *J. Asia-Pac. Entomol.* 2014, *17*, 459–466.
26. Chu, S.S.; Liu, S.L.; Liu, Q.Z.; Liu, Z.L.; Du, S.S. Composition and toxicity of Chinese *Dracocephalum moldavica*(Labiatae) essential oil against two grain storage insects. *J. Med. Plants Res.* 2011, *5*, 4621–4626.
27. Zhao, N.N.; Zhou, L.G.; Liu, Z.L.; Du, S.S.; Deng, Z.W. Evaluation of the toxicity of the essential oils of some common Chinese species against *Liposcelis bostrychophila*. *Food Control* 2012, *26*, 486–490.
28. Liu, X.C.; Zhou, L.G.; Liu, Z.L.; Du, S.S. Identification of insecticidal constituents of the essential oil of *Acorus calamus*Rhizomes against *Liposcelis bostrychophila* Badonnel. *Molecules* 2013, *18*, 5684–5696.
29. Liu, Z.L.; Zhao, N.N.; Liu, C.M.; Zhou, L.G.; Du, S.S. Identification of insecticidal constituents of the essential oil of*Curcuma wenyujin* Rhizomes active against *Liposcelis bostrychophila* Badonnel. *Molecules* 2012, *17*, 12049–12060.
30. Du, S.S.; Yang, K.; Wang, C.F.; You, C.X.; Geng, Z.F.; Guo, S.S.; Deng, Z.W.; Liu, Z.L. Chemical constituents and activities of the essential oil from *Myristica fragrans* against cigarette beetle *Lasioderma serricorne*. *Chem. Biodivers.* 2014, *11*, 1449–1456.
31. Nerio, L.S.; Olivero-Verbel, J.; Stashenko, E. Repellent activity of essential oils: A review. *Bioresour. Technol.* 2010, *101*, 372–399.
32. Liu, Z.L.; Goh, S.H.; Ho, S.H. Screening of Chinese medicinal herbs for bioactivity against *Sitophilus zeamais*Mostchulsky and *Tribolium castaneum* (Herbst). *J. Stored Prod. Res.* 2007, *43*, 290–296.
33. Liu, Z.L.; Chu, S.S.; Jiang, G.H. Insecticidal activity and composition of essential oil of *Ostericum sieboldii* (Apiaceae) against *Sitophilus zeamais* and *Tribolium castaneum*. *Rec. Nat. Prod.* 2011, *5*, 74–81.
34. Jiang, Z.; Akhtar, Y.; Bradbury, R.; Zhang, X.; Isman, M.B. Comparative toxicity of essential oils of *Litsea pungens* and*Litsea cubeba* and blends of their major constituents against the cabbage looper, *Trichoplusia ni*. *J. Agric. Food Chem.*2009, *57*, 4833–4837.
35. Liu, Z.L.; Ho, S.H. Bioactivity of the essential oil extracted from *Evodia rutaecarpa* Hook f. et Thomas against the grain storage insects, *Sitophilus zeamais* Motsch. and *Tribolium castaneum* (Herbst). *J. Stored Prod. Res.* 1999, *35*, 317–328.
36. Sakuma, M. Probit analysis of preference data. *Appl. Entomol. Zool.* 1998, *33*, 339–347.
37. Chaubey, M.K. Insecticidal activity of *Trachyspermum ammi* (Umbelliferae), *Anethum graveolens* (Umbelliferae) and*Nigella sativa* (Ranunculaceae) essential oils against stored-product beetle *Tribolium castaneum* Herbst (Coleoptera: Tenebrionidae). *Afr. J. Agric. Res.* 2007, *2*, 596–600.
38. Adam, R.P. Identification of essential oil components by gas chromatography/quadrupole mass spectrometry. *J. Am. Soc. Mass Spectrom.* 2005, *16*, 1902–1903.

CHAPTER 15

CHEMICAL COMPOSITION AND BIOLOGICAL ACTIVITY OF STAR ANISE ILLICIUM VERUM EXTRACTS AGAINST MAIZE WEEVIL, SITOPHILUS ZEAMAIS ADULTS

Linlin Wei, Rimao Hua, Maoye Li, Yanzhang Huang, Shiguang Li, Yujie He, and Zonghai Shen

Keywords: Contact toxicity, gas chromatography-mass spectrometry, repellency

INTRODUCTION

Sitophilus zeamais Motschulsky (Coleoptera, Curculionidae) is one of the major pests of stored products. It infests rice, wheat, corn, lotus seed, *Rhizoma gastrodiae,* and Chinese herbs, causing heavy losses of both quantity and quality of stored products worldwide. It attacks intact grains and is quite destructive, as its larvae and adults feed and develop inside the kernel (Lopez et al. 2008;Ojo and Omoloye 2012).

Control of *S. zeamais* populations around the world is primarily dependent on continued applications of synthetic insecticides (Liu et al. 2007). Although effective, the long repeated use of synthetic pesticides has led many insect pests, including *S. zeamais,* to develop pesticide resistance (Isman 2006) and pest resurgence. This phenomenon contributes to the vicious cycle of increasing doses, increasing application costs, and enhancing resistance. Moreover, chemical pesticide residues have exceeded acceptable limits in food and have polluted the environment, directly affecting the health of users and consumers. Therefore, investigating and using effective low- to nontoxic alternatives to synthetic pesticides to control stored-product pests is essential in ensuring food safety and human health.

Botanical pesticides have many advantages over synthetic pesticides, such as high selectivity, low or non-toxicity to nontarget organisms and the environment, rapid degradation, low residue, local availability, little cross-resistance due to their natural complex agents, and novel modes of action against insects (Isman 2006, 2008; Kudom et al. 2011; Lv et al. 2012; Ladhari et al. 2013) Therefore, increasing attention has been given to natural products from plants constituting a rich source of bioactive chemicals. Fennel, black pepper, menthol, camphor, eucalyptus oil, pyrethrum, and other vegetable matter have different degrees of effects on *S. zeamais* (Ukeh et al. 2009; Li et al. 2010).

Illicium verum Hook. f(Austrobaileyales: Schisandraceae), a fruit commonly known as star anise, is native to southwest China and Vietnam and is mainly distributed in the tropical and subtropical areas of Asia. *Illicium verum* was considered as one of the things "both food and medicine" by the Ministry of Health of the People's Republic of China (2002), implying its low or non-toxicity to humans. At present, the research focus on *I. verum* has been mainly on food and medical fields (Ohira et al. 2009). The fruits are commonly used as an ingredient of the traditional "five-spice" powder of Chinese cooking, and the essential oil of *I. verum* can be used as a flavoring. The extraction from *I. verum* has carminative, stomachic, stimulant, and diuretic properties, and is used as a pharmaceutical supplement (Lee et al. 2003). Shikimic acid extracted from *I. verum* is one of the main ingredients in the antiviral drug Tamiflu used to fight avian influenza (Ohira et al. 2009). It has also been reported to possess antimicrobial (Singh et al. 2006;

Citation: Linlin Wei, Rimao Hua, Maoye Li, Yanzhang Huang, Shiguang Li, Yujie He, and Zonghai Shen. Chemical composition and biological activity of star anise Illicium verum extracts against maize weevil, Sitophilus zeamais adults, Journal of Insect Science, Jun 2014; 14 (1) 80. DOI: 10.1093/jis/14.1.80.

Yang et al. 2010) andanti-oxidative properties (Padmashree et al. 2007) as well as significant anticancer potential (Shu et al. 2010).

In a previous screening program for new agrochemicals from Chinese medicinal herbs, *I. verum*powder has been shown to possess insecticidal activity against *S. zeamais* and *Cryptolestes pusillus* Schönherr (Li et al. 2013). Other studies have indicated that the essential oil of *I. verum*has repellent and fumigant actions on *S. Zeamais, Blattella germanica* (Chang and Ahn 2002),*Lasioderma serricorne, Sitophilus oryzae, Callosobruchus chinensis* (Kim et al. 2003a, b), and*Aedes aegypti* (Dana and Wej 2006). However, detailed studies on the biological activities of *I. verum* extracts against *S. zeamais* are lacking. Thus, in the current study, compounds of *I. verum*were extracted from the dried fruits with three types of organic solvents. The extract constituents were then determined by gas chromatography-mass spectrometry (GC-MS), and the repellency and contact toxicity of the extracts against *S. zeamais* adults were evaluated. The present study aims to provide additional data in support of its utilization and development as a new green storage protectant for control of stored-product insect pests.

MATERIALS AND METHODS

Insects

Adults of *S. zeamais* were originally collected from rice stores at Anhui Zhonggu State Grain Storage (Hefei, Anhui, China). The species were kept in the insectarium of the Plant Protection College of Anhui Agricultural University for more than five years. The insects were reared on whole wheat grains that were disinfected in the oven for 2 hr at 100°C; moisture content was adjusted to 13 ± 1%. The insects were placed in a glass jar (250 mL) containing 150 g of sterilized wheat grains (covered with a 60-mesh cloth) without exposure to any insecticide and then incubated at a temperature of 28 ± 1°C, a relative humidity of 70 ± 5%, and a 12:12 L:D photoperiod. The adults were left for seven days for oviposition and then sieved out. Adults (seven to 14 days old) were then used for the succeeding experiments.

Plant Extract Preparation

Fruits of *I. verum* were produced in Guangxi, China, but were purchased from a local supermarket in Hefei, China. *Illicium verum* fruits were dried in an oven (GRX-9071B; www.chem17.com) for two days at 40°C, ground to powder using an electric grinding mill (QE-100; Zhejiang, www.wyyili.com), and then sifted through a 40-mesh sieve. The *I. verum* dry powder (150 g) was placed in a 1.0 L round-bottomed flask. Methyl alcohol (MA; polarity, 5.1; highly polar), ethyl acetate (EA; polarity, 4.4; weakly polar), and petroleum ether (PE; boiling point range, 60°C to 90°C; polarity, 0.0; non-polar) were respectively added in a ratio of 1:5 (w/v) at room temperature (25°C), incubated in the dark for 48 hr, and then filtered (Whatman No. 2). The samples were leached twice using the same procedure. The final filtrates were collected for each solvent to obtain the respective crude extracts. The combined filtrate was concentrated to dryness using a vacuum rotary evaporator (Buchi Rotavapor R-124; www.buchi.com) and weighed using an electronic balance (FA2104; www.testmart.cn). The extraction rates were then calculated. All samples were stored in air-tight brown bottles at 4°C in a refrigerator for use in the subsequent experiments.

GC-MS

The composition of the extracts was investigated by GC-MS performed on a Varian Saturn 2200 GC system equipped with a capillary column with CP-Sil8CB-MS (30 m × 0.25 mm × 0.25 um) (Agilent Technologies, www.chem.agilent.com). The GC settings were as follows: the initial oven temperature was held at 50°C for 3 min and ramped at 20 °C/min to 120°C without reserve and then ramped at 10°C/min to 250°C for 5 min. The temperatures of the injector, interface, and gasification chamber were maintained at 260, 250, and 250°C, respectively. The samples (0.4 μL) were injected neat, with a split ratio of 1:100. The carrier gas was helium (99.999%). Mass spectra were obtained at 70 eV by electron impact ionization source. The temperatures of the interface and iron trap were 250 and 150°C. The electron multiplier voltage was 2.4 kV. The mass range analyzed was 20 amu to 650 amu. Most constituents were identified using GC by comparison of their retention indices based on literature data and MS data obtained from Saturn and NIST libraries. Component relative percentages were calculated based on GC peak areas without using correction factors (Gholivand et al. 2009; Chu et al. 2011).

Bioassays

Repellency Assay

The repellent activities of the extracts were assessed using a previously described filter-paper drug-film area preference method (Tapondjou et al. 2005) with slight modifications. Test areas consisted of 9 cm Whatman No. 2 filter paper cut in half (31.8 cm^2). Each extract (50.00 mg) was diluted with acetone to produce the following concentrations: 0.50, 1.00, 2.00, 4.00, and 8.00 mg/mL. Each diluent (0.5 mL) was uniformly dropped on the half filter paper disc with a pipette, corresponding to doses of 7.86, 15.72, 31.45, 62.89, and 125.79 μg/cm^2, whereas the other half of the paper was treated with 0.5 mL acetone alone as a control. The treated and control half discs were air dried at room temperature for 5 min to evaporate the solvents. The treated and untreated halves were attached to their corresponding halves using adhesive tape and then placed in 9 cm glass Petri dishes. The bottom of the dish was daubed with solid glue to prevent insects from going under the filter paper. Thirty mixed-sex *S. zeamais* adults were released at the center of each filter paper disc without food. Glass loops (8.8 cm i.d. × 6.0 cm) coated with polytetrafluoroethylene were then set on the interior walls of the Petri dishes to prevent insects from climbing upward. The dishes were kept in continual darkness. Each treatment was repeated three times. The numbers of *S. zeamais* present on the treated and untreated portions of the experimental paper halves were recorded after 24, 48, and 72 hr of exposure.

Contact Toxicity Assay

The Petri dish drug-film contact method was adopted in the contact toxicity examination with some adjustments (Huang et al. 2011). Each extract (50.00 mg) was respectively diluted with acetone to the following concentrations: 0.50,

1.00, 2.00, 4.00, and 8.00 mg/mL. Subsequently, 1.00 mL of each solution was poured into clean Petri dishes (9.0 cm i.d. × 1.0 cm), corresponding to doses of 7.86, 15.72, 31.45, 62.89, and 125.79 μg/cm². The dishes were swirled gently until the solutions formed a thin film. After the solutions evaporated for 1 min, glass loops (8.8 cm i.d. × 6.0 cm) coated with polytetrafluoroethylene were set on the interior walls of the Petri dishes to prevent insects from climbing upward. Thirty adult *S. zeamais* were placed into each Petri dish. The dishes were covered with a ventilated net lid to prevent insects from flying away. Each treatment was replicated three times, with acetone as control. The dishes were placed in a constant-temperature incubator (temperature, 27°C to 29°C; relative humidity, 70% to 80%). Mortalities were measured at 24, 48, and 72 hr after treatment. Test insects were considered dead if appendages did not move when prodded with a pin.

Statistical Analysis

Percentage repellency (PR) was calculated as follows:

$$PR = \left[(Nc - Nt) / (Nc + Nt) \right] \times 100$$

where Nc is the number of insects on the untreated area after the exposure interval, and Nt is the number of insects on the treated area after the exposure interval. Results are presented as the mean of PR ± SE. Mean PRs were assigned to repellency classes (Malik and Mujtaba 1984) from 0 to V, where class 0 < 0.1%, class I = 0.1% to 20%, class II = 20.01% to 40%, class III = 40.01% to 60%, class IV = 60.01% to 80%, and class V = 80.01% to 100%. The mean PRs of each dose were compared and separated by Duncan›s new multiple range tests with a significance level at α = 0.05.

The percentage mortalities were determined in contact toxicity examination and corrected using Abbott's formula (Abbott 1925). Treatment means were compared by Duncan's new multiple range tests with the significance level at α = 0.05. Data were transformed to arcsine square-root values for analysis of variance (ANOVA). The relationship between mortality and concentration was modeled by the DPS program (Tang and Feng 2002). Means (±SE) of untransformed data are presented.

RESULTS

Effects of Solvents on Extraction Yield from *Illicium Verum*

The highest extraction yield from *I. verum* was obtained with MA, followed by EA and PE (Table 1). The polarity values of MA, EA, and PE were 5.1, 4.4, and 0.0, respectively. The extraction yields were dependent on the similarity of the chemical characteristics of the solvent and polarity of the plant material. The result indicated that the compounds present in *I. verum* were mostly of high polarity.

Table 1. Extraction yields of Illicium verum using three types of organic solvents (%).

Solvents	Dry weight of fruit (g)	Volume of solvents (mL)	Dry weight of extracts (g)	Extraction Yield (%)[a]
MA	150.00	750.00	35.50	23.67
EA	150.00	750.00	21.32	14.21
PE	150.00	750.00	21.00	14.00

[a]Extraction Yield = (Dry weight of extract / Dry weight of star anise fruit) × 100.

Results of GC-MS analysis showed that compositions in the three extracts were almost the same (Figures 1–3). However, some minor components were different. For example, the MA extract contained more 4-ethyl benzaldehyde (3.4%) and 1-(4-methoxyphenyl)-2-propanone (3.7%); the EA extract contained more 1-(3-methyl-2-butenoxy)-4-(1-propenyl) benzene (6.2%), cis-3,5-dimethoxy-β-methyl-β-nitrostyrene (4.2%), hexadecanoic acid (4.1%), and Benzyl alcohol(4.0%); whereas the PE extract contained more 1-(3-methyl-2-butenoxy)-4-(1-propenyl) benzene (4.7%) and benzyl alcohol (2.8%).

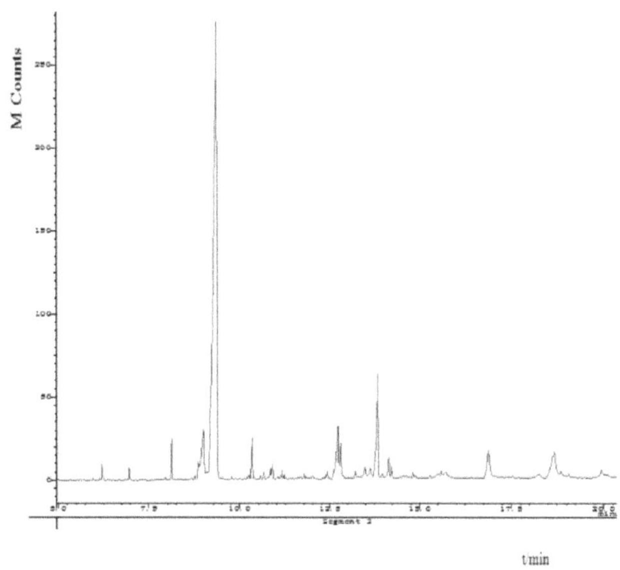

Figure 1. Total ion current chromatogram of compounds from *Illicium verum* by methyl alcohol extraction. High quality figures are available online.

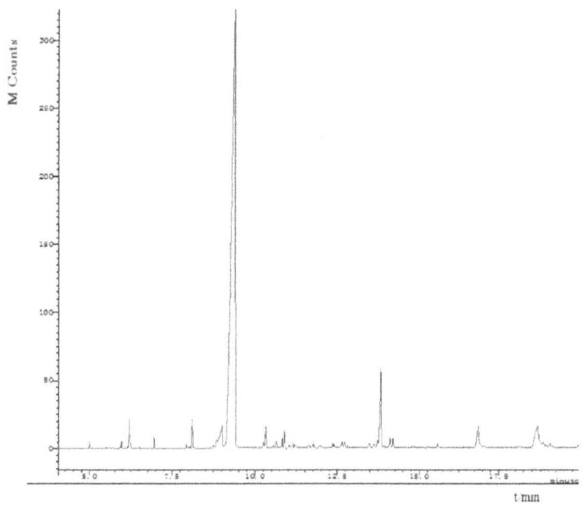

Figure 2. Total ion current chromatogram of compounds from *Illicium verum* by ethyl acetate extraction. High quality figures are available online.

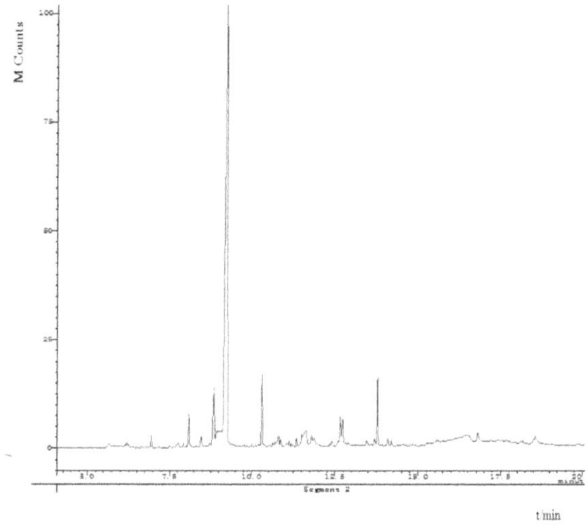

Figure 3. Total ion current chromatogram of compounds from *Illicium verum* by petroleum ether extraction. High quality figures are available online.

Chemical Composition of Extracts

Forty-four compounds from MA, EA, and PE extracts were separated and identified by GC-MS (Table 2), representing 70.5%, 82.9%, and 92.7%, respectively, of the whole composition of the extracts. The most abundant component was trans-anethole.

Table 2. Chemical composition of extracts from Illicium verum dried fruit using three types of organic solvents.

No.	Compound	RT (min)	Molecular formula	Relative area (%)		
				MA	EA	PE
1.00	Phenol	5.66	C_6H_6O	0.34		
2.00	Carene	5.96	$C_{10}H_{16}$		0.04	0.21
3.00	Terpinolene	6.20	$C_{10}H_{16}$		0.51	1.07
4.00	Benzoic acid	7.76	$C_7H_6O_2$	0.43		
5.00	Terpineol	7.93	$C_{10}H_{18}O$	0.24	0.09	0.12
6.00	Anethole	8.75	$C_{10}H_{12}O$	0.24	0.12	0.16
7.00	4-Ethyl benzaldehyde	8.86	$C_9H_{10}O$	3.40	0.67	0.53
8.00	Benzyl alcohol	9.00	C_7H_8O		4.04	2.81
9.00	Trans-anethole	9.38	$C_{10}H_{12}O$	41.14	52.54	72.25
10.00	β-Copaene	10.26	$C_{15}H_{24}$	0.06	0.12	0.21
11.00	1-(4-Methoxyphenyl)-2-propanone	10.32	$C_{10}H_{12}O$	3.68	1.57	0.80
12.00	Bergamotene	10.65	$C_{15}H_{24}$	0.11	0.23	0.36
13.00	Benzoic acid,4-methoxy	10.80	$C_8H_8O_3$	0.87		
14.00	Longifolene	10.83	$C_{15}H_{24}$	0.08	0.49	0.32
15.00	Bisabolene	10.89	$C_{15}H_{24}$	0.41	0.58	0.54
16.00	Acetaldehyde	11.14	C_2H_4O	0.24	0.32	0.22
17.00	4-Methoxy-benzaldehyde-oxime	11.44	C_8H_9NO2	0.35	0.03	0.04
18.00	O-Nitrosobenzoic acid	11.53	$C_7H_5O_3N$	0.92	0.06	0.05
19.00	4-Hydroxybenzoic acid	11.64	$C_7H_6O_3$	2.28		
20.00	β-Humulene	11.77	$C_{15}H_{24}$		0.17	0.24
21.00	3,6-Dimethyl-4H-furo[3,2-C]pyran-4-one	11.80	$C_9H_8O_3$	0.69		
22.00	3-Hydroxybenzoic acid	11.88	$C_7H_6O_3$	0.54		
23.00	3-Hydroxy-1,2-benzisoxazole	12.40	$C_7H_5NO_2$	0.37	0.26	0.11
24.00	4-Methoxy cinnamaldehyd	12.62	$C_{10}H_{10}O_2$		0.41	
25.00	4 -Ethyl - α- methyl benzyl alcohol	12.64	$C_{10}H_{14}O$	1.41		0.45
26.00	*cis*-3,5-Dimethoxy-β-methyl-β-nitrostyrene	12.73	$C_{11}H_{16}NO_4$	1.57	4.24	0.30
27.00	(s)-2,5,5,8-Tetramethyl-1,2,3,6-tetrahydro-azulen-4(5H)-one	13.19	$C_{14}H_{20}O$	0.39		0.32
28.00	δ- Cadinene	13.62	$C_{15}H_{24}$			0.23
29.00	9-Methyl-9H-fluorene	13.69	$C_{14}H_{12}$	0.41		0.33
30.00	1-(3- Methyl-2-butenoxy)-4-(1-propenyl) benzene	13.82	$C_{14}H_{18}O$	2.52	6.22	4.73
31.00	Bendazol	14.10	$C_{14}H_{12}N_2$	0.47	1.21	0.47
32.00	Trans-2-ethoxy-β-methyl-β-nitrostyrene	14.20	$C_{11}H_{15}NO_3$	0.29	0.44	0.41
33.00	N-(4-hydroxyphenyl)-2-methylbenzamide	14.81	$C_{14}H_{13}NO_2$		0.23	
34.00	Phenol-3-[2-(2-phenylethy)amino]ethyl]	15.52	$C_{17}H_{21}NO$		0.65	
35.00	1H-Benzimidazol-4-ol.5-ethoxy-1(methylphenyl)	15.57	$C_{16}H_{16}N_2O2$	0.77		
36.00	2,2-Diisobutyl-1,3-benzodioxole	15.59	$C_{15}H_{22}O_2$		0.37	0.27
37.00	Hexadecanoic acid, methyl ester	16.27	$C_{17}H_{34}O_2$	2.26	0.07	0.07
38.00	1,1'-(1-Ethyl-2methyl-1,2-ethenediyl)bis(4-methoxybenzene)	16.80	$C_{18}H_{20}O_2$	2.05	0.77	
39.00	Hexadecanoic acid	16.88	$C_{16}H_{32}O_2$		4.07	2.15
40.00	Octadeca-9C,12C dienoic acid(z,z)	18.51	$C_{18}H_{32}O_2$	0.71		
41.00	9,12-Octadecadienoic acid(z,z)-methyl ester	18.56	$C_{19}H_{34}O_2$	1.21		
42.00	1-Acetoxy-6-(1-hydroxy-1-methylethyl)-3,9-dimethyl-2z,8-deca-diene	18.66	$C_{18}H_{34}O_3$			2.15
43.00	(E,E)-Methyl linolelaidate	18.85	$C_{19}H_{34}O_2$		0.90	0.64
44.00	1-Methoxy-4-(1-propenyl)benzene	19.94	$C_{10}H_{12}O$		1.45	0.14
Total	-	-	-	70.45	82.87	92.70

Biological Activities of the Extracts from *I. Verum* against *S. Zeamais* Adults Repellency Activity

The repellent effects to *S. zeamais* adults increased with increasing dose (Table 3). At the highest dose (125.79 μg/cm²), EA had the highest average percentage repellency value, followed by PE and MA. The repellency of an extract at a dose declined as the exposure time was prolonged. This result may be attributed to solution volatilization that decreases extract concentration. In addition, the extracts from *I. verum* by three solvents to *S. zeamais* adults showed a certain paralysis effect.

Table 3. Percentage repellency (PR) of *Illicium verum* fruit extracts against *Sitophilus zeamais* adults.

Solvents	Dose (μg/cm²)	PR (%) 24 hr	48 hr	72 hr	PR mean values ± SE[a]	Repellecy level
MA	125.79	57.14	54.84	47.46	53.15 ± 2.92 a	III
	62. 89	47.46	42.11	33.33	40.97 ± 4.12 ab	III
	31.45	44.83	26.92	30.19	33.98 ± 5.51 b	II
	15.72	33.33	26.92	26.92	29.06 ± 2.14 b	II
	7.86	26.92	12.50	0.00	13.14 ± 7.78 c	I
EA	125.79	86.08	75.00	69.57	76.88 ± 4.86 a	IV
	62. 89	71.43	67.65	57.14	65.41 ± 4.27 ab	IV
	31.45	54.84	42.11	44.83	47.26 ± 3.87 b	III
	15.72	39.29	30.19	12.50	27.32 ± 7.86 c	II
	7.86	30.19	20.00	0.00	16.73 ± 8.87 c	I
PE	125.79	73.24	69.57	65.67	69.49 ± 2.18 a	IV
	62. 89	59.38	59.38	47.46	55.40 ± 3.97 b	III
	31.45	47.46	42.11	30.19	39.92 ± 5.10 c	II
	15.72	42.11	30.19	20.00	30.77 ± 6.38 cd	II
	7.86	23.53	12.50	16.33	17.45 ± 3.23 d	I

[a]The data marked with different small letters in the same column mean significant difference ($P < 0.05$) based on Duncan's new multiple range test.

Contact Toxicity

The extracts from *I. verum* by the three solvents showed strong contact toxicity against *S. zeamais*adults. The contact activities of the extracts by the three solvents increased with increasing treatment doses (Table 4). With the highest dose of 125.79 μg/cm² at 72 hr after treatment, the mortality of *S. zeamais* adults was highest for EA, followed by PE and MA. No mortality was observed in each control. For a solvent extract at lower doses, contact killing activity increased as the processing time was prolonged. However, at higher doses, the prolonged treatment time did not translate to an increase in contact toxicity.

Table 4. Contact toxicity of *Illicium verum* fruit extracts against *Sitophilus zeamais* adults.

Solvents	Dose (μg/cm²)	Corrected morality (%) 24 hr	48 hr	72 hr
MA	125.79	69.70 ± 1.21 a	83.24 ± 1.50 a	85.43 ± 0.90 a
	62. 89	58.45 ± 1.31 b	72.00 ± 1.88 b	74.21 ± 1.45 b
	31.45	35.94 ± 1.16 c	46.06 ± 1.12 c	48.31 ± 1.71 c
	15.72	16.72 ± 1.78 d	25.79 ± 1.28 d	28.04 ± 1.42 d
	7.86	7.79 ± 1.24 e	13.45 ± 0.93 e	14.57 ± 0.93 e
EA	125.79	76.46 ± 13.2 a	89.89 ± 0.00 a	94.50 ± 1.49 a
	62. 89	65.17 ± 0.67 b	79.85 ± 1.39 b	82.05 ± 0.83 b
	31.45	44.93 ± 1.72 c	61.80 ± 0.66 c	65.17 ± 0.67 c
	15.72	26.90 ± 1.47 d	38.20 ± 0.66 d	41.57 ± 0.65 d
	7.86	13.31 ± 1.97 e	25.79 ± 1.28 e	29.09 ± 2.17 e
PE	125.79	74.17 ± 0.74 a	87.69 ± 1.00 a	91.08 ± 1.16 a
	62. 89	61.80 ± 0.66 b	77.61 ± 1.52 b	79.85 ± 1.39 b
	31.45	42.69 ± 1.13 c	57.31 ± 0.65 c	60.68 ± 0.66 c
	15.72	22.41 ± 1.34 d	33.70 ± 0.68 d	37.08 ± 0.66 d
	7.86	11.18 ± 1.05 e	21.19 ± 2.12 e	25.68 ± 2.27 e

[a]The dose is the weight of drug residue per unit area of Petri dishes. The data marked with different small letters in the same column mean significant difference ($P < 0.05$) based on Duncan's new multiple range test.

The regression equations and LD_{50} (Table 5) for the contact toxicities of the three extracts against*S. zeamais* adults were obtained by linear regression analysis of the correlation between the treatment doses and the arcsine square-root values of average mortalities. The contact toxicities of the extracts by all three types of solvent were enhanced with increasing treatment doses. The results indicated that EA extract had the highest contact toxicity, followed by PE extract, and finally MA extract.

Table 5. Regression analysis of contact toxicities of *Illicium verum* fruit extracts against *Sitophilus zeamais* adults at 72 hr after treatment.

Solvent	Regression equation of toxicity	LD_{50} ($\mu g/cm^2$)	Relative coefficient (r)	95% confidence interval ($\mu g/cm^2$)	Chi square (χ^2)
MA	y = 1.80x + 2.20	36.11	1.00	35.52–42.64	0.80
EA	y = 1.74x + 2.69	21.19	0.99	16.83–25.54	1.41
PE	y = 1.69x + 2.65	24.65	1.00	19.94–29.52	1.12

DISCUSSION

In the current study, according to the extraction yields, yields of trans-anethole from *I. verum* in MA, EA, and PE solvents were 9.7%, 7.5%, and 10.1%, respectively, which are higher than those obtained by hydrodistillation. Moreover, the extracts obtained from MA contained more high-polar materials, whereas the extracts obtained from PE contained more non-polar materials. The polarity of the material is related to its size and molecular structure, and the polarity of target active components should be assessed to ensure that the appropriate solvent can be selected to separate them. The toxicity and reactivity of solvents with target components should also be considered.

Some minor compounds in the three extracts of *I. verum* have antibacterial activities. For example, benzyl alcohol is used as a preservative in skincare products because it is less irritating to the skin than other preservatives (Mashayekhi et al. 2011). These different minor constituents provide the possibility for different applications and targeted development.

The repellency of the extracts against *S. zeamais* adults increased with increasing dose. The repellency of the extracts from the three solvents to *S. zeamais* adults was in the order of EA extract > PE extract > MA extract (Table 3). The oils isolated from *Cupressus sempervirens* and*Eucalyptus saligna* are repellent against *S. zeamais* (Tapondjou et al. 2005). Repellency is crucial in protecting stored products from pests (Luz et al. 2009; Ukeh et al. 2009). In recent years, commercial repellents have developed rapidly. The current study suggests that the extracts from *I. verum* can be developed as a promising *S. zeamais* repellent applied on stored grain.

According to the mortalities and the LD50 values of MA, EA, and PE extracts to *S. zeamais* adults assayed in the current study, the EA extract had the highest contact toxicity to *S. zeamais.*Although the content of active constituent (trans-anethole) in the EA extract was the least among the three extracts, this extract showed the highest repellency and contact toxicity. This conflicting observation indicates the presence of other active constituents in the EA extract. Therefore, further studies on other insecticidal active constituents in the EA extract against *S. zeamais* and other economically important pests need to be undertaken.

Natural products can be used by small-scale farmers to protect stored grains from insect infestation. Several aromatic plants, such as *Lamiaceae,* have been used in developing countries to protect stored grain and legumes from pests (Ngamo et al. 2007; Li et al. 2013). In West Africa, farmers traditionally introduce *Hyptis spicigera* Lamarck leaves in their granaries to protect cowpea seeds against bruchids damages (Sanon et al. 2006). China has a long and rich tradition of using vegetable matter for pest control. For example, toosendanin, a good botanical insecticide, was derived from the bark of Chinaberry and is used against a broad spectrum of fruit and vegetable pests in China. *Illicium verum* is a potential grain protectant (Ho et al. 1995). In the present study, the extracts from *I. verum* exhibited pest-combating activities, including both repellency and toxicity. We conclude that the *I. verum* fruit-derived materials might be useful products for managing populations of stored-product pests. *Illicium verum* has been widely cultivated in Guangxi, Yunnan, Guangdong, Fujian, and Guizhou Provinces of China. In addition, the planting area and output of China's *I. verum* account for > 80% of the planting area and output worldwide. Thus, *I. verum* is readily available in China. In integrated stored-product protection, the coarse extracts of *I. verum* may be directly used for pest prevention, repelling pests from foods, and pest control by its toxic compounds.

In recent years, considerable effort has been exerted on the potential of plant extracts or phytochemicals as sources of commercial insect-control agents or as new leads for designing target-specific molecules (Ateyyat et al. 2009; Gaikwad et al. 2010; Li et al. 2011). Previous studies found that *I. verum* dried fruit contains 8% to 12% essential oil. A total of 49 compounds were separated and identified in the essential oil, including trans-anethole (81.4%), limonene (6.5%), chavicol (2.1%), and anisaldehyde (1.8%) (Gholivand et al. 2009). Our future studies will focus on isolating and purifying biologically active constituents derived from the EA extract of *I. verum* to select the compound with high biological activity and relatively simple structure for the development of a new type of repellent or insecticide. Moreover, an in-depth study on the relationship between the chemical composition and insecticidal activity of the extracts need to be performed with a biological activity tracking method to provide a clearer understanding of which compounds were responsible for the extracts' bioactivities.

ACKNOWLEDGMENTS

This research was supported by the grants from the Planning Subject of the "Twelfth Five-year-plan" from the Ministry of Science and Technology of China (2011BAD12B04-02), the National High Technology Research and Development Program of China (863 Program) (2011AA10A205), and the Research and Development Project of the Hefei Institute of Agricultural Products Processing and its Cooperating Company (03082317).

REFERENCES

1. Abbott WS. 1925. A method of computing the effectiveness of an insecticide. Journal of Economic Entomology 18: 265–267.
2. Ateyyat MA, Al-Mazra'awi M, Abu-Rjai T, Shatnawi

MA. 2009. Aqueous extracts of some medicinal plants are as toxic as Imidacloprid to the sweet potato whitefly, *Bemisia tabaci*. Journal of Insect Science 9(15). Available online: www.insectscience.org/9.15

3. Chang KS, Ahn YJ. 2001. Fumigant activity of (E)-anethole identified in *Illicium verum* fruit against *Blattella germanica*. Pest Management Science 58: 161–166.
4. Chu SS, Hu JF, Liu ZL. 2011. Composition of essential oil of Chinese Chenopodium ambrosioides and insecticidal activity against maize weevil, *Sitophilus zeamais*. Pest Management Science 67: 714– 718.
5. Dana C, Wej C, 2006. Essential oils as potential adulticides against two populations of *Aedes aegypti*, the laboratory and natural field strains, in Chiang Mai province, northern Thailand. Parasitology Research 99:715– 721.
6. Gaikwad YB, Gaikwad SM, Bhawane GP. 2010. Effect of induced oxidative stress and herbal extracts on acid phosphatase activity in lysosomal and microsomal fractions of midgut tissue of the silkworm, *Bombyx mori*. Journal of Insect Science 10(113). Available online: www.insectscience.org/10.113.
7. Gholivand MB, Rahimi-Nasrabadi M, Chalabi H. 2009. Determination of essential oil components of star anise (*Illicium verum*) using simultaneous hydrodistillation-static headspace liquid-phase microextraction-gas chromatography mass spectrometry. Analytical Letters 42: 1382–1397.
8. Ho SH, Ma Y, Goh PM, Sim KY. 1995. Star anise, *Illicium verum* Hook f as a potential grain protectant against*Tribolium castaneum* (Herbst) and *Sitophilus zeamais* Motsch. Postharvest Biology and Technology 6: 341–347.
9. Huang YZ, Hua HX, Li SG, Yang CJ. 2011. Contact and fumigant toxicities of calamusenone isolated from*Acorus gramineus* rhizome against adults of *Sitophilus zeamais* and *Rhizopertha dominica*. Insect Science18: 181–188.
10. Isman MB. 2006. Botanical insecticides, deterrents, and repellents in modern agriculture and an increasingly regulated world. Annual Review of Entomology 51: 45–66.
11. Isman MB. 2008. Perspective botanical insecticides: For richer, for poorer. Pest Management Science 64:8–11.
12. Kim SI, Park C, Ohh MH, Cho HC, Ahn YJ. 2003a. Contact and fumigant activities of aromatic plant extracts and essential oils against *Lasioderma serricorne* (Coleoptera: Anobiidae). Journal of Stored Products Research 39: 11–19.
13. Kim SI, Roh JY, Kim DH, Lee HS, Ahn YJ. 2003b. Insecticidal activities of aromatic plant extracts and essential oils against *Sitophilus oryzae* and *Callosobruchus chinensis*. Journal of Stored Products Research 39:293–303.
14. Kudom AA, Mensah BA, Botchey MA. 2011. Aqueous neem extract versus neem powder on *Culex quinquefasciatus*: Implications for control in anthropogenic habitats. Journal of Insect Science 11(142).www.insectscience.org/11.142.
15. Ladhari A, Laarif A, Omezzine F, Haouala R. 2013. Effect of the extracts of the spiderflower, *Cleome arabica,*on feeding and survival of larvae of the cotton leafworm, *Spodoptera littoralis. Journal of Insect Science* 13(61). Available online: www.insectscience.org/13.61
16. Lee SW, Li G, Lee KS, Jung JS, Xu ML, Seo CS, Chang HW, Kim SK, Song DK, Son JK. 2003. Preventive agents against sepsis and new phenylpropanoid glucosides from the fruits of Illicium verum. Planta Medica69: 861-864.
17. Li HQ, Bai CQ, Chu SS, Zhou LG, Du SS, Liu ZL. 2011. Chemical composition and toxicities of the essential oil derived from *Kadsura heteroclita* stems against *Sitophilus zeamais* and Meloidogyne incognita. Journal of Medicinal Plants Research 5: 4943-4948.
18. Li SG, Li MY, Huang YZ, Hua RM, Lin HF, He YJ, Wei LL, Liu ZQ. 2013. Fumigant activity *ofIllicium verum*fruit extracts and their effects on the acetylcholinesterase and glutathione S-transferase activities in adult *Sitophilus zeamais*. Journal of Pest Science 86: 677-683.
19. Li WQ, Jiang CH, Chu SS, Zuo MX, Liu ZL. 2010. Chemical composition and toxicity against *Sitophilus zeamais* and *Tribolium castaneum* of the essential oil *of Murraya exotica* aerial parts. Molecules 15: 5831-5839.
20. Liu ZL, Goh SH, Ho SH. 2007. Screening of Chinese medicinal herbs for bioactivity against *Sitophilus zeamais* Motschulsky and *Tribolium castaneum* (Herbst). Journal of Stored Products Research 43: 290-296.
21. López MD, Jordán MJ, Pascual-Villalobos MJ. 2008. Toxic compounds in essential oils of coriander, caraway and basil active against stored rice pests. Journal of Stored Products Research 44: 273–278.
22. Luz SN, Jesus OV, Elena ES. 2009. Repellent activity of essential oils from seven aromatic plants grown in Colombia against *Sitophilus zeamais* Motschulsky (Coleoptera). Journal of Stored Products Research 45:212–214.
23. Lv C, Zhong B, Zhong G, Weng Q, Chen S, Hu M, Sun X, Qin W. 2012. Four botanical extracts are toxic to the hispine beetle, *Brontispa longissima*, in laboratory and semi–field trials. Journal of Insect Science 12(58). Available online: www.insectscience.org/12.58.
24. Malik MM, Muiutaba NSH. 1984. Screening of some indigenous plants as repellents of antifeedants for stored grain insect. Journal of Stored Products Research 20: 45–48.
25. Mashayekhi HA, Rezaee M, Garmaroudi SS, Montazeri N, Ahmadi SJ. 2011. Rapid and sensitive determination of benzaldehyde arising from benzyl alcohol used as preservative in an injectable formulation solution using dispersive liquid-liquid microextraction followed by gas chromatography. Analytical Sciences27: 865– 868.
26. Ngamo TSL, Goudoum A, Ngassoum MB, Mapongmestsem PM, Lognay G, Malaisse F, Hance T. 2007.Chronic toxicity of essential oil of 3 local aromatic plants towards *Sitophilus zeamais* Motsch. (Coleoptera: Curculionidae). African Journal of Agricultural Research 2: 164–167.
27. Ohira H, Torii N, Aida TM, Watanabe M, Smith RLJ. 2009. Rapid separation of shikimic acid from Chinese

star anise *Illicium verum* (Hook f) with hot water extraction. Separation and Purification Technology 69: 102-108.

28. Ojo JA, Omoloye AA. 2012. Rearing the maize weevil, *Sitophilus zeamais,* on an artificial maize-cassava diet. Journal of Insect Science 12(69). Available online: www.insectscience.org/12.69

29. Padmashree A, Roopa N, Semwal AD, Sharma GK, Agathian G, Bawa AS. 2007. Star anise *(Illicium verum)* and black caraway *(Carum nigrum)* as natural antioxidants. Food Chemistry 104: 59-66.

30. Sanon A, Ilboudo Z, Dabire C, Nebie R, Dicko I, Monge JP. 2006. Effects *of Hyptis spicigera* Lam (Labiatae) on the behaviour and development of *Callosobruchus maculatus F.* (Coleoptera: Bruchidae), a pest of stored cowpeas. International Journal of Pest Management 52: 117-123.

31. Singh G, Maurya S, de Lampasona MP, Catalan C. 2006. Chemical constituents, antimicrobial investigations and antioxidative potential of volatile oil and acetone extract of star anise fruits. Journal of the Science of Food and Agriculture 86: 111-121.

32. Shu X, Liu XM, Fu CL, Liang QX. 2010. *Extraction, characterization and antitumor effect of the polysaccharides from star anise Illicium verum (Hook f).*Journal of Medicinal Plants Research 4: 2666-2673.

33. Tang QY, Feng MG. 2002. *DPS data processing system for practical statistics.*pp. 417-452. Science Press. (in Chinese)

34. Tapondjou A, Adler C, Fontem D, Bouda H, Reichmuth C. 2005. Bioactivities of cymol and essential oils of *Cupressus sempervirens* and *Eucalyptus saligna* against *Sitophilus zeamais* Motschulsky and *Tribolium confusum* du Val. Journal of Stored Products Research 41: 91–102.

35. Ukeh DA, Birkett MA, Pickett JA, Bowman AS, Luntz AJM. 2009. Repellent activity of alligator pepper,*Aframomum melegueta*, and ginger, *Zingiber officinale*, against the maize weevil, *Sitophilus zeamais.* Phytochemistry 70: 751–758.

36. Yang JF, Yang CH, Chang HW, Yang CS, Wang SM, Hsieh MC, Chuang LY. 2010. Chemical composition and antibacterial activities of *Illicium verum* against antibiotic-resistant pathogens. Journal of Medicinal Food 13:1254–1262.

CHAPTER 16

TYPICAL MONOTERPENES AS INSECTICIDES AND REPELLENTS AGAINST STORED GRAIN PESTS

Suelen L. Reis, Anieli G. Mantello, Jeferson M. Macedo, Erica A. Gelfuso, Cássio P. da Silva, Ana L. Fachin, Alexandre M. Cardoso, and Rene O. Beleboni

Keywords: Essential oils, Callosobruchus maculatus, fumigant, pest management, Sitophilus zeamais

INTRODUCTION

Crop pests are one of the most important factors contributing to the decrease in productivity of different agricultural crops. These pests can attack during different stages of the food production process, affecting crops in the field or stored grains and causing major economic losses [1]. Therefore, pest management is crucial for the longevity and quality maintenance of food and related agricultural products, especially stored products since cumulative production costs are higher [1,2].

One of the most important pests attacking stored grains in Brazil and in the world are the cowpea seed beetle *Callosobruchus maculatus* (Fabr.) (Coleoptera: Bruchidae) and the maize weevil *Sitophilus zeamais* Motschulsky (Coleoptera: Curculionidae). These pests are particularly important due to wide range of crops attacked, higher economic losses and problems in their management. The management of these pests can be difficult since infestation with adult insects occurs throughout the entire crop growth period and is characterized by a high degree of inbreeding between different insect populations [3,4]. Today, *C. maculatus* and *S. zeamais* are the most important pests associated with stored cowpea (*Vigna unguiculata* L.) and maize (*Zea mays* L.), respectively [5,6]. Both insect species are commonly used as model organisms in biological research, especially to screen for new insecticides and repellents, because of their short generation interval and effortless maintenance [3,4].

Botanical pesticides are considered interesting alternatives to synthetic pesticides since they are potentially safer due to their rapid biodegradation and consequent short persistence in the environment. Furthermore, these pesticides are usually less expensive and less likely to induce pest resistance, especially when used as phytocomplexes in which a mixture of active compounds has different complementary modes of action on different molecular targets and insect stages of development [7,8].

Within this context, essential oils of plants are an interesting source of new botanical pesticides, especially because they are rich in monoterpenes [9]. Indeed, several monoterpenes have been shown to have remarkable insecticidal and repellent activities, including against *C. maculatus* and *S. zeamais*. Several studies have demonstrated these properties for monoterpenes such as cineol, limonene, terpinolene, and thymol, among others [10,11]. However, no direct insecticidal or repellent activity against *C. maculatus* and *S. zeamais* has been reported for other monoterpenes. Further investigation of these activities involving a larger number of different chemical structures is important, especially when considering the critical contribution of structure-activity relationship studies to the design and development of new biotechnological compounds such as safer and less expensive pesticides.

Citation: Suelen L. Reis, Anieli G. Mantello, Jeferson M. Macedo, Erica A. Gelfuso, Cássio P. da Silva, Ana L. Fachin, Alexandre M. Cardoso, and Rene O. Beleboni. Typical monoterpenes as insecticides and repellents against stored grain pests, Molecules (2016); 21(3), 258. DOI:10.3390/molecules21030258.

The aim of the present study was to evaluate the insecticidal and repellent activity of typical monoterpenes (geraniol, geranial, (±) citronellal, citronellol and eugenol) against the stored grain pests *C. maculatus* and *S. zeamais*. The monoterpenes were chosen based on their insecticidal and repellent potential, scientific originality, easy availability, and wide occurrence in essential oils of plants.

RESULTS AND DISCUSSION

The results of the mortality tests show considerable effectiveness of the different monoterpenes against the insects tested, especially against *C. maculatus*. With respect to the number of dead *C. maculatus* after treatment with geranial (LD_{50} = 0.9832 μL), significant differences were observed for all doses tested when compared to the control group [F(7.71) = 56.22; $p < 0.0001$] (Figure 1A). The same was observed for the treatments with geraniol (LD_{50} = 0.7140 μL) [F(7.71) = 53.79; $p < 0.0001$] (Figure 1B), citronellal (LD_{50} = 2.261 μL) [F(7.71) = 217.8; $p < 0.0001$] (Figure 1C) (exception only for the dose of 1 μL), citronellol (LD_{50} = 1.534 μL) [F(7.71) = 76.54; $p <$ 0.0001] (Figure 1D), and eugenol (LD_{50} = 0.9473 μL) [F(7.71) = 53.85; $p < 0.0001$] (Figure 1E).

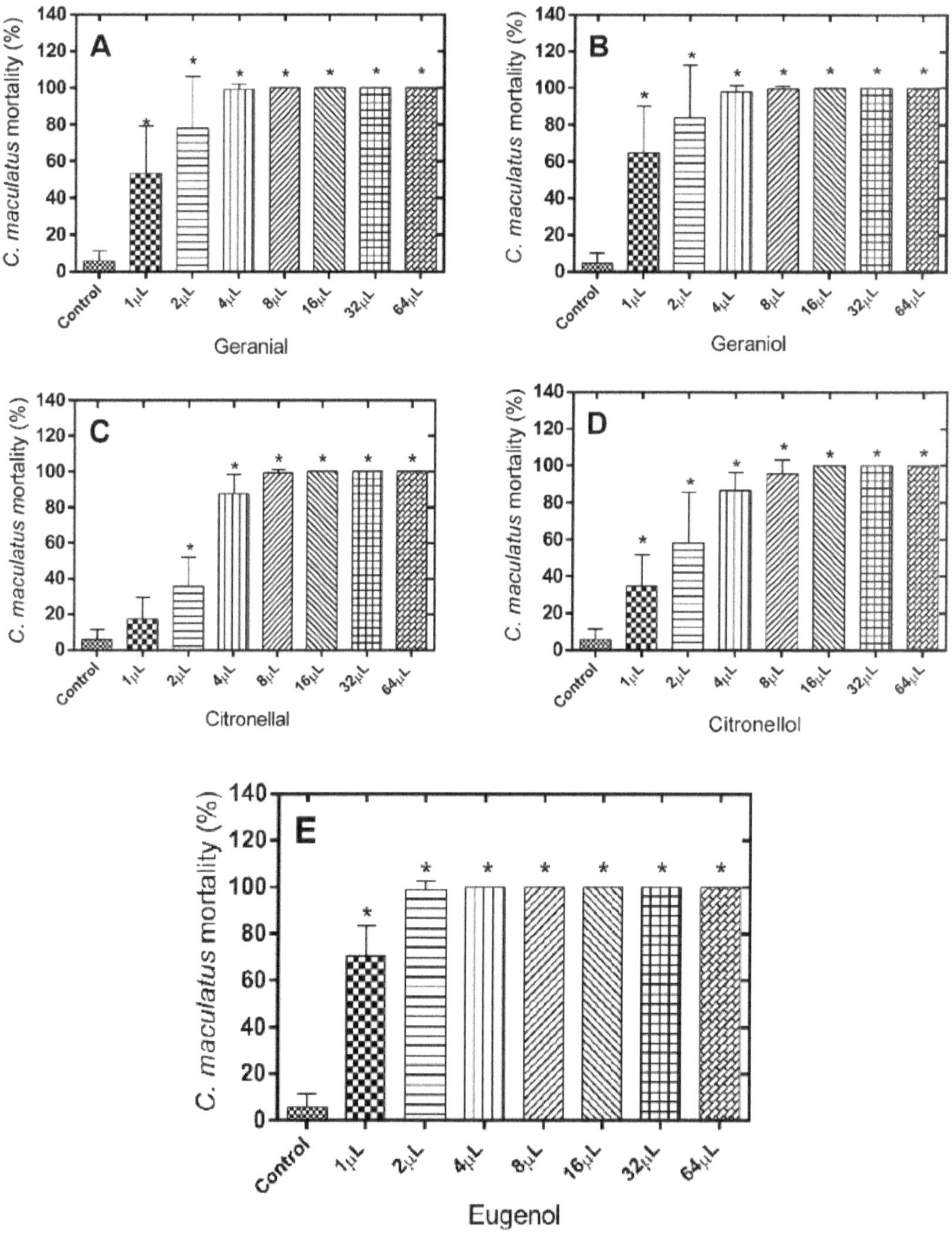

Figure 1. Mortality (%) of *C. maculatus* 24 h after exposure to the alimentary substrate treated with different monoterpene doses. (A) Geranial; (B) Geraniol; (C) Citronellal; (D) Citronellol; (E) Eugenol. * Significantly different compared to control, test ($p < 0.05$).

Regarding the number of dead *S. zeamais* after treatment with geranial (LD_{50} = 14.26 μL), only doses higher than 8 μL resulted in significant differences when compared to the control group [F(7.71) = 127,2; $p < 0.0001$] (Figure 2A). For geraniol (LD_{50} = 22.60 μL), only doses higher than 16 μL [F(7.71) = 39.92; $p < 0.0001$] (Figure 2B) was effective.

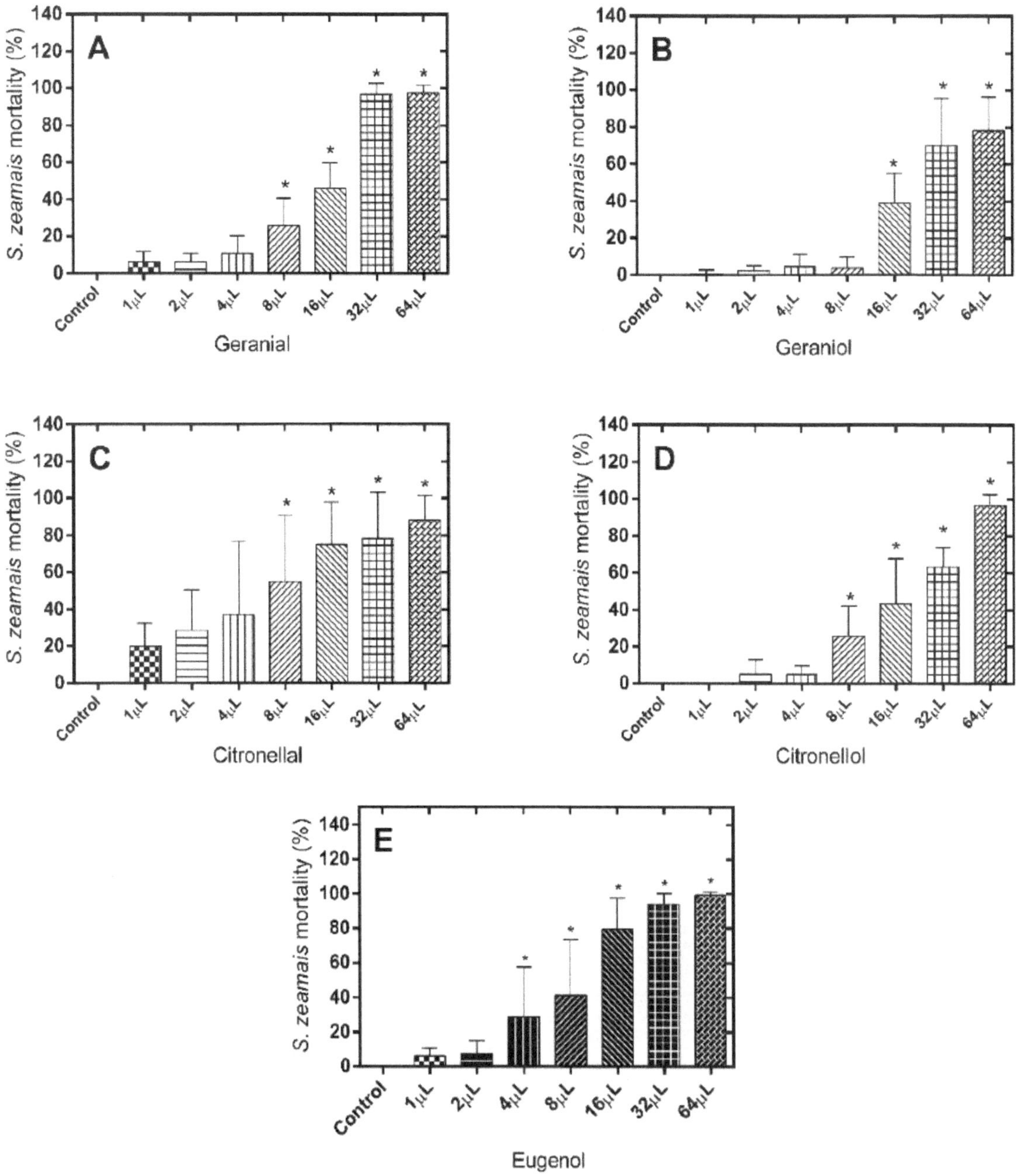

Figure 2. Mortality (%) of *S. zeamais* 24 h after exposure to the alimentary substrate treated with different monoterpene doses. (A) Geranial; (B) Geraniol; (C) Citronellal; (D) Citronellol; (E) Eugenol. * Significantly different compared to control, test ($p < 0.05$).

Similar results to those observed for geranial were obtained for the citronellal/citronellol pair (LD_{50} = 6.112 µL and 18.67 µL, respectively), with the occurrence of a significant effect only for doses higher than 8 µL (Figure 2C,D). In the case of eugenol (LD_{50} = 8.172 µL), the fumigant effect was already observed at a dose of 4 µL (Figure 2E). Treatment with the different monoterpenes exhibited a clearer dose-dependent profile of fumigant effect against *S. zeamais* than in case of *C. maculatus*.

Despite the proven effectiveness of the different monoterpenes against the two insect species tested; indistinctly considering LD_{50} and the generic profile of results, it can be stated that the terpenes tested clearly exhibited a greater fumigant effect against *C. maculatus* than against *S. zeamais*. These differences in effect may be due to the type of alimentary substrate used for maintenance of the test. In this respect, *C. maculatus* uses cowpea (*V. unguiculata* L.) as a substrate, while*S. zeamais* requires alimentary paste. These substrates differ in terms of the levels of humidity, rugosity and porosity, factors that can affect adhesion between the monoterpene/grain and monoterpene volatility. Nevertheless, these differences in fumigant potential can be better explained by differences in the sensitivity/resistance of the insects to the compounds tested as a result of their strict biological differences. They differ completely in genera, differing then in feeding parameters, exoskeleton, enzymatic systems, neurotransmission and olfactory reception [12,13,14]. Within this context, all compounds are effective against both insects, but are more selective for *C. maculatus* than *S. zeamais*. Considering the two species, eugenol is one of the most effective insecticides despite a slightly greater effectiveness of geraniol against *C. maculatus*which, on the other hand, is compensated for by its lower effectiveness against *S. zeamais*.

The functional isomerism of aldehyde and alcohol groups between the geranial/geraniol and citronellal/citronellol pairs does not seem to exert an important influence on the fumigant potential of the monoterpenes, since the LD_{50} values and especially the general profile of the results did not differ so much within each pair against *S. zeamais* or, especially, against*C. maculatus* (Figure 1 and Figure 2). The exception here is about the pair citronellal/citronellol against *S. zeamais*, in which the LD_{50} for citronellal was about three-fold lower compared with citronellol. On the other hand, positional isomerism appears to have a more important influence, since when considered specifically the values of LD_{50}, geranial and geraniol presented LD_{50} about two-fold lower than in case of citronellal and citronellol, respectively, against *C. maculatus*. It is similar for the pair geranial and citronellal in which the LD_{50} is two-fold higher for the first compound against *S. zeamais*. (Figure 1and Figure 2).

Table 1 and Table 2 show the repellent effects of the different monoterpene doses against *C. maculatus* and *S. zeamais*, respectively. A clear dose-dependent profile of the repellent effect could not be established for either insect or any of the monoterpenes tested. This fact permits comparison of the different doses of the compounds based on overall mean percent repellency.

Geranial exerted a significant repellent effect against *C. maculatus* at doses higher than 4 µL, with preference indices of −0.34 to −0.51. No dose-dependent profile was also observed for geraniol, which exhibited greater repellency at doses of 8 µL (−0.61), 1 µL (−0.57), and 64 µL (−0.48). Citronellal exerted a significant repellent effect against *C. maculatus* at all doses tested, with preference indices of −0.16 to −0.60. Similar results were obtained for citronellol, with repellency indices ranging from −0.13 to −0.58, except for the dose of 2 µL which was inefficient as a repellent (Table 1).

It is possible to assume that functional and positional isomerism exerts no influence on monoterpenes repellency effectiveness. This statement is based on the general profile of results, on the repellency performance exhibited by each isomer pairs (function and position) at a single dose and on the overall mean percent of repellency exhibited by them along the range of doses. The overall mean percent repellency at all doses of each compound ranged from 60 to 70% for the geranial (63.33% ± 9.5%)/geraniol (69.50% ± 8.7%), citronellal (67.50% ± 7.0%)/citronellol (66.50% ± 9.7%), geranial (63.33% ± 9.5%)/citronellal (67.50% ± 7.0%), and geraniol (69.50% ± 8.7%)/citronellol (66.50% ± 9.7%) pairs, with no significant differences between results ($p < 0.05$) (Table 1). When pairs (function or position) were compared at the same dose, the only significant difference for functional pairs was observed between geranial/geraniol and geranial/citronellal at dose of 2 µL (Table 1) and between geraniol and citronellol at the doses of 2 and 8 µL (Table 1). It is a very discrete correlation in view of the enormous universe of possible comparison between pairs. Eugenol exhibited significant effectiveness as a repellent at all doses tested (preference indices ranging from −0.30 to −0.76) showing the highest overall mean percent repellency at all doses tested (75.63% ± 7.0%) compared to the other monoterpenes (Table 1).

Table 1. Repellent effect of different monoterpene doses against *C. maculatus* according to the percentage of repellency and preference index (PI).

Treatment	Adults Repelled (%) (*C. maculatus*)	PI
Geranial/Geraniol (1 µL)	51.67 ± 17.85/78.33 ± 15	−0.03 N/−0.57 R
Geranial/Geraniol (2 µL)	50.00 ± 10.38 [a,b]/58.89 ± 8.85 [a,b]	−0.00 N/−0.18 R
Geranial/Geraniol (4 µL)	67.22 ± 5.96/67.78 ± 14.16	−0.34 R/−0.36 R
Geranial/Geraniol (8 µL)	70.56 ± 14.24/80.56 ± 12.61 [b]	−0.41 R/−0.61 R
Geranial/Geraniol (16 µL)	61.11 ± 19.32/58.20 ± 18.67	−0.22 R/−0.16 R
Geranial/Geraniol (32 µL)	67.22 ± 13.01/68.89 ± 13.64	−0.34 R/−0.38 R
Geranial/Geraniol (64 µL)	75.56 ± 18.44/73.89 ± 9.93	−0.51 R/−0.48 R
Mean between all doses of each compound	63.33 ± 9.5/69.5 ± 8.7 *	--------------------
Citronellal/Citronellol (1 µL)	57.78 ± 17.69/66.11 ± 14.6	−0.16 R/−0.32 R
Citronellal/Citronellol (2 µL)	80.00 ± 11.91 [b]/52.22 ± 9.93 [b]	−0.60 R/−0.04 N
Citronellal/Citronellol (4 µL)	68.33 ± 11.99/63.33 ± 16.0	−0.37 R/−0.27 R
Citronellal/Citronellol (8 µL)	68.33 ± 11.45/56.67 ± 15.81 [b]	−0.37 R/−0.13 R
Citronellal/Citronellol (16 µL)	63.89 ± 10.54/78.89 ± 8.58	−0.28 R/−0.58 R
Citronellal/Citronellol (32 µL)	71.11 ± 11.93/73.89 ± 13.41	−0.42 R/−0.48 R
Citronellal/Citronellol (64 µL)	63.33 ± 13.46/75.00 ± 7.9	−0.27 R/−0.50 R
Mean between all doses of each compound	67.53 ± 7.0/66.7 ± 9.7 *	--------------------
Eugenol (1 µL)	77.22 ± 11.75	−0.54 R
Eugenol (2 µL)	72.78 ± 12.52	−0.46 R
Eugenol (4 µL)	73.89 ± 12.93	−0.48 R
Eugenol (8 µL)	65.00 ± 12.74 *	−0.30 R
Eugenol (16 µL)	79.44 ± 12.85	−0.59 R

Eugenol (32 μL)	73.33 ± 7.5	−0.47 R
Eugenol (64 μL)	87.78 ± 12.0	−0.76 R
Mean between all doses of eugenol	75.63 ± 7.0 *	--------------------

* Significantly different compared to control, test ($p < 0.05$); [a] compared to differ significantly between function isomers test ($p < 0.05$); [b] significantly different compared between position isomers test ($p < 0.05$); PI = Preference Index; Rating: R = repellent; N = neutral.

Geranial and geraniol exerted a significant repellent effect against *S. zeamais* at all doses tested. The preference indices ranged from −0.33 to −0.92 for geranial and from −0.17 to −0.83 for geraniol, without the observation of a clear dose-response profile effect. Citronellal was also effective at all doses, with preferences indices of −0.72 to −0.95, while citronellol was ineffective at the lowest dose (1 μL) and showed indices of −0.23 to −0.68 after the second dose (Table 2). Despite the slight discrepancy between the citronellal/citronellol pair, with a mean percent repellency of 92.16% ± 3.9% and 72.99% ± 12.5%, respectively, functional or positional isomerism does not seem to influence the repellency results reported here, as seen in the interpretations above. Indeed, the functional pair geranial/geraniol exhibited overall mean percent repellency of 87.75% ± 10.3% and 83.5% ± 12.1%, respectively. The positional pair geranial/citronellal exhibited overall mean percent repellency of 87.75% ± 10.3% *vs.* 92.16% ± 3.9% and geraniol/citronellol of 83.50% ± 12.1% *vs.* 72.99% ± 12.5% (Table 2). Comparison according to isomer pair (function or position) revealed results similar to those obtained for *C. maculatus*, with a very discrete or zero correlation between functional or positional isomers and the repellent effect observed (Table 2). The few significant differences between functional isomers were exactly observed between citronellal and citronellol at the lower doses. Some significant differences between positional isomers occurred between the geranial and citronellal pair only at the doses of 1 and 16 μL and between the geraniol and citronellol pair only at the dose of 2 μL (Table 2). The mean percentage of repellency of eugenol was only 63.81% ± 13.14%. Eugenol did not exhibited repellency for all dose tested. In fact, eugenol is not repellent at both the lowest and the highest doses, while it is repellent at intermediate doses. In this point is important to consider the well-known attractant effect of eugenol for some pests, particularly flies. For *S. zeamais*, citronellal and geranial exhibit a better repellent effectiveness than the other monoterpenes (Table 2).

Table 2. Repellent effect of different monoterpene doses against *S. zeamais* according to the percentage of repellency and preference index (PI).

Treatment	Adults Repelled (%) *(S. zeamais)*	PI
Geranial/Geraniol (1 μL)	66.67 ± 14.0 [b]/58.33 ± 8.1	−0.33 R/−0.17 R
Geranial/Geraniol (2 μL)	96.67 ± 2.58/91.67 ± 5.16 [b]	−0.93 R/−0.83 R
Geranial/Geraniol (4 μL)	95.83 ± 4.91/77.50 ± 14.95	−0.92 R/−0.55 R
Geranial/Geraniol (8 μL)	91.67 ± 5.16/89.17 ± 9.17	−0.83 R/−0.78 R
Geranial/Geraniol (16 μL)	92.50 ± 8.16 [b]/90.00 ± 4.47	−0.85 R/−0.80 R
Geranial/Geraniol (32 μL)	84.17 ± 10.68/90.83 ± 7.36	−0.68 R/−0.82 R
Geranial/Geraniol (64 μL)	86.67 ± 9.3/87.50 ± 15.0	−0.73 R/−0.75 R
Mean between all doses of each compound	87.74 ± 10.3 */83.57 ± 12.1 *	--------------------
Citronellal/Citronellol (1 μL)	85.83 ± 5.48 [a,b]/50.00 ± 21.21 [a]	−0.72 R/−0.00 N
Citronellal/Citronellol (2 μL)	90.98 ± 8.2 [a]/61.67 ± 8.7 [a,b]	−0.82 R/−0.23 R
Citronellal/Citronellol (4 μL)	93.33 ± 12.5/80.00 ± 10.5	−0.87 R/−0.60 R
Citronellal/Citronellol (8 μL)	97.50 ± 4.18/80.00 ± 12.65	−0.95 R/−0.60 R
Citronellal/Citronellol (16 μL)	95.83 ± 5.48 [b]/80.83 ± 9.7	−0.92 R/−0.62 R
Citronellal/Citronellol (32 μL)	90.83 ± 10.2/74.26 ± 4.91	−0.82 R/−0.49 R
Citronellal/Citronellol (64 μL)	90.83 ± 5.48/84.17 ± 11.43	−0.82 R/−0.68 R
Mean between all doses of each compound	92.12 ± 3.9 [a,*]/72,99 ± 12.5 [a,*]	--------------------
Eugenol (1 μL)	43.33 ± 19.66	0.13 A
Eugenol (2 μL)	75.00 ± 14.83	−0.50 R
Eugenol (4 μL)	54.17 ± 13.93	−0.08 N
Eugenol (8 μL)	73.33 ± 15.38	−0.47 R
Eugenol (16 μL)	69.17 ± 26.15	−0.38 R
Eugenol (32 μL)	77.50 ± 2.73 *	−0.55 R
Eugenol (64 μL)	54.17 ± 22.23	−0.08 N
Mean between all doses of each compound	63.81 ± 13.14 *	--------------------

* Significantly different compared to control, test ($p < 0.05$); [a] compared to differ significantly between function isomers test ($p < 0.05$); [b] significantly different compared between position isomers test ($p < 0.05$); PI= Preference Index; Rating: R = repellent; N = neutral; A = attractive.

Despite the suggestion that functional or positional isomerism exerts little or no influence especially in case of repellent potential of the monoterpenes tested, some technical limitations of this descriptive-preliminary interpretation should be considered, including the need for a wider dose range and a greater representativeness of the functional or positional isomer pairs, represented here only by the geranial/geraniol, citronellal/citronellol, geranial/citronellal and geraniol/ citronellol pairs.

Considering the general profile of the results, as well as LD_{50} values, preference indices and percent repellency, it can be suggested that most of the monoterpenes were more effective fumigants against *C. maculatus* than against *S. zeamais*, but in general more effective repellents for *S. zeamais* than for *C. maculatus*. Of course, since *S. zemais* is more repelled it avoids these insects to progress to the death when in comparison to *C. maculatus*. Again, these differences between the results obtained for *S. zeamais* and *C. maculatus* can be explained by the biological differences of the insects which belong to completely different genera and species, although both are pests of stored grains. Differences in feeding patterns, exoskeleton and enzymatic and neurotransmission systems, as well as marked differences in the olfactory systems/ receptors used to process volatile chemical information from the environment (odorant receptors, specific gustatory receptor proteins and ionotropic receptors, as well as peripheral sensory reception and signal transduction), may at least in part explain the different results. Although ubiquitous in some cases, many of these biological systems cited above are species specific [13,14,15].

Monoterpenes and other compounds are found at different concentrations and proportions in different essential oils. The synergistic or complementary activities of different compounds present in the same oil play an extremely important role in the final insecticidal and/or repellent activity. Within this context, minor compounds can support or regulate the activity of major compounds or, less commonly, vice-versa [12,16]. In some cases, this fact can favor the choice of the intact essential oil of a plant for the final insecticide/repellent formulation instead of an isolated compound, with different mechanisms of action contributing to the final effect on the target organism. The choice of the oil can be guided by the knowledge of the insecticidal and/or repellent activity of the isolated compounds, as provided in the present study for the typical monoterpenes tested. Thus, the validation of the isolated active ingredients for different effects, including insecticidal and repellent activity, may provide a more advantageous, rapid, economic, effective and optimized approach to the identification of promising oils for commercial formulations than the conventional strategy. The latter consists of prospecting an enormous number of oils for pharmacological activities of interest and the subsequent study of the active ingredients responsible for the action observed. In other words, once the action of a given active ingredient is validated, oils rich in this compound can become potentially optimized candidates for new studies and for the elaboration of commercial formulations, including insecticides and repellents.

The main families of essential oil-producing plants studied for insecticidal and/or repellent activity include Lamiaceae, Labiatae, Lauraceae, Asteraceae, Myrtaceae, and Cupressaceae. These plants may be optimal sources for the choice of pharmaceutical phytocomplexes to be used as insecticides/repellents based on the knowledge of isolated active ingredients with activity against defined target insects [10,17,18]. Once the insecticidal activity of certain active ingredients is known, it is possible to search for these substances in essential oils from different plants for the production of an effective pharmaceutical phytocomplex. Plants of the genera *Cymbopogon* spp., *Eucalyptus* spp. and *Ocimum* spp. are the most widely studied as sources of insecticidal and/or repellent oils and are also the ones most commonly found in commercial formulations for this purpose [18,19]. Characteristically, these genera are rich in monoterpenes, including eugenol, citronellal and citronellol also investigated in this study [18].

Indeed, essential oils of *Eucalyptus benthamii*, *E. dunnii* and *E. saligna* exerted insecticidal activity against *S. zeamais*, with LD_{50} values of 25.03, 37.93 and 121.09 µL, respectively, confirming the popular use of eucalyptus as an insecticide for different purposes. All three oils were also significantly effective as repellents for *S. zeamais* [20]. Tests against *S. zeamais*using plants of the genus *Ocimum* ssp. also showed good activity. For example, the oil of *Ocimum gratissimum* killed 98% of adult weevils after 24 h of exposure in fumigation tests [21]. Curiously, there is no or only very little information about tests involving species of the genus *Cymbopogon* against *S. zeamais*, which could be an attractive alternative considering the great potential of the oil as insecticide and repellent, particularly because it is rich in monoterpenes [18]. Tests against *C. maculatus*using plants of the genus *Eucalyptus* ssp. have also shown good activity, with the oil of *Eucalyptus citriodora* killing 100% of adults of this insect after 48 h of exposure in fumigation tests [22]. Izakmehri *et al.* [23] reported LD_{50} values of 56.7 and 26.1 µL/L air, respectively, after 12 and 24 h of exposure of *C. maculatus* to the oil of *E. camaldulensis*. Tests involving *C. maculatus*treated with different doses of essential oils of *Ocimum basilicum* L. and *Ocimum gratissimum* L. for the control of adults also yielded excellent results [24]. These studies highlight the importance of the potential use of essential oils and/or isolated monoterpenes to management of these insects.

Considering the potential use of monoterpenes and oils as insecticides/repellents, as well as the advantages of botanical pesticides and the importance of *C. maculatus* and *S. zeamais*, the present study contributes to the identification of new agents of biotechnological interest which could be used in commercial formulations that are less expensive, safer and more effective than currently available ones.

EXPERIMENTAL SECTION

Chemicals

All monoterpenes tested (geraniol, geranial, (±) citronellal, citronellol and eugenol) were purchased from Sigma-Aldrich (São Paulo, Brazil), with their respective datasheet containing the certificate of analysis attesting for purity and chemical identity.

Insects

The bruchid beetle *C. maculatus* and the maize weevil *S. zeamais* were maintained in stock cultures as described by Toscano *et al.* [25], with minor modifications. The culture conditions, which were the same for all experiments, were: 25 ± 2 °C, light-dark cycle of 12 h, and 60%–80% relative humidity. Unsexed weevils or beetles used in the fumigation and repellency assays were about 48–72 h post-eclosion.

Fumigant Toxicity Assay

The fumigant toxicity tests were carried out according to the procedures described by Mossi [20], with modifications. Briefly, 20 adult insects were placed in the screw cap of a glass vial (6.1 cm in diameter and height of 2.1 cm) containing 10 g of small pieces (homogeneous in weight and length) of alimentary paste for *S. zeamais* or cowpea (*V. unguiculata* L.) for *C. maculatus*. For the different treatments, different volumes of each monoterpene tested (1–64 μL) were uniformly dispersed in the respective alimentary substrate. The vials were then sealed until counting of the insects at the end of the experiment. The same experimental arrangement was used for the control groups, except for the addition of monoterpenes. The percentage of insect mortality was determined after 24 h of the experiment using the formula of Abbott [26]. LD_{50} values were calculated by non-linear regression. Each experiment was performed in triplicate and repeated at least two (*C. maculatus*) and three times (*S. zeamais*).

Repellency Assay

These experiments were carried out as described by Mossi [20] and Pettersson [27], with modifications. The repellent effects of the monoterpenes were assessed using an arena apparatus (30 × 30 cm) composed of five circular plastic boxes referred to as A, B, C, D and E (6.1 cm in diameter and 2.1 cm in height). The central box (A) was interconnected to the other boxes (B, C, D and E) by plastic tubes (0.5 cm in open diameter and 10 cm in length) for free choice of the 20 unsexed insects to access treated or untreated alimentary substrate. In all boxes, 10 g of small pieces (homogeneous in weight and length) of alimentary paste for *S. zeamais* or cowpea (*V. unguiculata* L.) for *C. maculatus* was added, except for the central box (A). For the different treatments, different volumes of each monoterpene tested (1–64 μL) were uniformly dispersed in the respective alimentary substrate in boxes B and D arranged in a diagonal orientation, while boxes C and E, also in a diagonal orientation, served as controls (without monoterpenes). The number of insects per box was determined 24 h after the beginning of the experiment. The preference index (PI) was determined as follows [28]: PI = It − Ic/It + Ic, where *It* is the percentage of insects in the boxes treated with the monoterpenes and Ic is the percentage of insects in the control boxes (without monoterpene). The preference index was classified as follows: −1.00 to −0.10, repellent monoterpene; −0.10 to +0.10, neutral monoterpene, and +0.10 to +1.00, attractant monoterpene. Each experiment was performed in triplicate and repeated at least two (*C. maculatus*) and three times (*S. zeamais*). The percentage of repellency per treatment at each dose was calculated and used for the comparison between the different function and position isomers. The overall mean percentages of repellency obtained for all doses of each compound were also compared.

Statistical Analysis

All statistical analysis was performed using ANOVA followed by the Tukey's test ($p < 0.05$) using the GraphPad Prism 5.1 program (GraphPad Software, San Diego, CA, USA, version 5.01).

ACKNOWLEDGMENTS

The authors are thankful to the CAPES, CNPq, FAPESP (13/13738-8), University of Ribeirão Preto and Instituto Federal de Educação, Ciência e Tecnologia de São Paulo (IFSP), Barretos-SP for financial support.

AUTHOR CONTRIBUTIONS

S.L. Reis, A.G. Mantello, E.A. Gelfuso work at execution of biological assays; J.M. Macedo works at maintenance and in providing of stored grains pests; C.P. da Silva, A.L Fachim, A.M. Cardoso and Rene O. Beleboni work at interpretation of results, writing the paper, students supervisioning and had different intellectual contributions at several phases of work.

REFERENCES

1. Oerke, E.C. Crop losses to pests. *J. Agric. Sci.* 2006, *144*, 31–43.
2. Brooker, D.B.; Bakker-Arkema, F.W.; Hall, C.W. *Drying and Storage of Grains and Oilseeds*, 1st ed.; Van Nostrand Reinold: New York, NY, USA, 1992; p. 450.
3. Lorini, I.; Scheneider, A. *Controle Integrado de Pragas de Grãos Armazenados*; EMBRAPA-CNPT: Passo Fundo, Brazil, 1994; pp. 8–10.
4. Lorini, I. *Manual Técnico Para o Manejo Integrado de Pragas de Grãos de Cereais Armazenados*, 1st ed.; Embrapa Trigo: Passo Fundo, Brazil, 2003; pp. 16–27.
5. Mound, L. *Common Insect Pest of Stored Food Products*, 7th ed.; British Museum (Natural History): London, UK, 1989; p. 68.
6. Loeck, A.E. *Principais Pragas que Atacam Produtos Armazenados*, 1st ed.; EGUFPEL: Pelotas, Brazil, 2002; pp. 35–59.
7. Isman, M.B. Botanical insecticides, deterrents, and repellents in modern agriculture and an increasingly regulated world. *Annu. Rev. Entomol.* 2006, *51*, 45–66.
8. Isman, M.B. Botanical insecticides: For richer, for poorer. *Pest Manag. Sci.* 2008, *64*, 8–11.
9. Isman, M.B. Plant essential oils as green pesticides for pest and disease management. *ACS Sym. Ser.* 2004, *887*, 41–51.
10. Nerio, L.S.; Olivero-Verbel, J.; Stashenko, E. Repellent activity of essential oils: a review. *Bioresource Technol.* 2010, *101*, 372–378.
11. Regnault-Roger, C.; Vincent, C.; Arnason, J.T. Essential oils in insect control: Low-risk products in a high-stakes world. *Annu. Rev. Entomol.* 2012, *57*, 405–424.
12. Berenbaum, M. Allelochemical interactions in plants. *Rec. Adv. Phytochem.* 1985, *19*, 139–169.
13. Suh, E.; Bohbot, J.; Zwiebel, L.J. Peripheral olfactory signaling in insects. *Curr. Opin. Insect Sci.* 2014, *6*, 86–92.
14. Wicher, D. Olfactory signaling in insects. *Prog. Mol. Biol. Transl. Sci.* 2015, *130*, 37–54.

15. Lutz, A.; Sfara, V.; Alzogaray, R.A. Repellence produced by monoterpenes on *Rhodnius prolixus* (Hemiptera: Reduviidae) decreases after continuous exposure to these compounds. *J. Insect Sci.* 2014, *14*, 1–6.

16. Gillij, Y.G.; Gleiser, R.M.; Zygadlo, J.A. Mosquito repellent activity of essential oils of aromatic plants growing in Argentina. *Bioresource Technol.* 2008, *99*, 2507–2515.

17. Adeyemi, M.M.H. The potential of secondary metabolites in plant material as deterrents against insect pests: A review. *Afr. J. Pure Appl. Chem.* 2010, *4*, 243–246.

18. Rehman, J.U.; Ali, A.; Khan, I.A. Plant based products: Use and development as repellents against mosquitoes: A review. *Fitoterapia* 2014, *95*, 65–74.

19. Padilha, J.P.; Gomes-Carneiro, M.R.; Paumgartten, F.J.R. Chemical composition, toxicity and mosquito repellency of*Ocimum selloi* oil. *J. Ethnopharmacol.* 2003, *88*, 253–260.

20. Mossi, A.J.; Astolfi, V.; Kubiak, G.; Lerin, L.; Zanella, C.; Toniazzo, G.; de Oliveira, D.; Treichel, H.; Devilla, I.A.; Cansian, R.; Restello, R. Insecticidal and repellent activity of essential oil of *Eucalyptus* sp. against *Sitophilus zeamais*Motschulsky (Coleoptera, Curculionidae). *J. Sci. Food Agric.* 2011, *91*, 273–277.

21. Kouninki, H.; Haubruge, E.; Noudjou, F.E.; Lognay, G.; Malaisse, F.; Ngassoum, M.B.; Goudoum, A.; Mapongmetsem, P.M.; Ngamo, L.S.; Hance, T. Potential use of essential oils from Cameroon applied as fumigant or contact insecticides against *Sitophilus zeamais* Motsch. (Coleoptera: Curculionidae). *Commun. Agric. Appl. Biol. Sci.* 2005, *70*, 787–792.

22. Brito, J.P.; Oliveira, J.E.M.; Bortoli, S.A. Toxicidade de óleos essenciais de *Eucalyptus* spp. sobre *Callosobruchus maculatus* (Fabr., 1775) (Coleoptera: Bruchidae). *Rev. Biol. Ciênc. Terra* 2006, *6*, 96–103.

23. Izakmehri, K.; Saber, M.; Mehrvar, A.; Hassanpouraghdam, M.B. Vojoudi S, Lethal and sublethal effects of essential oils from *Eucalyptus camaldulensis* and *Heracleum persicum* against the adults of *Callosobruchus maculatus*. *J. Insect Sci.*2013, *13*, 152.

24. Keita, S.M.; Vincent, C.; Schimit, J.P.; Arnason, J.T.; Bélanger, A. Efficacy of essential oil of *Ocimum basilicum* L. and *O. gratissimum* L. applied as an insecticidal fumigant and powder to control *Callosobruchus maculates* (Fab.) (Coleoptera: Bruchidae). *J. Stored Prod. Res.* 2001, *37*, 339–349.

25. Toscano, L.C.; Boiça Junior, A.L.; Lara, F.M.; Waquil, J.M. Resistência e mecanismos envolvidos em genótipos de milho em relação ao ataque do gorgulho, *Sitophilus zeamais* Mots (Coleoptera, Curculionidae). *An. Soc. Entomol. Bras.*1999, *28*, 141–147.

26. Abbott, W.W. A method for computing the effectiveness of an insecticide. *J. Econ. Entomol.* 1925, *18*, 265–267.

27. Pettersson, J. An aphid sex attractant. Biological studies. *Entomol. Scand.* 1970, *1*, 63–73.

28. Procopio, S.O.; Vendramin, J.D.; Ribeiro, J.I.; Santos, J.B. Bioatividade de diversos pós de origem vegetal em relaçáo a*Sitophilus zeamais* Mots. (Coleoptera: Curculionidae). *Cienc. Agrotec.* 2003, *27*, 1231–1236.

CHAPTER 17

IDENTIFICATION OF GENES RELEVANT TO PESTICIDES AND BIOLOGY FROM GLOBAL TRANSCRIPTOME DATA OF MONOCHAMUS ALTERNATUS HOPE (COLEOPTERA: CERAMBYCIDAE) LARVAE

Songqing Wu, Xiaoli Zhu, Zhaoxia Liu, Ensi Shao, CarballarLejarazú Rebeca, Yajie Guo, Yueting Xiong, Yani Mou, Runxue Xu, Xia Hu, Guanghong Liang, Shuangquan Zou, Xiong Guan, and Feiping Zhang

INTRODUCTION

Pine Wilt Disease is a devastating disease in pine trees caused by the infection of*Bursaphelenchus xylophilus* and it is commonly known as the cancer of pine trees [1]. Since the discovery of *B. xylophilus* in Japanese black pines in the Sun Yat-sen Mausoleum in Nanjing City (Jiangsu Province in China) in 1982, Pine Wilt Disease has occurred in a total of 275 county-level administrative regions (excluding Hong Kong and Taiwan) of 17 provinces (autonomous regions and municipalities), causing immense damage to forest resources and having impact in China's ecological environment [2]. In China, the principal vector for Pine Wilt Disease is the beetle *Monochamus alternatus* Hope (*M. alternatus*) larvae as carriers; after emergence, t larvae use pine trees as food and oviposition sources, therefore they are considered the invasive stage of the insect. Effective control of *M.alternatus* plays an important role in the prophylaxis and treatment of this disease [3].

At present, the principal strategies to control *M. alternatus* include: trap trees, biological control, silvicultural control and chemical prevention [3, 4]. Among these, the biological control presents unique advantages: (1) It is difficult for pests to become resistant as microorganisms have adapted to the immune systems of insects during the process of concurrent evolution. Therefore insect immunity to pathogenic microorganisms has been kept at a low level; (2) It has high environmental security; microorganisms typically have strong specificity for their targets and they are harmless to vertebrates, which protects the natural predators of those hosts; (3) insects are an ideal medium for various types of pathogens; the proliferation of insect pathogens *in vivo* can be spread by diseases and pests or the insect's body; (4) It's easy to obtain strains that are strongly pathogenic using genetic engineering and transformation techniques [5, 6]. Current biological control techniques for *M. alternatus* have progressed, including the spreading of effective natural enemies, creation of black lights and trap-trees. Application of the above techniques has successfully controlled Pine Wilt Disease at test locations [5]. Among the methods of natural enemies are the parasitoid beetles *Dastarcus helophoroides* and *Sclerodermus* spp. (Hymenoptera: Bethylidae), showing up to 90% of effectiveness [7, 8]. In general, biological control has brought new insights to control *M. alternatus* forest infestations. Therefore, microbial control of *M. alternatus* has increasingly gained attention [4]. However, there is currently a lack of knowledge regarding *Monochamus alternatus* Hope transcripts, gene expression and gene function in this insect vector.

We used next generation sequencing technology to sequence the whole fourth instar larvae transcriptome of *M. alternatus* and successfully built a *M. alternatus* Hope

Citation: Wu S, Zhu X, Liu Z, Shao E, Rebeca C-L, Guo Y, et al. (2016) Identification of Genes Relevant to Pesticides and Biology from Global Transcriptome Data of Monochamus alternatus Hope (Coleoptera: Cerambycidae) Larvae. PLoS ONE 11(1): e0147855. DOI:10.1371/journal.pone.0147855.

transcriptome database. In addition, our data describe genes related to putative insecticide resistance, intestinal digestive enzymes, possible future insect control targets and immune-related molecules. This study provides valuable basic information that can be used as a key point to develop new molecular tools for *M. alternatus* Hope control strategies.

RESULTS AND DISCUSSION

Sequencing and De Novo Assembly

Illumina sequencing produced 46, 761 and 743 clean reads for the larvae samples, which is equivalent to an accumulated length of 11, 777, 133 and 130 bp (Table 1).

Table 1. Sequence statistics of the Illumina sequencing assembly.

	Reads	Contig	Transcript	Unigene
Number of sequences	94,545,777	11,433,166	107,259	73,090
Mean length (bp)	252	42	942	693
Total length (bp)	23,805,690,981	475,935,825	101,013,380	50,668,837

doi:10.1371/journal.pone.0147855.t001

The average of raw reads length was 252 bp. First, the sequencing reads were broken into K-mers using Trinity software [9]; the small fragments were assembled into 11,433,166 contigs with a mean length of 41.63 bp. The length of the contigs mainly ranged from 0 to 3000 bp, representing 99.61% of the reads, although some contigs were longer than 3000 bp. Size distribution of the contigs. Finally, using the method of De Bruijn graphing and sequencing read information, we identified 107,259 transcripts with a mean length of 941.77 bp, transcripts ranged in length from ~200–3000 bp, identifying 49,615 transcripts with a length larger than 500 bp. We obtained 73,090 unigenes with a mean length of 693.24 bp. The lengths of 25,718 and 13,668 unigenes were larger than 500 bp and 1000 bp, respectively, while 64.82% of the unigenes had lengths between 0 to 500 bp (Fig 1). This result indicates that the length distribution of the transcripts and unigenes were represented in majority by short sequences with relatively little redundancy, which is similar to transcriptome analysis reported for other insect species using the same technology [10–14]. Importantly the longer sequences contribute only for 7.09% of the *M. alternatus* transcriptome, the majority of transcripts and unigenes were still less than 500 bp after assembly; this is probably due to the capacity of shorter sequences and low coverage of the transcriptome [5, 15]. A large number of assembled sequential data could provide a more deeply transcriptome information for future research, allowing rapid characterization for most of the transcripts and a reference for the genes of interest [15].

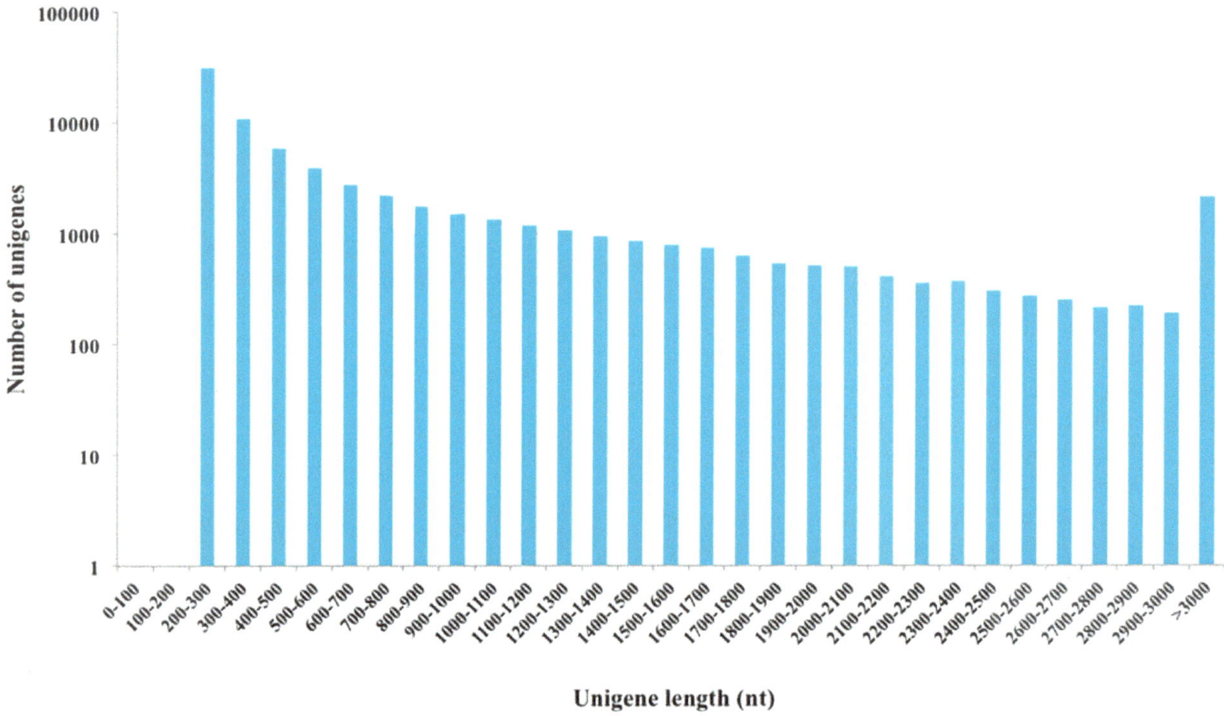

Figure 1. Unigenes length distribution. The y-axis number has been converted into logarithmic scale.

Annotation of Predicted Proteins

All assembled unigenes were used as an input for NR, Swiss-Prot, Gene Oontology (GO), Clusters of Orthologous Groups (COG), KOG and KEGG databases. BLAST and HMMER parameter E-values were set at 10^{-5} and 10^{-10} respectively, we were able to obtain annotated information for 36,828 unigenes, representing 50.38% of the unigenes. The rest of the unigene sequences (49.62%) had no significant matches in the existing databases. Unigenes comparison with the NR database produced 34,702 hits, the distribution of E-values demonstrated that 26.91% of the mapped sequences have strong homology (smaller than $1.0E^{-49}$) with an annotated sequence, and 62.70% of the homolog sequences ranged from $1.0E^{-5}$ to $1.0E^{-49}$ (Fig 2A). Based on the best species match, we found that *M.alternatus* sequences have 30.59% and 8.89% matches with sequences from the *Tribolium castaneum* and *Dendroctonus ponderosae*, both belonging to Coleopteran order, while only <6% matched to other insects (Fig 2B). Therefore, *M. alternatus* have the closest evolutionary distance with *T. castaneum*.

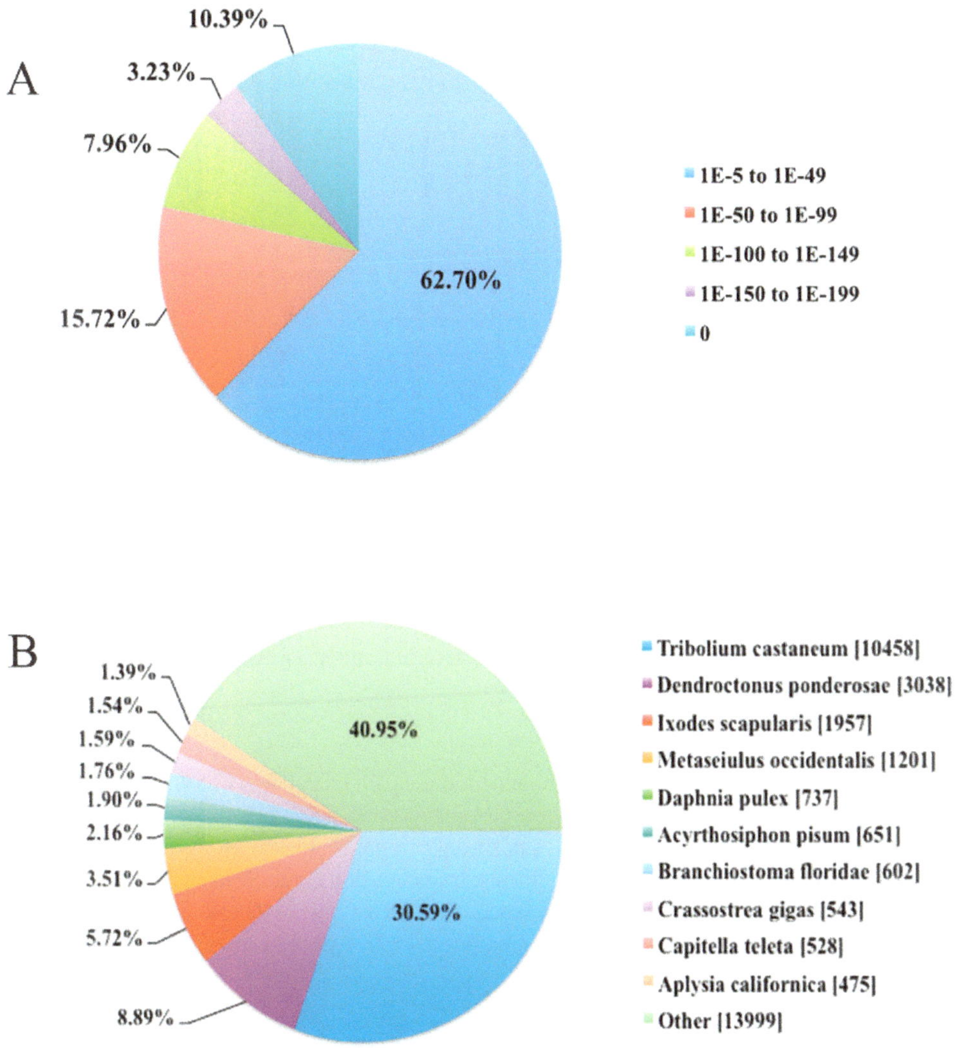

Figure 2. Characteristics of the homology search of Illumina sequences against the NR database. (A) E-value distribution of the BLAST hits for each unique sequence with a cut-off E-value of $1.0E^{-5}$. (B) Species distribution of the BLASTX results. The first hit of each sequence was used for further *in silico* analysis.

GO Assignments

GO database is an internationally standardized gene function classification system, which provides a suitably updated standard vocabulary to describe the functional attributes of genes and gene products in an organism [16]. Transcript sequences were used as an input in the GO database; using BlastX a total of 105,612 unigenes were assigned to GO terms (Fig 3). According to the standard GO terms and 60 subcategories, differential gene expression and all unigenes from *M. alternatus* larva were statistically classified into three main GO categories: biological process, cellular component, and molecular function. Biological process represented the majority of GO annotations (52,883, 50.07%), followed by cellular component (29,970, 28.38%) and molecular function (22,759, 21.55%). The metabolic process (20.48%) and cellular process (18.55%) were predominant within the biological process category, indicating that the analyzed tissue has a high degree of metabolic activity; with the following subcategories: single-organism process (15.34%), biological regulation (8.05%), developmental process (5.59%), response to stimulus (5.49%), localization (5.38%), multicellular organismal process (5.12%), cellular component organization or biogenesis (4.26%) and signaling (3.16%). Biological processes contain most major cellular processes, from transportation and cell formation to transcription, translation and supersession. According to the classification of the cellular components, cell part (21.55%), cell (21.39%), and organelle (14.85%) are the most representative subcategories. Most of the annotated unigenes from the cellular component category, correspond to plastids and mitochondria. We also identified genes involved in the synthesis of secondary metabolites, and were grouped into catalytic activity (41.61%), binding (39.45%), transporter activity (5.63%), structural molecule activity (2.86%), molecular transducer activity (2.76%), receptor activity (2.43%) and nucleic acid binding transcription factor activity (2.02%), etc. A previous study reported similar classifications related to metabolic processes for *Tomicus yunnanensis* transcriptome [15]. GO annotations describe the contour features of the overall gene expression of *M. alternatus*, and revealed expressed genes encoding diverse structural, stress and regulatory proteins.

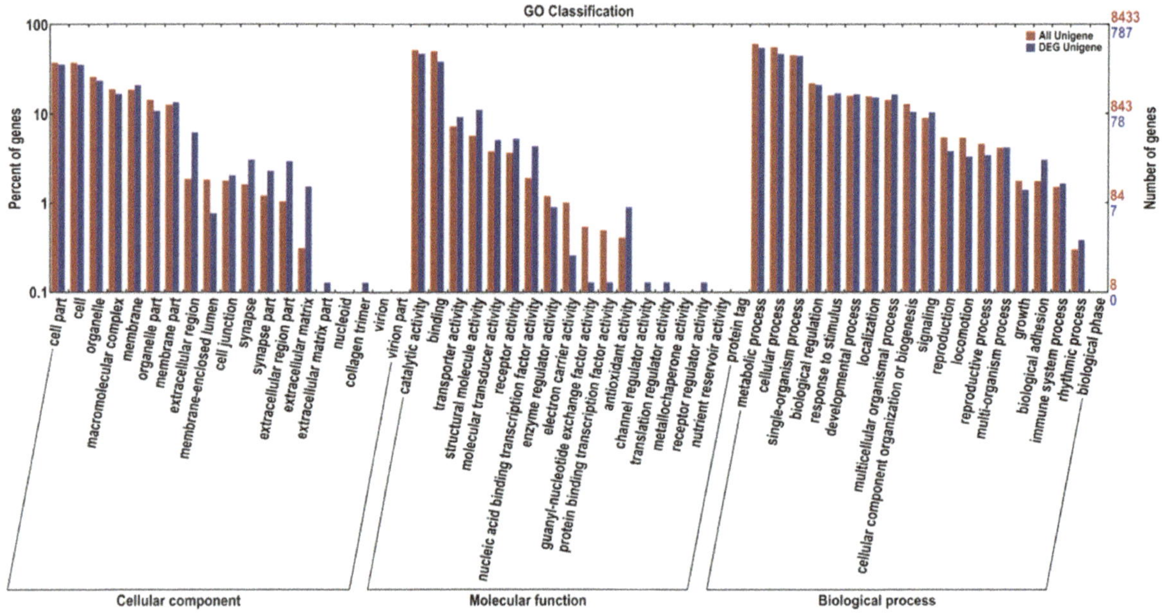

Figure 3. Distribution of second level GO of *Monochamus alternatus* Hope transcriptome. Distribution of GO categories assigned to the *Monochamus alternatus* Hope transcriptome. Unigenes were annotated in three categories: cellular components, molecular functions, and biological process. Right y-axis indicates the number of genes in a category; left y-axis indicates the genes percentage in a specific.

COG Classification

In total, information for 16,730 classified unigenes was obtained in the COG database (Fig 4). COG classifications were divided into 25 functional categories. Among these categories, general function prediction (19.90%) was the largest group, followed by translation, ribosomal structure and biogenesis (10.19%), posttranslational modification, protein turnover, chaperones (8.29%) and transcription (7.57%). Secondary metabolites biosynthesis, transport and catabolism represented 2.71%, given the relative importance of secondary metabolic activity for insect resistance. To some extent, COG classifications further reveal the potential specific reactions and the functional participation in molecular processes for genes expressed in *M. alternatus*.

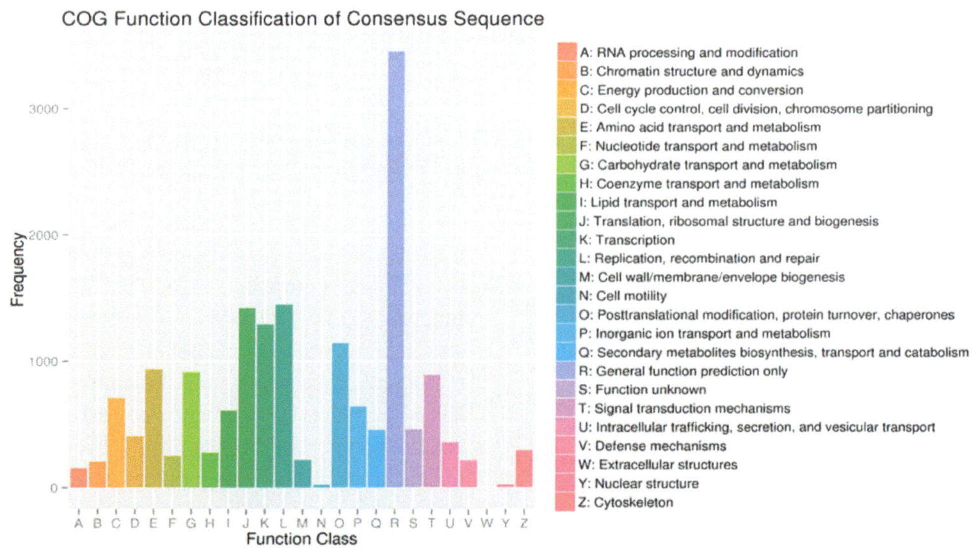

Figure 4. COG function classification of *Monochamus alternatus* Hope transcriptome. All putative proteins were analyzed using the COG database. COG classifications were divided into 25 functional categories; 16,730 classified unigenes were assigned to 25 COG classifications.

KEGG Analysis

The KEGG Pathway database, is a collection of graphical maps representing different cellular processes, to systematically analyze metabolic pathways and functions of gene products in a cell [17]. To identify the represented biological pathways in *M. alternatus*, 34,302 annotated unigenes were analyzed using KEGG database. The results indicated that 13,024 unigenes matched with 224 KEGG pathways. The top 10 pathways were Protein processing in endoplasmic reticulum (503 members), Ribosome (486 members), Spliceosome (444 members), RNA transport (415 members), Purine metabolism (410 members), Pyrimidine metabolism (321 members), Ribosome biogenesis in eukaryotes (312 members), Endocytosis (305 members), Oxidative phosphorylation (302 members) and Ubiquitin mediated proteolysis (288 members). These annotations provide powerful information for our research related specific biological processes and pathways of *M. alternatus*.

Putative Insecticide Resistance-Related Genes

Cytochrome P450 (P450)

Research over the past 10 years has provided clear evidence that in insects, P450 is involved in a number of physiological functions, such as hormone metabolism [18–20], adaptability of parasitic plants [21, 22] and resistance to insecticides [23, 24]. Approximately 222 types of P450-related unigenes were identified in *M. alternatus* transcriptome. We identified 103 P450-related sequences with a length bigger than 600 bp (46.40% of the global pests in the database); approximately half of the unigene sequences were long sequences (600 bp). In addition, the length of 30 P450-related sequences was larger than 1800 bp (13.51%) (Table 2). P450 genes identified from the *M. alternatus* transcriptome was comparable in number to those from *T. yunnanensis* transcriptome, although the final number of genes still needs to be verified via gene cloning [15]. Ai Junwen *et al.* classified the P450 enzyme system into four large families: CYP2, CYP3, CYP4 and mitochondrial P450, based on the phylogenetic analysis of *Drosophila* and silkworm P450 gene homology [25].

Table 2. Putative P450 genes identified in *Monochamus alternatus* Hope.

#Gene ID	Length	TFPKM	E_value	Identity (%)	Annotation
c40573.graph_c0	1814	1.16	0	69.31	PREDICTED: similar to cytochrome P450 CYP6BK17 [Tribolium castaneum]
c40753.graph_c0	1816.61	4.14	4.47E-179	56.8	cytochrome P450 [Leptinotarsa decemlineata]
c30543.graph_c0	1820	62.62	0	67.41	cytochrome P450 CYP4g56 [Dendroctonus ponderosae]
c31302.graph_c0	1838	6.33	4.52E-140	54.2	PREDICTED: similar to Cytochrome P450 315a1, mitochondrial precursor (CYPCCCXVA1) (Protein shadow) [Tribolium castaneum]
c42017.graph_c0	1863.46	94.8	0	58.7	cytochrome P450 [Leptinotarsa decemlineata]
c40340.graph_c0	1890	6.08	4.37E-146	53.67	cytochrome P450-like protein [Tribolium castaneum]
c41067.graph_c0	1893.65	52.99	7.25E-173	57.23	cytochrome P450 [Leptinotarsa decemlineata]
c41279.graph_c0	1927.2	5.72	2.73E-143	53.14	cytochrome P450-like protein [Tribolium castaneum]
c36280.graph_c0	1929	0.98	2.88E-177	58.92	hypothetical protein YQE_06277, partial [Dendroctonus ponderosae]
c31219.graph_c0	1933	450.22	0	86.12	cytochrome P450 [Leptinotarsa decemlineata]
c36851.graph_c0	1934.48	10.03	4.08E-153	49.5	PREDICTED: similar to antennae-rich cytochrome P450 [Tribolium castaneum]
c41193.graph_c0	1938.57	3.47	0	77.51	PREDICTED: similar to Cyp49a1 [Tribolium castaneum]
c33148.graph_c0	1948.26	7.19	1.00E-108	45.96	PREDICTED: cytochrome P450 4c3-like [Metaseiulus occidentalis]
c40989.graph_c0	1969	2.07	9.89E-157	50.39	cytochrome P450 6BQ7 [Tribolium castaneum]
c32887.graph_c0	1976	2.59	8.79E-121	40.76	cytochrome P450 monooxygenase [Panonychus citri]
c40370.graph_c0	1980.74	108.63	1.61E-170	53.98	cytochrome P450 9Z4 [Tribolium castaneum]
c37941.graph_c0	1993.46	28.81	3.61E-154	53.13	cytochrome P450 [Leptinotarsa decemlineata]
c39854.graph_c0	2006.71	796.85	0	64.19	cytochrome P450 9Z4 [Tribolium castaneum]
c30479.graph_c0	2050	14.46	1.07E-177	60.12	cytochrome P450 CYP314A1 [Tribolium castaneum]
c40839.graph_c0	2080	0.59	2.97E-159	53.91	cytochrome P450 6BQ7 [Tribolium castaneum]
c32624.graph_c0	2097.79	8.71	5.61E-93	54.24	hypothetical protein TcasGA2_TC006764 [Tribolium castaneum]
c41739.graph_c1	2128.53	108.31	6.44E-173	53.73	cytochrome P450 [Leptinotarsa decemlineata]
c37224.graph_c0	2131	15.25	1.13E-153	51.11	PREDICTED: similar to antennae-rich cytochrome P450 [Tribolium castaneum]
c40116.graph_c0	2135	0.35	2.85E-173	56.86	cytochrome P450 [Leptinotarsa decemlineata]
c38458.graph_c0	2434	36.65	3.33E-173	55.34	cytochrome P450 [Leptinotarsa decemlineata]
c29697.graph_c0	2490.42	5.88	0	64.56	cytochrome P450, putative [Ixodes scapularis]
c42120.graph_c0	2513.73	25.61	2.28E-147	50.1	cytochrome P450 CYP6CR2 [Dendroctonus ponderosae]
c40510.graph_c0	2515.24	10.43	4.67E-100	41.25	PREDICTED: similar to cytochrome P450 monooxygenase [Tribolium castaneum]
c39155.graph_c0	2599.72	266.53	0	84.85	PREDICTED: similar to nadph cytochrome P450 [Tribolium castaneum]
c40802.graph_c0	2692.71	0.6	7.00E-146	56.42	cytochrome P450 353A1 [Tribolium castaneum]

doi:10.1371/journal.pone.0147855.t002

Previous studies have reported that 17 P450 genes from CYP3 and CYP4 families are associated with plant toxins and pesticide resistance [25,26]. Based in phylogenic analyses; we found evidence that CYP1, CYP3, CYP4 and mitochondrial families are present in *M.alternatus* transcriptome. We also identified CYP1A1, which belongs to the CYP1 family, and CYP3A4, CYP3A5 and CYP3A7, which belong to the CYP3 family. Interestingly, the P450 genes identified in *M. alternatus* differ from those reported in *T. castaneum* and other insect systems.

Recent studies have shown that P450 has an increased expression in insecticide resistant insects; moreover, there is also evidence for P450 gene duplication and amplification in four types of insects *in vivo* [27]. However, there is no experimental data to support the importance and biological role of P450 complex related to insecticide resistance in *M.alternatus*.

Glutathione S-transferase (GST)

GSTs specifically catalyze glutathione thiol and interact with other electrophilic groups [28,29]; GST is one of the main detoxification enzymes in insecticide metabolism. GST high levels of expression are related to insecticide resistance mechanisms in insects [27]. We identified 96 GST unigenes in *M. alternatus* transcriptome. The length of 19 GST-related sequences was greater than 1000 bp (19.79%) (Table 3). GSTs are divided into three major categories according to their cellular location: cytosolic, microsomal, and mitochondrial [30]. Cytosolic matrix GSTs in insects are further divided into at least six classes (delta, epsilon, omega, sigma, theta, and zeta) based on sequence homology of the N-terminus, substrate specificity, immunoreactivity, and sensitivity to different inhibitors [31–33]. Delta and epsilon classes are unique in insects [34]. We found delta, omega, and theta types of GST in *M.alternatus* transcriptome, however we could not identify any gene from the epsilon class. Previous studies have reported the identification of sigma, delta and theta GST classes in Nasonia vitripennis; epsilon, sigma, omega, and delta in *T. castaneum*; and one delta unigene in *T. yunnanensis* [35, 36]. The GST genes identified in *M. alternatus* transcriptome can contribute to a greater understanding of the relationship between GSTs and insecticide resistance in insects.

Table 3. Putative identified GST genes in *Monochamus alternatus* Hope.

#Gene ID	Length (bp)	FPKM	E_value	Identity (%)	Annotation
c10463.graph_c0	1125.51	3	2.16E-06	30.98	unnamed protein product, partial [Leishmania mexicana MHOM/GT/2001/U1103]
c8990.graph_c0	1175.51	2.25	6.02E-15	31.75	predicted protein [Nematostella vectensis]
c44708.graph_c0	1277.51	1.85	1.99E-67	44.01	elongation factor 1-gamma [Cryptosporidium hominis TU502]
c36023.graph_c0	1307.03	4.98	1.15E-47	44.93	glutathione transferase delta-like Yv4019D08 [Sarcoptes scabiei type hominis]
c27703.graph_c0	1919.09	18.56	3.85E-106	49.22	failed axon connections, putative [Ixodes scapularis]
c21401.graph_c0	1080.51	6.46	7.11E-84	51.77	PREDICTED: similar to metaxin 1 [Tribolium castaneum]
c29135.graph_c0	3037.97	2.24	2.39E-102	54.39	PREDICTED: lachesin-like [Metaseiulus occidentalis]
c41606.graph_c0	2305.51	4.56	0	60.52	glutathione S-transferase C-terminal domain-containing protein [Tribolium castaneum]
c39852.graph_c0	3260.51	8.74	0	62.51	PREDICTED: similar to AAEL014709-PA [Tribolium castaneum]
c41863.graph_c0	1835.51	0.67	6.55E-150	66.07	hypothetical protein YQE_09889, partial [Dendroctonus ponderosae]
c33923.graph_c0	1133.51	109.27	9.20E-98	67.9	unknown [Dendroctonus ponderosae]
c35423.graph_c0	1148.51	48.56	3.94E-96	68.46	unknown [Dendroctonus ponderosae]
c30871.graph_c0	1594.51	10.77	1.21E-88	69.49	glutathione transferase mu class Yv5001D03 [Sarcoptes scabiei type hominis]
c10168.graph_c0	1176.51	55.03	7.72E-87	71.5	glutathione-S-transferase theta, GST, putative [Pediculus humanus corporis]
c39366.graph_c0	1552.51	10.74	3.40E-120	79.54	PREDICTED: similar to metaxin 2 [Tribolium castaneum]
c10006.graph_c1	1441.51	2940.6	0	82.28	PREDICTED: similar to Elongation factor 1-gamma (EF-1-gamma) (eEF-1B gamma) [Tribolium castaneum]
c35156.graph_c0	2321.46	37.05	2.41E-102	82.95	PREDICTED: similar to AGAP008106-PA [Tribolium castaneum]
c33364.graph_c0	2144.51	273.41	0	84.07	PREDICTED: similar to failed axon connections protein [Tribolium castaneum]
c40777.graph_c0	1429.64	32.41	0	86.15	PREDICTED: similar to SVOP protein [Tribolium castaneum]

doi:10.1371/journal.pone.0147855.t003

Insecticide Receptors and Resistance-Related Genes

In addition to the identified genes described above, we identified 384 unigenes that represent potential pesticide receptors and insecticide resistance-related genes; including FigCys-loop ligand-gated ion channel (Cys-loop LGIC), carboxylesterase, superoxide dismutase, acetyl-CoA carboxylase, acetylcholinesterase, c-aminobutyric, acid (GABA) receptors, nicotinic acetylcholine receptors, sodium channels, chloride channels and ryanodine receptors (Fig 5). Further study of these genes could uncover potential insecticide receptors and provide the bases to test whether these genes play functional roles in insecticide resistance.

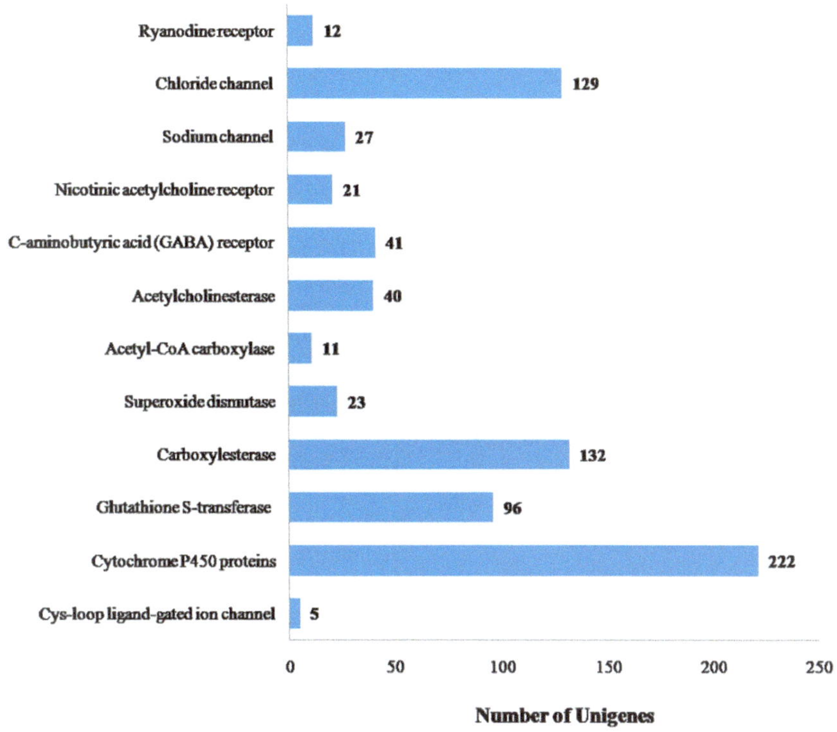

Figure 5. Number of unigenes related to pesticide receptors and resistance-related genes. x-axis indicates the number of unigenes, y-axis indicates the specific unigenes related to pesticide receptors and resistance-related genes. Seven hundred and fifty-nine insecticide receptors and resistance-related unigenes were identified in *M.alternatus* transcriptome.

RNA Interference-Genes

RNA interference (RNAi) is an important pathway that is used in many organisims to regulate gene expression [37]. RNAi pathways have been indentified in *Drosophila melanogaster* [38–40], *T. castaneum* [41–46] and *Bombyx mori* [47–49].

Forty-two unigenes related to RNAi pathways were identified (Fig 6); we identified two SID-1, 34 scavenger receptors, Figone RNA-dependent RNA polymerase, and five vacuolar H^{+}ATPase unigenes in *M. alternatus* transcriptome. These components have previously been reported as part of the RNAi uptake mechanisms in insects [50]. Interestingly, we did not identify any RSD-3 unigenes in the transcriptome, although its participation in RNAi pathways. Among the identified genes, 26 unigenes were larger than 600 bp (61.90%) and 23 were more than 1 kb (54.76%). Moreover, two SID-1-related unigenes were both represented by longer sequences in the transcriptome, at c38753.graph_c0 (2731.42 bp) and c41338.graph_c0 (3066.51 bp), respectively.

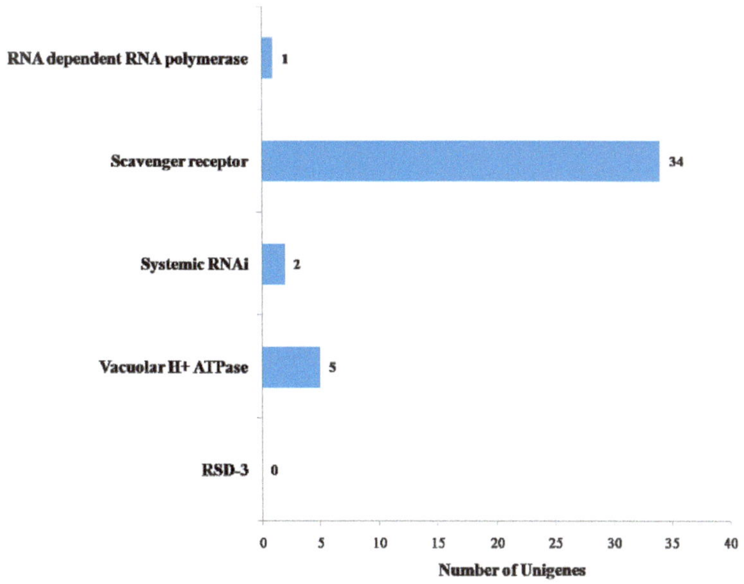

Figure 6. Number of unigenes related to RNAi. x-axis indicates the corresponding number of unigenes and y-axis indicates the specific unigenes related to RNAi. Forty-two unigenes coding for RNAi identified in the database, 26 of them were larger than 600 bp (61.90%) and 23 were more than 1 kb (54.76%).

SID-1 is a multispan transmembrane protein that is crucial for systemic RNAi pathways in*Caenorhabditis elegans*, it delivers dsRNAs to cells [50]. However, SID-1 has not been described in *D. melanogaster*, indicating that the existence or absence of *sid-1* gene could play an important part in determining whether systemic RNAi is present in the biome [51]. Although a sid-1 ortholog has also been found in the cotton aphid *Aphis gossypii* [52], further research is still needed to provide evidence for the molecular basis of systemic RNAi in *M. alternatus*.

Scavenger receptors can recognize extensive polyanionic ligands, and they play key roles mediating phagocytosis of pathogens in *Drosophila* [50]. Among the 34 identified unigenes related to scavenger receptors, nine of them had lengths above 1 kb, while the rest ranged from 100 bp to 1 kb. However, we were not able to identify important relevant sequences homologous sequences to the scavenger receptors of mammals [53, 54], but we were able to identify them by homology to the flesh fly and C. elegans.

Finally, we identified five vacuolar H^+ ATPase unigenes and one RNA-dependent RNA polymerase-1 unigene. Previous studies have found that vacuolar H+ ATPase-deficient S2 *Drosophila* cells accumulate dsRNA in endocytic vesicles and showing an endocytosis-based mechanism as a way to disrupt vacuolar H^+ ATPase in the S2 cells to induce RNAi silencing [55].

The study of SID-1, scavenger receptors, vacuolar H^+ ATPase, RSD-3 and RNA-dependent RNA polymerase genes can deepen our understanding of the biology of defense against parasitic endogenous nucleic acids and exogenous pathogen nucleic acids and provides a basis for the expression of regulatory protein coding genes [56, 57].

Potential *Bacillus Thuringensis* (Bt) Receptors

Receptor molecules related to B. thuringensis Cry toxin mechanisms, have been characterized in the the insect midgut; they have been widely studied, most notably: aminopeptidase (APN), alkaline phosphatase (ALP) and cadherin [58, 59]. We identified a total of 448 Bt receptor unigenes in *M. alternatus* transcriptome, including: ALP, APN, cadherin and the ABC transporter, which are currently the four described Bt insect receptor molecules (Fig 7).

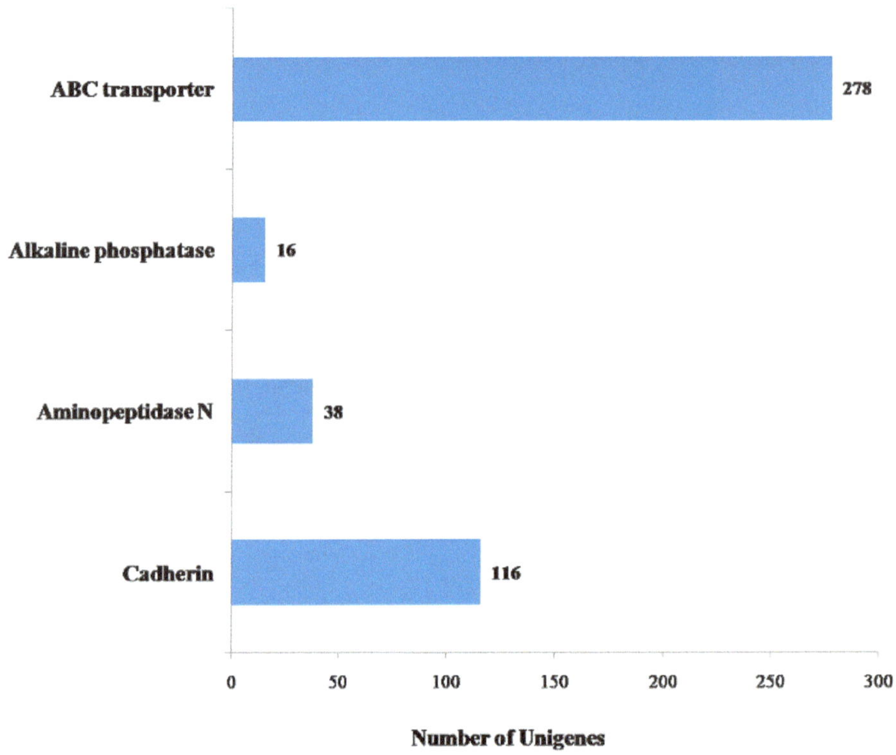

Figure 7. Number of unigenes related to Bt receptors. x-axis indicates the corresponding number of unigenes and y-axis indicates the specific unigenes related to Bt receptors. Four hundred and forty-eight Bt receptor unigenes were confirmed in the *Monochamus alternatus* Hope transcriptome through*in silico* comparisons.

APN [60–62] and cadherin [63] are considered the most important type of receptors within the putative insect Cry receptors. *M. alternatus* transcriptome revealed 38 APN unigenes and 116 cadherin unigenes, representing both approximately 8% of the total number of related Bti-receptor unigenes. APNs belong to the zinc-binding metalloprotease/peptidase enzymes and are anchored to the midgut membrane through GPI anchors [64]. Crava *et al.* clustered APNs into seven classes based on phylogenetic analyses of lepidopteran, Hughes found that in intestinal tissue, APN1 class was the most highly expressed within a complex mix of APN expression data [65, 66]. Cadherins are mainly localized in adherens junctions and are involved in cell adhesion [59]; they have been found anchoring midgut epithelial cells of *Manduca sexta* and *Lymantria dispar* larvae. We found that the number of identified cadherin unigenes contributed in nearly 26% of the total number of Bt receptor molecules in *M. alternatus*, second only to ABC transporters (278, ~62%). It has previously been reported that APN and cadherin-like (CAD-like) midgut proteins in lepidotera can interact with Cry1 toxins. In Diptera, homologous APN, CAD-like, and alkaline phosphatase proteins of mosquitoes are also considered as Cry11 and Cry4 receptor proteins. In the longicorn of coleopterous, a cadherin-like protein acts as Cry3Aa receptor; finally, gene silencing has confirmed that APN and CAD-like proteins are the most representative Cry receptors [67]. In insects, ALPs are a major group of Cry-binding proteins; their roles as receptor molecules have extensively studied in Lepidoptera, Coleoptera and Diptera larvae [68–71]. Data suggest that the interaction between Cry1Ac and ALP affects the midgut protease activity during the incubation period in *Heliothis virescens* and *M. sexta* larvae [62,68]. *Monochanus alternatus* transcriptome revealed 16 ALP unigenes, which contribute to only ~3% of the Bt receptor-related unigenes. Using RNAi silencing of APN and ALP genes in *M. sexta* larvae, Flores-Escobar *et al.* found that for Cry1Ab toxicity, binding to ALP was more important than APN, however for Cry1Ac, APN was more important, these results suggest that Cry binding receptors have specific affinity for different Cry toxins [72].

In addition to ANP, ALP and cadherins, *M. alternatus* transcriptome revealed 278 ABC transporter-related unigenes, contributing for 62.05% of the total number of Bt receptor-related unigenes. The ABC family in insects is related to multi-drug resistance [73]. Recently, ATP-binding cassette transporter subfamily C member 2 (ABCC2) was identified as a Cry1 toxin receptor in *B. mori* larvae [74], with a consistent role in the mechanism of Cry1 toxin resistance [75, 76].

Intestinal Digestive Enzymes

Midgut proteases play an important role in activating Cry toxins, producing the toxin's core 3D structure in the insect midgut [77] and the type and abundance of insect proteases are important for toxin specificity [78]. Changes in the content and activity of proteases can lead to resistance [78]. Frederick S. Walters *et al.* found that mCry3A toxicity to corn rootworm larvae was attributed to a chymotrypsin/cathepsin G site, which enhances cleavage and subsequent binding of the activated toxin to midgut cells [79].

In Lepidoptera and Diptera, serine proteases are the main type of intestinal protease [80,81]; meanwhile in Coleoptera, cysteine and aspartic proteases are the principal class of digestive enzymes. Cathepsin G serine protease has been considered as the principal enzyme because its activity is a key step towards Cry toxicity. It has been demonstrated that cathepsin activity can be Cry toxin specific, for example, *Acyrthosiphon pisum* mainly expresses cathepsin L and cathepsin B types [82], but different serine proteases have activity for Cry4A and Cry4B in sensitive Diptera. [80]. Moreover, cathepsins B, L and serine peptidases such as trypsins and chymotrypsins are the major enzyme components in the larval midgut of tenebrionid beetles [83]. In general, the digestive protease activity of Coleopteran insects depends mainly on cysteine proteases [84, 85]. We found 394 protease-related unigenes in *M. alternatus* transcriptome (Fig 8). Among these unigenes, serine proteases were the most represented group (232, 58.88%), followed by cysteine proteases (79, 20.05%), metalloproteases (77, 19.54%) and aspartic acid proteases (61.52%). Threonine proteases and glutamic acid proteases were not found in our data. Interestingly, we found 212 serine proteases genes (Fig 8). Only 82 serine protease genes had lengths above 1 kb. The unigenes with lengths less than 1 kb were represented by 95 trypsin and 28 chymotrypsin genes. Serine proteases are the main protein digestive enzymes, accounting for 95% of the digestive activity in Lepidoptera [86]. In insects, genes encoding serine proteases (SP) and serine protease homologs (SPH) comprise a large family of proteins involved in digestion, immune defense, development and other process [87]. Cysteine proteases play an important role as virulence factors in *Entamoeba histolytica*. Furthermore, from invasive trophozoites to parasites in dormant infective cyst stages, cysteine proteases have an important role in parasite morphology [88]. It has been reported that Colorado potato beetle (CPB) midgut membrane metalloproteases participate in the proteolytic processing of Cry3Aa toxin [89]. C. Rausell *et al.* found that Brush border membrane vesicles (BBMV)-associated metalloproteases can cut Cry3Aa toxin specificity, thus significantly reducing the activity of pore formation [90]. In vertebrates, the four main aspartic proteases are pepsins, cathepsin D, cathepsin E and renins [91–93], these enzymes mainly degrade endogenous proteins. Our data confirmed the abundance of protease-related unigenes in *M. alternatus* transcriptome, which has significant implications for future research regarding the mechanism of action of Cry toxin in *M. alternatus*. The information provided by insect transcriptomes can contribute in understanding of digestive enzyme functions, in addition to generate new tools to improve the activity of proteases.

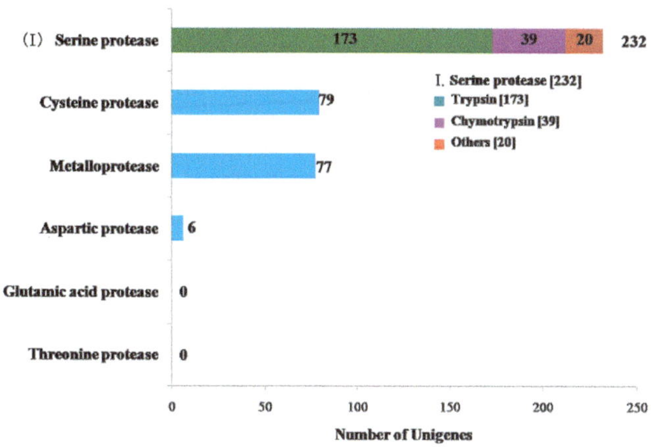

Figure 8. Number of unigenes related to intestinal digestive enzymes. x-axis indicates the corresponding number of unigenes and y-axis indicates the specific unigenes related to intestinal digestive enzymes. Three hundred and ninety-four protease-related unigenes were identified in *M. alternatus* Hope transcriptome. (I) Number of specific unigenes related to serine proteases in intestinal digestive enzymes. The icon indicates specific unigenes. The number in brackets indicates the corresponding number of unigenes.

Possible Future Insect Control Targets

In addition to the diverse type of genes described above (Fig 9), we also identified candidate unigenes as targets for future insect control strategies, such as cathepsin B, cysteine peptidases, neuropeptides and serine peptidases. Among them, we found serine carboxypeptidase and serine-type endopeptidase, belonging to the serine peptidases (Fig 9). Of these, 117 unigenes had a length above 600 bp, 87 of which were above 1 kb. Moreover, we found 96 neuropeptide unigenes, corresponding to insect neurohormones that signal via G-protein-coupled receptors (GPCRs) to control growth, reproduction, behavior, breeding and other physiological processes [94]. Identifying important molecule-related unigenes for these basic physiological processes in *M. alternatus* will provide a reference for possible future research into targets for insect control and other applications.

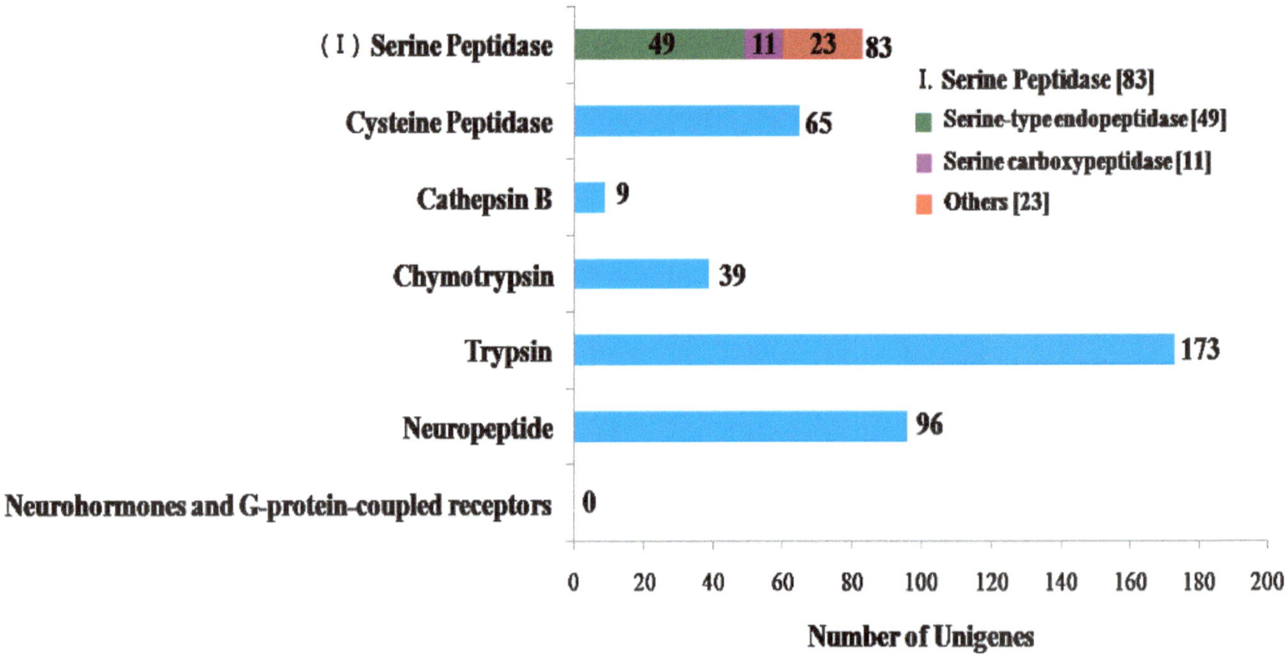

Figure 9. Number of unigenes related to possible future insect control targets. x-axis indicates the corresponding number of unigenes and y-axis indicates the specific unigenes related to possible future insect control targets. 465 possible future insect control targets unigenes were found in *M. alternatus* Hope transcriptome. Among these, 117 unigenes had a length above 600 bp and 87 above 1 kb. (I) Number of specific unigenes related to Serine Peptidase in possible future insect control targets. The icon indicates specific unigenes. The number in brackets indicates the corresponding number of unigenes. Serine carboxypeptidase, serine-type endopeptidase and others belongs to the serine peptidases.

Immune-Related Molecules

Organisms have unique immune mechanisms against pathogens in the environment. Compared to vertebrates, invertebrates lack of an acquired immune system and rely on an innate immune system; adaptive immunity in insects has yet to be identified [95]. The innate immune system of insects is divided into cellular immunity and humoral immunity. Cellular immunity comprises phagocytosis, melanization and encapsulation. Humoral immunity has three steps:1) pattern recognition protein receptors (PRRs) recognize and bind to pathogen-associated molecular patterns (PAMPs); 2) a series of innate immune responses are activated, and 3) finally triggering the generation of innate effector activities (phagocytosis) and effector molecules (antimicrobial peptides (AMPs)) [96, 97]. RNAi machinery also plays an important role in regulating the innate immune response in insects and other organisms [98]. We identified 478 unigenes related to immune molecules and receptors related to immune activities (Fig 10). This group contains 20 widely recognized immune factors, including serine protease inhibitors (94, ~20%) and being the most abundant, followed by peroxidases (75, 16%). Interestingly, MD2(an extracellular binding partner of Toll-like receptor 4 (TLR4),)-like proteins and ML (MD-2-related lipid-recognition) -related unigenes were not found, although MD2-like gene family encodes for secretory proteins [99]. Therefore, molecules directly related to MLs may not be present in *M. alternatus* immune system.

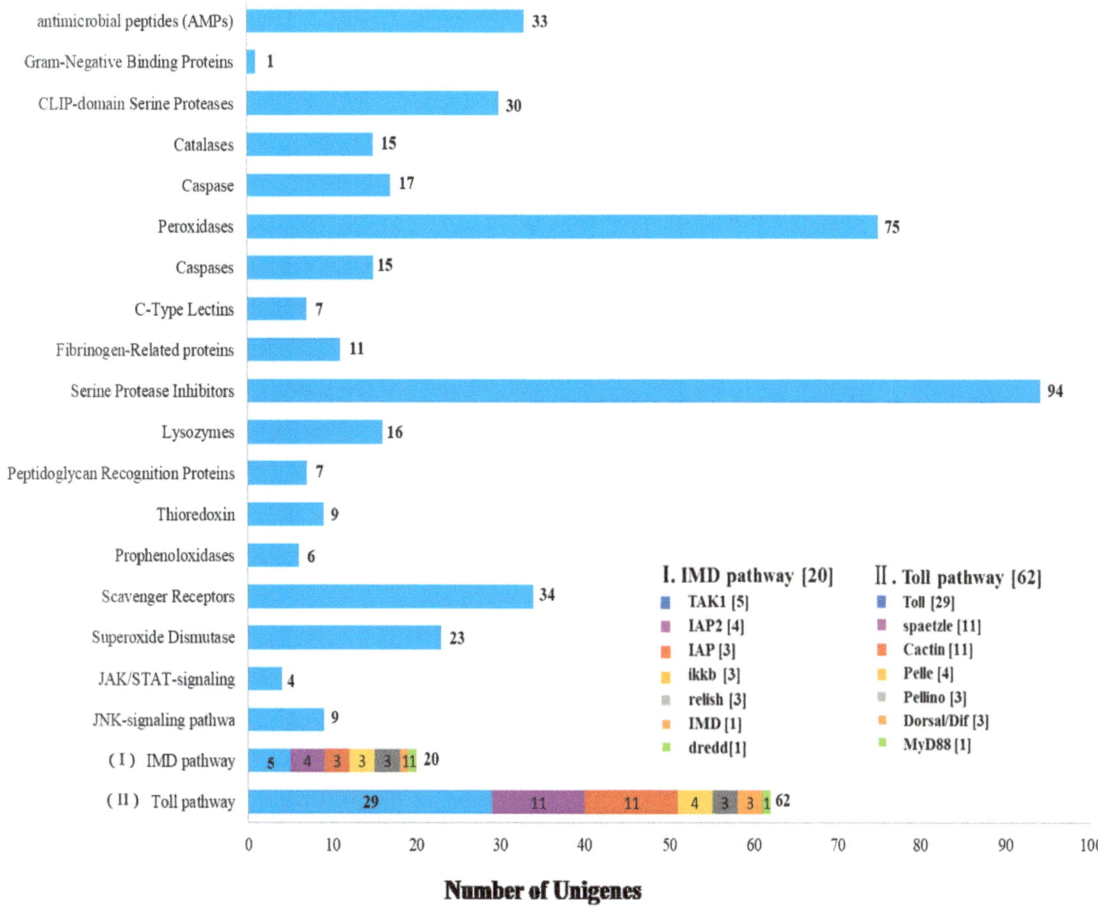

Figure 10. Number of unigenes related to immune-related molecules. x-axis indicates the corresponding number of unigenes and y-axis indicates the specific unigenes related to immune-related molecules. 478 unigenes were identified related to immune molecules and receptors in the transcriptome. This group contains 20 widely recognized immune factors. (I) Number of specific unigenes related to IMD pathway in immune-related molecules. (II) Number of specific unigenes related to Toll pathway in immune-related molecules. The icon indicates specific unigenes. The number in brackets indicates the corresponding number of unigenes.

Antimicrobial Peptides

AMP-related unigenes were represented by approximately 7% of the total identified immune factors. AMPs play important roles in the innate immune system of invertebrates and possess significantly broad antimicrobial activity. AMPs are widespread in nature and mainly distributed in Lepidoptera, Coleoptera, Diptera, Hymenoptera, Hemiptera, Isoptera, Homoptera and Odonata orders. Antimicrobial peptides such as metchnikowin, drosocin, defensin, diptericins, attacins, cecropins, drosomycins, have been characterized in *D.melanogaster* [100, 101]. We identified 33 antimicrobial peptide-related coding unigenes: antifungal peptides were the most abundant (15, ~ 45%), while attacins, cecropins and defensins represented 15%, 9% and 6%, respectively. Iijima R *et al.* isolated an anti-fungal peptide (AFP) from *Sarcophaga peregrina* as the first reported AFP [102], it is composed of 67 amino acid residues and is rich in glycine and histidine. We identified 5 attacin-related unigenes. Attacins only participate in bacteriostasis for some gram-negative bacteria [103]. At present, more than 20 cecropin analogues have been isolated from Lepidoptera and Diptera insects [104]. Cecropines form a voltage-dependent channel that changes the permeability of the bacterial cell membrane. Its antibacterial spectrum includes gram-negative and gram-positive bacteria. Moreover, we found two defensin unigenes (~6%), insect defensins were first isolated in *Phormia terranovae*, due to their high homology with mammalian defensins and later identified in *S. peregrina* (sapecins) [105]. Members of insect defensins family were also identified in *A. aegypti* and *A. gambiae* [106]. Although fewer in number defensin activities cannot be ignored in *M. alternatus* immune system. Dobson *et al.* hypothesized that the highly diverse function of antibacterial peptides have implications in the evolution of pathogen inhibition mechanisms in the host [107].

Immune Enzymes and Inhibitors

We found 17 unigenes related to caspases (CASPs), a class of aspartic acid proteolytic enzymes containing cysteine and responsible for apoptosis initiation; they play a critical role in regulating cell death in the growth process of an organism. CASPs 4, 5 and 11 have been reported as cytoplasmic receptors for gram-negative lipopolysaccharides (LPS) [30,108]. LPS are important structural components of gram-negative bacterial outer cell membranes and can activate the innate immune response through TLR4.

Ninety-four serine protease inhibitors (SRPN) unigenes were identified in our data, contributing for ~20% of the total number of immune factor unigenes; which may possess important active defense immune functions in the process of pathogenic microorganism infection. SRPNs can participate in a variety of physiological reactions and are present higher eukaryotes and viruses [109, 110].

Oxidative Stress Molecules

In insect immune responses, a high level of reactive oxygen species (ROS) is produced. Peroxidase-related unigenes identified in *M. alternatus* contributed for 16% of the total number of immune factor-related unigenes, we identified 75 peroxidase-related unigenes, 21 unigenes corresponding to including glutathione [GPXs], 8 corresponding to thioredoxin [TPXs], and 30 unigenes corresponding to Haem. Furthermore, we identified 15 catalase-related unigenes (CAT), 16 Lysozyme-related unigenes (LYS), and 6 prophenoloxidases-related unigenes (PPO). CATs can effectively convert H_2O_2 into water and oxygen, removing hydrogen peroxide from the cell to prevent H_2O_2 poisoning. Therefore, CATs are key enzymes in the biological defense system. Studies have shown that CAT gene disruption in*D. melanogaster* quickly leads to death [111]. Moreover, we also identifed superoxide dismutases (SODs) unigenes (5%). SODs can convert $\bullet O_2^{-}$ into hydrogen peroxide (H_2O_2), which is less toxic.

Lectins and Galectins

Seven unigenes related to C-Type lectins (CTLs) and five galectin (GALEs) genes were identified in our data. CTLs are secreted proteins or membrane proteins that depend on calcium ions for their function, they are the largest and most diverse family of animal lectins. Several invertebrate immune responses involving CTLs include opsonization [112, 113], pathogen elimination, hemocytes biosynthesis and activating prophendoxidase to produce melaninization [114–116]. GALEs are β-galactoside binding proteins that depend on mercaptans. There is evidence that galectins are related to congenital immunity in*Drosophila* and *An. gambiae* [117].

Pattern Recognition Proteins

Peptidoglycan recognition proteins (PGRP) are pattern recognition receptors that can recognize bacteria peptidoglycans and they play an important role in recognition and adaption of innate immunity. We found 7 PGRP-related unigenes and interestingly, 34 pattern recognition scavenger receptor (SCR), SCR mediates the recognition of pathogen LPS and LTA [118–120] and "clean up" after apoptosis [53, 54, 121–125]. We hypothesize that SCR-associated molecules in *M. alternatus* may have similar actions. In addition, we identified 11 fibrinogen-related proteins (FREP) and 30 CLIP-domain serine proteases, contributing for 2% and 6% respectively. CLIP proteases have regulatory roles for immune responses in insects [111, 126–132]. We also discovered one unigene that correlated with a Gram-Negative Binding Protein (GNBP). Based on the genetics and biochemistry analysis GNBP1 has been identified as a co-receptor for gram positive microbes and PGRP-SA [133,134]. Finally, we identified 9 thioester-containing (TEP) protein-related unigenes (~2%). The TEP family is mainly composed of vertebrate complementary factors (C3, C4, C5) and α2-macroglobulin. *Anopheles gambiae* TEP1 is a typical TEP in insects, which can bind to the surface of pathogenic bacteria and promote the haemocytes of mosquitoes to phagocyte the bacteria or eliminate *Plasmodium* [135–137].

IMD, Toll, JAK/STAT and JNK-signaling Pathway

We identifed unigenes that are closely related to the IMD, Toll, JAK/STAT and JNK-signaling pathway, represented by 20, 62, 4 and 9 unigenes, respectively.

Imd Pathway

The Imd pathway is the primary immune pathway that acts against Gram-negative bacteria. We identified 20 unigenes, such as IMD, TAK1, IAP, IAP2, dredd, ikkb and relish (Fig 10). Imd pathway regulates the activity of a third *Drosophila* NF-kB protein named Relish, controls the expression of most of the *Drosophila* AMPs and, thus is indispensable for normal immunity [138]. TAK1 is ubiquitin-dependent kinase of IKK [68,139]. IAP2 participates in Relish nuclear localization in *Drosophila* [140]. IKK enhanced the phosphorylation of Relish, after Relish cleavage by DREDD, causing translocation of the N-terminal end to the nucleus [141].

Toll Pathway

ixty-two identified unigenes were related to the Toll pathway, including spatzle, Toll, MyD88, Pelle, pellino, Cactin and Dorsal/Dif (Fig 10). In Drosophila the Toll receptors are essential for embryonic development and immunity. The induction of the Toll pathway by Gram-positive bacteria or fungi leads to the activation od cellular immunity and the systemic production of AMPs. The Toll receptor is activated when the proteolytically cleaved ligand Spatzle binds the receptor leading to the activation of the NF-kB factors Dorsal-related immunity factor [141, 142]. Cactus is phosphorylated by a complex consisting of MyD88, tube and pelle, causing degradation of cactus and release of Dorsal/Dif [143, 144]. Besides, pellino and Cactin genes showed high expression levels in *T. castaneum* transcriptome [145].

JAK/STAT-signaling Pathway/ JNK-signaling Pathway

two additional signaling pathways have been shown to have immune functions in insects: the JNK and JAK-STAT pathways involved in cell stress or wound response as other immune pathways in insects [139]. Only two genes involved in JAK-STAT signaling pathway were identified in our transcriptome sequencing data, including 3 unigenes related to STAT92E and 1 unigenes related to hopscotch. We could not identified the reaining components of this pathway (unpaired, STAT, JAK, and the receptor Domeless/Master of marelle (DOME/MOM)) [146, 147]. We also identified 9 unigenes related to the JNK-signaling pathway. Five unigenes were related to Kay and 4 to hemipterous.

CONCLUSION

M. alternatus is known as the cancer of pine trees and is considered a devastating disease, causing serious environmental and economic losses. Vector control strategies are needed to stop and/or control the spreading of this disease [5]. In this work, we sequenced and characterized the transcriptome in the insect vector *M. alternatus* using Illumina sequencing. We identified a large set of genes related to putative insecticide resistance, intestinal digestive enzymes, possible future insect control targets and immune-related molecules. This study provides valuable information that may serve as key point to develop new control strategies for Pine Wilt Disease.

MATERIALS AND METHODS

Ethics Statement

There are no specific permits for insect collection in the selected locations. The chosen locations are not privately-owned or natural protected areas. Insects used for the experiments are not considered endangered or protected species, and its collection is legal in China.

Insects

Pinus massoniana (*P. massoniana*) trunks infested with *M. alternatus* were selected from trees withered during the first year in the town of GuanTou, LianJiang county and FuJian province (N 26.15046°;E 119.59261°). Trunks were cut into 1 m size with a chain saw in the open field and transported to the isolated laboratory of FuJian Agriculture and Forestry University in sealed canvas bags. Trucks were kept in a rearing cage (1.5 m x 1.0 m in length/width) with 1mm iron mesh to prevent insect scaping. Insects were maintained using an artificial diet and kept for two generations. Twenty-five whole larvae (fourth instar) were collected for RNA extraction.

cDNA Library and Illumina Sequencing

Total RNA was extracted from 25 whole larvae (fourth instar) using TRIzol Reagent (Invitrogen). Extracted RNA was processed using the E.Z.N.A. ® HP Total RNA Kit (OMEGA RNA, Invitrogen) to eliminate polysaccharides. RNA purity, quality and concentration were determined using Nanodrop, Qubit 2.0 and Agilent 2100 methods. Messenger RNA was extracted from 6 μg (50 ng/ul) of total RNA, using oligo (dT) magnetic beads, fragmentation buffer was added to the beads coated with mRNA and mRNA was broken randomly. mRNA was used to synthetize the first cDNA chain and then subjected to a second amplification to obtain double stranded cDNA. Double stranded cDNA was purified using AMPure XP. Finally, the cDNA library was created by PCR enrichment using HiSeq2500 high-throughput sequencing, the read length for the sequencing was PE125.

Bioinformatic Analysis

The cDNA library was sequenced by High-throughput sequencing platform to produce a large number of high quality reads based on Sequencing by Synthesis (SBS) technology. Raw data were cleaned from joint sequence and low-quality reads. Depurated reads were assembled into contigs using Trinity software. The transcript sequences were identified in each fragment collection using the De Bruijn method of graphing the sequencing read information [9]. The BLAST parameter E-value was set at 10^{-5}, and the HMMER parameter E-value was set at 10^{-10}. All unigenes were compared with NR [148], Swiss-Prot [149], GO [150], COG [150], KOG [151] and KEGG [152] databases using BLAST [153] software. GO database was used to determine the function of the identified transcript and to assign Gene Ontology (GO) terms. In addition, metabolic pathways were predicted using the COG and KEGG databases. Amino acid sequence of the unigenes was analyzed by HMMER [154] software and the outputs were searched in the Pfam [155] database to obtain annotated information for the unigenes.

ACKNOWLEDGMENTS

We wish to thank Dr. Brian McGarvey and Dr. Rebecca Carballar for proofreading this paper. This study was supported by the Special Fund for Forestry Research in the Public Interest of China [Grant No. 201304401], Project of Fujian-Taiwan Joint Center for Ecological Control of Crop Pests [Minjiaoke (2013) 51], the United Fujian Provincial Health and Education Project for Tackling Key Research [Grant No. WKJ-FJ-25], National Natural Science Foundation of China [Grant No. 31301724, No. 31201574 and No. 31401802], Education Department Foundation of Fujian Province Distinguished Young Scholars in Universities of Fujian Province [Grant No.JA12092], Science Fund for Distinguished Young Scholars of Fujian Agriculture and Forestry University [Grant No. xjq201203], Key Projects of Fujian Provincial Department of Science and Technology [No. 2011|N5003 and No. 2013N0003], Natural Science Foundation of Fujian Province [Grant No. 2013J01079] and Programme for New Century Excellent Talents in Fujian Province [JA13089], the Special Fund for Grain Research in the Public Interest of China [Grant No. 201313002-3], the Special Fund for Key Construction Project of Fujian Agriculture and Forestry University [Grant No. 6112C035003], National Undergraduate Training Programme for Innovation and Entrepreneurship [Grant No. 201510389013, No.201510389018, No. 201510389023, No. 201510389025, No. 201510389028, No. 201510389101 and No. 201510389173]. Science and Technology Project in Fujian Province Department of Education [JA15161], Central Finance Forestry Science and Technology Demonstration Project [Min (2015) TG017].

AUTHOR CONTRIBUTIONS

Conceived and designed the experiments: FPZ XG SQW XLZ. Performed the experiments: SQW XLZ ZXL. Analyzed the data: SQW XLZ ZXL ESS CLR YJG YTX RXX YNM XH GHL SQZ. Contributed reagents/materials/analysis tools: SQW XLZ ZXL ESS CLR YJG YTX RXX YNM XH GHL SQZ. Wrote the paper: SQW XLZ.

REFERENCES

1. Wu XQ, Yuan WM, Tian XJ, Fan B, Fang X, Ye JR, et al. Specific and functional diversity of endophytic bacteria from pine wood nematode *Bursaphelenchus xylophilus* with different virulence. Int J Biol Sci. 2013; 9(1):34–44. doi: 10. 7150/ijbs. 5071 pmid:23289015; PubMed Central PMCID: PMC3535532.
2. State Forestry Bureau of China (2014) The announcement of Bursaphelenchus xylophilus infected area in 2014. On February 12, 2014. http://wwwforestrygovcn/main/3600/content-657977html.
3. Zhao BG, Futai K, Sutherland JR, Takeuchi Y. Pine wilt disease: Springer; 2008.
4. Hajek AE, Glare TR, O'Callaghan M. Use of microbes for control and eradication of invasive arthropods:

Springer Science & Business Media; 2008.

5. Yang ZQ, Wang XY, Zhang YN. Recent advances in biological control of important native and invasive forest pests in China. Biol Control. 2014; 68:117–28. doi: 10. 1016/J. Biocontrol. 2013. 06. 010. WOS:000327821500011.

6. YanLong Z, ZhongQi Y, XiaoYi W, YiNan Z, ChengJin W, ShengFu M, et al. Functional response of the parasitoid *Sclerodermus sp*. (Hymenoptera: Bethylidae) to the third instar larvae of host *Monochamus alternatus* (Coleoptera: Cerambycidae). Acta Entomologica Sinica. 2012; 55(4):426–34.

7. Wang Y-H, Lai Y-X, Cen D-H, Wang J-Y, Yang Z-Q. Behavior and numbers of Dastarcus helophoroides beetle's laying egg and its first-instar larvae parasitism on larvae of Monochamus alternatus. Journal of Environmental Entomology. 2011; 4.

8. Yuanliang Y, Yang Zq, Wang Xy, Yu Jx, Yan Xw. Biological control of Monochamus alternatus (Coleoptera: Cerambycidae) by releasing eggs and adults of Dastarcus helophoroides (Coleoptera: Bothrideridae). Scientia Sivae Sinicae. 2013; 49(3):103–9.

9. Grabherr MG, Haas BJ, Yassour M, Levin JZ, Thompson DA, Amit I, et al. Full-length transcriptome assembly from RNA-Seq data without a reference genome. Nature Biotechnology. 2011; 29(7):644–U130. doi: 10. 1038/ Nbt. 1883. WOS:000292595200023. pmid:21572440

10. Bai X, Zhang W, Orantes L, Jun T-H, Mittapalli O, Mian MAR, et al. Combining next-generation sequencing strategies for rapid molecular resource development from an invasive aphid species, Aphis glycines. PloS one. 2010; 5(6):e11370. doi: 10. 1371/journal. pone. 0011370. pmid:20614011

11. Crawford JE, Guelbeogo WM, Sanou A, Traoré A, Vernick KD, Sagnon NF, et al. De novo transcriptome sequencing in Anopheles funestus using Illumina RNA-seq technology. PloS one. 2010; 5(12):e14202. doi: 10. 1371/journal. pone. 0014202. pmid:21151993

12. Karatolos N, Pauchet Y, Wilkinson P, Chauhan R, Denholm I, Gorman K, et al. Pyrosequencing the transcriptome of the greenhouse whitefly, *Trialeurodes vaporariorum* reveals multiple transcripts encoding insecticide targets and detoxifying enzymes. BMC genomics. 2011; 12:56. doi: 10. 1186/1471-2164-12-56 pmid:21261962; PubMed Central PMCID: PMC3036619.

13. Wang XW, Luan JB, Li JM, Bao YY, Zhang CX, Liu SS. De novo characterization of a whitefly transcriptome and analysis of its gene expression during development. BMC genomics. 2010; 11:400. doi: 10. 1186/1471-2164-11-400 pmid:20573269; PubMed Central PMCID: PMC2898760.

14. Xue J, Bao YY, Li BL, Cheng YB, Peng ZY, Liu H, et al. Transcriptome analysis of the brown planthopper Nilaparvata lugens. PloS one. 2010; 5(12):e14233. doi: 10. 1371/journal. pone. 0014233 pmid:21151909; PubMed Central PMCID: PMC2997790.

15. Zhu JY, Zhao N, Yang B. Global Transcriptome Profiling of the Pine Shoot Beetle, Tomicus yunnanensis (Coleoptera: Scolytinae). PloS one. 2012; 7(2). ARTN e32291 doi: 10. 1371/journal. pone. 0032291. WOS:000302916100067. pmid:22384206

16. Gene Ontology C. The Gene Ontology project in 2008. Nucleic acids research. 2008; 36(Database issue):D440–4. doi: 10. 1093/nar/gkm883 pmid:17984083; PubMed Central PMCID: PMC2238979.

17. Cancino-Rodezno A, Lozano L, Oppert C, Castro JI, Lanz-Mendoza H, Encarnacion S, et al. Comparative proteomic analysis of *Aedes aegypti* larval midgut after intoxication with Cry11Aa toxin from *Bacillus thuringiensis*. PloS one. 2012; 7(5):e37034. doi: 10. 1371/journal. pone. 0037034 pmid:22615881; PubMed Central PMCID: PMC3353955.

18. Daimon T, Kozaki T, Niwa R, Kobayashi I, Furuta K, Namiki T, et al. Precocious metamorphosis in the juvenile hormone-deficient mutant of the silkworm, *Bombyx mori*. PLoS genetics. 2012; 8(3):e1002486. doi: 10. 1371/journal. pgen. 1002486 pmid:22412378; PubMed Central PMCID: PMC3297569.

19. Helvig C, Koener JF, Unnithan GC, Feyereisen R. CYP15A1, the cytochrome P450 that catalyzes epoxidation of methyl farnesoate to juvenile hormone III in cockroach corpora allata. Proceedings of the National Academy of Sciences of the United States of America. 2004; 101(12):4024–9. doi: 10. 1073/Pnas. 0306980101. WOS:000220472200009. pmid:15024118

20. Niwa R, Matsuda T, Yoshiyama T, Namiki T, Mita K, Fujimoto Y, et al. CYP306A1, a cytochrome P450 enzyme, is essential for ecdysteroid biosynthesis in the prothoracic glands of *Bombyx* and *Drosophila*. Journal of Biological Chemistry. 2004; 279(34):35942–9. doi: 10. 1074/Jbc. M404514200. WOS:000223303400095. pmid:15197185

21. Niu G, Rupasinghe SG, Zangerl AR, Siegel JP, Schuler MA, Berenbaum MR. A substrate-specific cytochrome P450 monooxygenase, CYP6AB11, from the polyphagous navel orangeworm (*Amyelois transitella*). Insect Biochem Mol Biol. 2011; 41(4):244–53. doi: 10. 1016/j. ibmb. 2010. 12. 009 pmid:21220011.

22. Schuler MA, Berenbaum MR. Structure and function of cytochrome P450S in insect adaptation to natural and synthetic toxins: insights gained from molecular modeling. J Chem Ecol. 2013; 39(9):1232–45. doi: 10. 1007/s10886-013-0335-7 pmid:24036972.

23. Daborn PJ, Lumb C, Boey A, Wong W, Ffrench-Constant RH, Batterham P. Evaluating the insecticide resistance potential of eight *Drosophila melanogaster*cytochrome P450 genes by transgenic over-expression. Insect Biochem Mol Biol. 2007; 37(5):512–9. doi: 10. 1016/j. ibmb. 2007. 02. 008 pmid:17456446.

24. Feyereisen R. Evolution of insect P450. Biochem Soc T. 2006; 34:1252–5. WOS:000243013500061. doi: 10. 1042/bst0341252

25. Ai JW, Zhu Y, Duan J, Yu QY, Zhang GJ, Wan F, et al. Genome-wide analysis of cytochrome P450 monooxygenase genes in the silkworm, Bombyx mori. Gene. 2011; 480(1–2):42–50. doi: 10. 1016/J. Gene. 2011. 03. 002. WOS:000291773600006. pmid:21440608

26. Xia Q, Cheng D, Duan J, Wang G, Cheng T, Zha X, et al. Microarray-based gene expression profiles in multiple tissues of the domesticated silkworm, *Bombyx mori*. Genome biology. 2007; 8(8):R162. pmid:17683582 doi: 10. 1186/gb-2007-8-8-r162

27. Bass C, Field LM. Gene amplification and insecticide resistance. Pest management science. 2011; 67(8):886–90. doi: 10. 1002/Ps. 2189. WOS:000293327300002. pmid:21538802

28. Mannervik B, Helena Danielson U, Ketterer B. Glutathione transferases-structure and catalytic activit. Critical Reviews in Biochemistry and Molecular Biology. 1988;23(3):283–337. doi: 10.3109/10409238809088226

29. Pickett CB, Lu AYH. Glutathione S-Transferases: Gene Structure, Regulation, and Biological Function. Annual Review of Biochemistry. 1989; 58(1):743–64. doi: 10. 1146/annurev. bi. 58. 070189. 003523 pmid:2673020.

30. Shi J, Zhao Y, Wang Y, Gao W, Ding J, Li P, et al. Inflammatory caspases are innate immune receptors for intracellular LPS. Nature. 2014; 514(7521):187–92. doi: 10. 1038/nature13683 pmid:25119034.

31. Ranson H, Claudianos C, Ortelli F, Abgrall C, Hemingway J, Sharakhova MV, et al. Evolution of supergene families associated with insecticide resistance. Science. 2002; 298(5591):179–81. doi: 10. 1126/Science. 1076781. WOS:000178370500053. pmid:12364796

32. Ding YC, Ortelli F, Rossiter LC, Hemingway J, Ranson H. The *Anopheles gambiae*glutathione transferase supergene family: annotation, phylogeny and expression profiles. BMC genomics. 2003; 4. Artn 35 doi: 10. 1186/1471-2164-4-35. WOS:000185828600001. pmid:12914673

33. Ranson H, Rossiter L, Ortelli F, Jensen B, Wang X, Roth CW, et al. Identification of a novel class of insect glutathione S-transferases involved in resistance to DDT in the malaria vector *Anopheles gambiae*. Biochem J. 2001; 359(2):295–304. doi: 10. 1042/bj3590295

34. Pemble SE, Taylor JB. An evolutionary perspective on glutathione transferases inferred from class-theta glutathione transferase cDNA sequences. Biochem J. 1992; 287 (Pt 3):957–63. pmid:1445253; PubMed Central PMCID: PMC1133100. doi: 10. 1042/bj2870957

35. Friedman R. Genomic organization of the glutathione S-transferase family in insects. Mol Phylogenet Evol. 2011; 61(3):924–32. doi: 10. 1016/j. ympev. 2011. 08. 027 pmid:21930223.

36. Gullipalli D, Arif A, Aparoy P. Identification of a developmentally and hormonally regulated Delta-Class glutathione S-transferase in rice moth Corcyra cephalonica. Comparative Biochemistry & Physiology Part B Biochemistry & Molecular Biology. 2010; 156(1):33–9. doi: 10. 1016/j. cbpb. 2010. 01. 016

37. Burand JP, Hunter WB. RNAi: future in insect management. J Invertebr Pathol. 2013; 112 Suppl:S68–74. doi: 10. 1016/j. jip. 2012. 07. 012 pmid:22841639.

38. Roignant JY, Carre C, Mugat B, Szymczak D, Lepesant JA, Antoniewski C. Absence of transitive and systemic pathways allows cell-specific and isoform-specific RNAi in*Drosophila*. RNA. 2003; 9(3):299–308. pmid:12592004; PubMed Central PMCID: PMC1370397. doi: 10. 1261/rna. 2154103

39. Bischoff V, Vignal C, Duvic B, Boneca IG, Hoffmann JA, Royet J. Downregulation of the *Drosophila* immune response by peptidoglycan-recognition proteins SC1 and SC2. PLoS Pathog. 2006; 2(2):e14. doi: 10. 1371/journal. ppat. 0020014 pmid:16518472; PubMed Central PMCID: PMC1383489.

40. Miller SC, Brown SJ, Tomoyasu Y. Larval RNAi in *Drosophila*? Development Genes And Evolution. 2008; 218(9):505–10. doi: 10. 1007/s00427-008-0238-8. WOS:000258717800005. pmid:18663472

41. Arakane Y, Li B, Muthukrishnan S, Beeman RW, Kramer KJ, Park Y. Functional analysis of four neuropeptides, EH, ETH, CCAP and bursicon, and their receptors in adult ecdysis behavior of the red flour beetle, *Tribolium castaneum*. Mechanisms of Development. 2008; 125(11–12):984–95. doi: 10. 1016/J. Mod. 2008. 09. 002. WOS:000262302100006. pmid:18835439

42. Fujita K, Shimomura K, Yamamoto K, Yamashita T, Suzuki K. A chitinase structurally related to the glycoside hydrolase family 48 is indispensable for the hormonally induced diapause termination in a beetle. Biochemical and Biophysical Research Communications. 2006; 345(1):502–7. doi: 10. 1016/J. Bbrc. 2006. 04. 126. WOS:000237877700068. pmid:16684504

43. Konopova B, Jindra M. Broad-complex acts downstream of Met in juvenile hormone signaling to coordinate primitive holometabolan metamorphosis. Development. 2008; 135(3):559–68. doi: 10. 1242/Dev. 016097. WOS:000252254900015. pmid:18171683

44. Minakuchi C, Namiki T, Shinoda T. Kruppel homolog 1, an early juvenile hormone-response gene downstream of Methoprene-tolerant, mediates its anti-metamorphic action in the red flour beetle *Tribolium castaneum*. Developmental Biology. 2009; 325(2):341–50. doi: 10. 1016/J. Ydbio. 2008. 10. 016. WOS:000262589400004. pmid:19013451

45. Parthasarathy R, Palli SR. Molecular Analysis of Juvenile Hormone Analog Action in Controlling the Metamorphosis of the *Red Flour Beetle*, *Tribolium Castaneum*. Archives of Insect Biochemistry and Physiology. 2009; 70(1):57–70. doi: 10. 1002/Arch. 20288. WOS:000262205000006. pmid:19072925

46. Tomoyasu Y, Denell RE. Larval RNAi in *Tribolium* (Coleoptera) for analyzing adult development. Development genes and evolution. 2004; 214(11):575–8. doi: 10. 1007/s00427-004-0434-0 pmid:15365833.

47. Quan GX, Kanda T, Tamura T. Induction of the white egg 3 mutant phenotype by injection of the double-stranded RNA of the silkworm white gene. Insect molecular biology. 2002; 11(3):217–22. pmid:12000640. doi: 10. 1046/j. 1365-2583. 2002. 00328. x

48. Ohnishi A, Hull JJ, Matsumoto S. Targeted disruption of genes in the *Bombyx mori*sex pheromone biosynthetic pathway. Proc Natl Acad Sci U S A. 2006; 103(12):4398–403. doi: 10. 1073/pnas. 0511270103 pmid:16537410; PubMed Central PMCID: PMC1450183.

49. Hossain M, Shimizu S, Matsuki M, Imamura M, Sakurai

S, Iwami M. Expression of 20-hydroxyecdysone-induced genes in the silkworm brain and their functional analysis in post-embryonic development. Insect Biochem Mol Biol. 2008; 38(11):1001–7. doi: 10. 1016/j. ibmb. 2008. 08. 006 pmid:18835445.

50. Huvenne H, Smagghe G. Mechanisms of dsRNA uptake in insects and potential of RNAi for pest control: a review. Journal of insect physiology. 2010; 56(3):227–35. doi: 10. 1016/j. jinsphys. 2009. 10. 004. pmid:19837076

51. Tribolium Genome Sequencing C, Richards S, Gibbs RA, Weinstock GM, Brown SJ, Denell R, et al. The genome of the model beetle and pest *Tribolium castaneum*. Nature. 2008; 452(7190):949–55. doi: 10. 1038/nature06784 pmid:18362917.

52. Xu W, Han Z. Cloning and phylogenetic analysis of sid-1-like genes from aphids. J Insect Sci. 2008; 8:1–6. doi: 10. 1673/031. 008. 3001 pmid:20302524; PubMed Central PMCID: PMC3061602.

53. Pearson A, Lux A, Krieger M. Expression cloning of dSR-CI, a class C macrophage-specific scavenger receptor from *Drosophila melanogaster*. Proc Natl Acad Sci U S A. 1995; 92(9):4056–60. pmid:7732030; PubMed Central PMCID: PMC42101. doi: 10. 1073/ pnas. 92. 9. 4056

54. Ramet M, Pearson A, Manfruelli P, Li X, Koziel H, Gobel V, et al. *Drosophila*scavenger receptor CI is a pattern recognition receptor for bacteria. Immunity. 2001; 15(6):1027–38. pmid:11754822. doi: 10. 1016/ s1074-7613(01)00249-7

55. Saleh MC, van Rij RP, Hekele A, Gillis A, Foley E, O'Farrell PH, et al. The endocytic pathway mediates cell entry of dsRNA to induce RNAi silencing. Nature Cell Biology. 2006; 8(8):793–802. doi: 10. 1038/ncb1439. WOS:000239423000009. pmid:16862146

56. Tomoyasu Y, Miller SC, Tomita S, Schoppmeier M, Grossmann D, Bucher G. Exploring systemic RNA interference in insects: a genome-wide survey for RNAi genes in *Tribolium*. Genome Biol. 2008; 9(1):R10. doi: 10. 1186/gb-2008-9-1-r10 pmid:18201385; PubMed Central PMCID: PMC2395250.

57. Sijen T, Fleenor J, Simmer F, Thijssen KL, Parrish S, Timmons L, et al. On the role of RNA amplification in dsRNA-triggered gene silencing. Cell. 2001; 107(4):465–76. pmid:11719187. doi: 10. 1016/s0092-8674(01)00576-1

58. Bravo A, Likitvivatanavong S, Gill SS, Soberon M. *Bacillus thuringiensis*: A story of a successful bioinsecticide. Insect Biochem Mol Biol. 2011; 41(7):423–31. doi: 10. 1016/j. ibmb. 2011. 02. 006 pmid:21376122.

59. Pigott CR, Ellar DJ. Role of receptors in *Bacillus thuringiensis* crystal toxin activity. Microbiol Mol Biol Rev. 2007; 71(2):255–81. doi: 10. 1128/MMBR. 00034-06 pmid:17554045; PubMed Central PMCID: PMC1899880.

60. Knight PJ, Knowles BH, Ellar DJ. Molecular cloning of an insect aminopeptidase N that serves as a receptor for *Bacillus thuringiensis* CryIA(c) toxin. J Biol Chem. 1995; 270(30):17765–70. pmid:7629076. doi: 10. 1074/ jbc. 270. 30. 17765

61. Knight PJ, Crickmore N, Ellar DJ. The receptor for *Bacillus thuringiensis* CrylA(c) delta-endotoxin in the brush border membrane of the lepidopteran *Manduca sexta*is aminopeptidase N. Mol Microbiol. 1994; 11(3):429–36. pmid:7908713. doi: 10. 1111/j. 1365-2958. 1994. tb00324. x

62. Sangadala S, Walters FS, English LH, Adang MJ. A mixture of *Manduca sexta*aminopeptidase and phosphatase enhances *Bacillus thuringiensis* insecticidal CryIA(c) toxin binding and 86Rb(+)-K+ efflux in vitro. J Biol Chem. 1994; 269(13):10088–92. pmid:8144508.

63. Vadlamudi RK, Weber E, Ji I, Ji TH, Bulla LA Jr. Cloning and expression of a receptor for an insecticidal toxin of *Bacillus thuringiensis*. J Biol Chem. 1995; 270(10):5490–4. pmid:7890666. doi: 10. 1074/jbc. 270. 10. 5490

64. Garczynski SF, Adang MJ. *Bacillus thuringiensis* CryIA (c) δ-endotoxin binding aminopeptidase in the *Manduca sexta* midgut has a glycosyl-phosphatidylinositol anchor. Insect biochemistry and molecular biology. 1995; 25(4):409–15. doi: 10. 1016/0965-1748(94)00072-7

65. Crava CM, Bel Y, Lee SF, Manachini B, Heckel DG, Escriche B. Study of the aminopeptidase N gene family in the lepidopterans *Ostrinia nubilalis* (Hubner) and*Bombyx mori* (L.): sequences, mapping and expression. Insect Biochem Mol Biol. 2010; 40(7):506–15. doi: 10. 1016/j. ibmb. 2010. 04. 010 pmid:20420910.

66. Hughes AL. Evolutionary diversification of aminopeptidase N in Lepidoptera by conserved clade-specific amino acid residues. Mol Phylogenet Evol. 2014; 76:127–33. doi: 10. 1016/j. ympev. 2014. 03. 014 pmid:24675701.

67. Contreras E, Schoppmeier M, Real MD, Rausell C. Sodium Solute Symporter and Cadherin Proteins Act as *Bacillus thuringiensis* Cry3Ba Toxin Functional Receptors in *Tribolium castaneum*. Journal of Biological Chemistry. 2013; 288(25):18013–21. doi: 10. 1074/Jbc. M113. 474445. WOS:000320721900008. pmid:23645668

68. English LH, Readdy TL. Delta endotoxin inhibits a phosphatase in midgut epithelial membranes of *Heliothis virescens*. Insect biochemistry. 1989; 19(2):145–52. doi: 10. 1016/0020-1790(89)90085-1

69. McNall RJ, Adang MJ. Identification of novel *Bacillus thuringiensis* Cry1Ac binding proteins in *Manduca sexta* midgut through proteomic analysis. Insect Biochemistry and Molecular Biology. 2003; 33(10):999–1010. doi: 10. 1016/S0965-1748(03)00114-0. WOS:000185958200005. pmid:14505693

70. Krishnamoorthy M, Jurat-Fuentes JL, McNall RJ, Andacht T, Adang MJ. Identification of novel Cry1Ac binding proteins in midgut membranes from *Heliothis virescens*using proteomic analyses. Insect Biochem Mol Biol. 2007; 37(3):189–201. doi: 10. 1016/j. ibmb. 2006. 10. 004 pmid:17296494.

71. Bayyareddy K, Andacht TM, Abdullah MA, Adang MJ. Proteomic identification of*Bacillus thuringiensis* subsp. israelensis toxin Cry4Ba binding proteins in midgut

membranes from *Aedes* (Stegomyia) *aegypti Linnaeus* (Diptera, Culicidae) larvae. Insect Biochem Mol Biol. 2009; 39(4):279–86. doi: 10. 1016/j. ibmb. 2009. 01. 002 pmid:19272330.

72. Flores-Escobar B, Rodriguez-Magadan H, Bravo A, Soberon M, Gomez I. Differential role of *Manduca sexta* aminopeptidase-N and alkaline phosphatase in the mode of action of Cry1Aa, Cry1Ab, and Cry1Ac toxins from *Bacillus thuringiensis*. Appl Environ Microbiol. 2013; 79(15):4543–50. doi: 10. 1128/AEM. 01062-13 pmid:23686267; PubMed Central PMCID: PMC3719532.

73. Heckel DG. Learning the ABCs of Bt: ABC transporters and insect resistance to*Bacillus thuringiensis* provide clues to a crucial step in toxin mode of action. Pesticide Biochemistry and Physiology. 2012; 104(2):103–10. doi: 10. 1016/J. Pestbp. 2012. 05. 007. WOS:000311068900005.

74. Tanaka S, Miyamoto K, Noda H, Jurat-Fuentes JL, Yoshizawa Y, Endo H, et al. The ATP-binding cassette transporter subfamily C member 2 in *Bombyx mori* larvae is a functional receptor for Cry toxins from *Bacillus thuringiensis*. FEBS J. 2013; 280(8):1782–94. doi: 10. 1111/febs. 12200 pmid:23432933.

75. Atsumi S, Miyamoto K, Yamamoto K, Narukawa J, Kawai S, Sezutsu H, et al. Single amino acid mutation in an ATP-binding cassette transporter gene causes resistance to Bt toxin Cry1Ab in the silkworm, *Bombyx mori*. Proc Natl Acad Sci U S A. 2012; 109(25):E1591–E8. doi: 10. 1073/pnas. 1120698109 pmid:22635270; PubMed Central PMCID: PMC3382473.

76. Gahan LJ, Pauchet Y, Vogel H, Heckel DG. An ABC transporter mutation is correlated with insect resistance to *Bacillus thuringiensis* Cry1Ac toxin. PLoS genetics. 2010; 6(12):e1001248. doi: 10. 1371/journal. pgen. 1001248 pmid:21187898; PubMed Central PMCID: PMC3002984.

77. Bravo A, Gomez I, Porta H, Garcia-Gomez BI, Rodriguez-Almazan C, Pardo L, et al. Evolution of Bacillus thuringiensis Cry toxins insecticidal activity. Microbial biotechnology. 2013; 6(1):17–26. doi: 10. 1111/j. 1751-7915. 2012. 00342. x pmid:22463726; PubMed Central PMCID: PMC3815381.

78. Deist BR, Rausch MA, Fernandez-Luna MT, Adang MJ, Bonning BC. Bt toxin modification for enhanced efficacy. Toxins. 2014; 6(10):3005–27. doi: 10. 3390/ toxins6103005 pmid:25340556; PubMed Central PMCID: PMC4210883.

79. Walters FS, Stacy CM, Lee MK, Palekar N, Chen JS. An engineered chymotrypsin/cathepsin G site in domain I renders Bacillus thuringiensis Cry3A active against western corn rootworm larvae. Applied and environmental microbiology. 2008; 74(2):367–74. pmid:18024675 doi: 10. 1128/aem. 02165-07

80. Borovsky D. Biosynthesis and control of mosquito gut proteases. IUBMB Life. 2003; 55(8):435–41. doi: 10. 1080/1521654031000159772l pmid:14609198.

81. Schnepf E, Crickmore N, Van Rie J, Lereclus D, Baum J, Feitelson J, et al. *Bacillus thuringiensis* and its pesticidal crystal proteins. Microbiol Mol Biol Rev. 1998; 62(3):775–806. pmid:9729609; PubMed Central PMCID: PMC98934.

82. Carrillo L, Martinez M, Alvarez-Alfageme F, Castanera P, Smagghe G, Diaz I, et al. A barley cysteine-proteinase inhibitor reduces the performance of two aphid species in artificial diets and transgenic *Arabidopsis plants*. Transgenic Res. 2011; 20(2):305–19. doi: 10. 1007/ s11248-010-9417-2 pmid:20567901.

83. Vinokurov KS, Elpidina EN, Oppert B, Prabhakar S, Zhuzhikov DP, Dunaevsky YE, et al. Diversity of digestive proteinases in *Tenebrio molitor* (Coleoptera: Tenebrionidae) larvae. Comp Biochem Physiol B Biochem Mol Biol. 2006; 145(2):126–37. doi: 10. 1016/j. cbpb. 2006. 05. 005 pmid:16859942.

84. Bown DP, Wilkinson HS, Jongsma MA, Gatehouse JA. Characterisation of cysteine proteinases responsible for digestive proteolysis in guts of larval western corn rootworm (*Diabrotica virgifera*) by expression in the yeast *Pichia pastoris*. Insect Biochem Mol Biol. 2004; 34(4):305–20. doi: 10. 1016/j. ibmb. 2003. 11. 005 pmid:15041015.

85. Murdock LL, Brookhart G, Dunn PE, Foard DE, Kelley S, Kitch L, et al. Cysteine digestive proteinases in Coleoptera. Comparative Biochemistry and Physiology Part B: Comparative Biochemistry. 1987; 87(4):783–7. doi: 10. 1016/0305-0491(87)90388-9

86. Srinivasan A, Giri AP, Gupta VS. Structural and functional diversities in lepidopteran serine proteases. Cellular & molecular biology letters. 2006; 11(1):132–54. doi: 10. 2478/s11658-006-0012-8 pmid:16847755.

87. Kuwar S, Pauchet Y, Vogel H, Heckel DG. Adaptive regulation of digestive serine proteases in the larval midgut of *Helicoverpa armigera* in response to a plant protease inhibitor. Insect Biochemistry and Molecular Biology. 2015. doi: 10. 1016/j. ibmb. 2015. 01. 016

88. De Cádiz AE, Jeelani G, Nakada-Tsukui K, Caler E, Nozaki T. Transcriptome analysis of encystation in Entamoeba invadens. PloS one. 2013; 8(9):e74840. doi: 10. 1371/journal. pone. 0074840. pmid:24040350

89. Rausell C, Ochoa-Campuzano C, Martínez-Ramírez AC, Bravo A, Real MD. A membrane associated metalloprotease cleaves Cry3Aa *Bacillus thuringiensis* toxin reducing pore formation in *Colorado potato beetle* brush border membrane vesicles. Biochimica et Biophysica Acta (BBA)-Biomembranes. 2007; 1768(9):2293–9. doi: 10. 1016/j. bbamem. 2007. 06. 014

90. Rausell C, Ochoa-Campuzano C, Martinez-Ramirez AC, Bravo A, Real MD. A membrane associated metalloprotease cleaves Cry3Aa *Bacillus thuringiensis* toxin reducing pore formation in *Colorado potato beetle* brush border membrane vesicles. Biochimica et biophysica acta. 2007; 1768(9):2293–9. doi: 10. 1016/j. bbamem. 2007. 06. 014 pmid:17643388.

91. Barrett AJ. Cathepsin D: the lysosomal aspartic proteinase. Protein degradation in health and disease. 1980:37–42. doi: 10. 1002/9780470720585. ch3

92. Hamilton KA, Nisbet AJ, Lehane MJ, Taylor MA, Billingsley PF. A physiological and biochemical model

for digestion in the ectoparasitic mite, *Psoroptes ovis* (Acari: Psoroptidae). International journal for parasitology. 2003; 33(8):773–85. pmid:12865077 doi: 10. 1016/s0020-7519(03)00089-4

93. Akov S. Blood digestion in ticks. Physiology of ticks. 1982:197–211. doi: 10. 1016/b978-0-08-024937-7. 50011-1

94. Richards S, Gibbs RA, Weinstock GM, Brown SJ, Denell R, Beeman RW, et al. The genome of the model beetle and pest Tribolium castaneum. Nature. 2008; 452(7190):949–55. doi: 10. 1038/nature06784. pmid:18362917

95. Brennan CA, Anderson KV. *Drosophila*: the genetics of innate immune recognition and response. Annual Review of Immunology. 2004; 22:457–83. doi: 10. 1146/ annurev. immunol. 22. 012703. 104626 pmid:15032585.

96. Vallet-Gely I, Lemaitre B, Boccard F. Bacterial strategies to overcome insect defences. Nature Reviews Microbiology. 2008; 6(4):302–13. doi: 10. 1038/ Nrmicro1870. WOS:000253961800012. pmid:18327270

97. Welchman DP, Aksoy S, Jiggins F, Lemaitre B. Insect immunity: from pattern recognition to symbiont-mediated host defense. Cell Host Microbe. 2009; 6(2):107–14. doi: 10. 1016/j. chom. 2009. 07. 008 pmid:19683677.

98. Lindsay MA. microRNAs and the immune response. Trends in immunology. 2008; 29(7):343–51. doi: 10. 1016/j. it. 2008. 04. 004 pmid:18515182.

99. Inohara N, Nunez G. ML—a conserved domain involved in innate immunity and lipid metabolism. Trends Biochem Sci. 2002; 27(5):219–21. pmid:12076526. doi: 10. 1016/s0968-0004(02)02084-4

100. Hetru C, Troxler L, Hoffmann JA. *Drosophila melanogaster* antimicrobial defense. J Infect Dis. 2003; 187 Suppl 2:S327–S34. doi: 10. 1086/374758 pmid:12792847.

101. Bulet P, Stocklin R, Menin L. Anti-microbial peptides: from invertebrates to vertebrates. Immunol Rev. 2004; 198:169–84. pmid:15199962. doi: 10. 1111/j. 0105-2896. 2004. 0124. x

102. Iijima R, Kurata S, Natori S. Purification, characterization, and cDNA cloning of an antifungal protein from the hemolymph of *Sarcophaga peregrina* (flesh fly) larvae. J Biol Chem. 1993; 268(16):12055–61. pmid:8505329.

103. Carlsson A, Engstrom P, Palva ET, Bennich H. Attacin, an antibacterial protein from*Hyalophora cecropia*, inhibits synthesis of outer membrane proteins in *Escherichia coli* by interfering with omp gene transcription. Infect Immun. 1991; 59(9):3040–5. pmid:1715318; PubMed Central PMCID: PMC258132.

104. Hoffmann JA, Kafatos FC, Janeway CA, Ezekowitz RAB. Phylogenetic perspectives in innate immunity. Science. 1999; 284(5418):1313–8. pmid:10334979 doi: 10. 1126/science. 284. 5418. 1313

105. Matsuyama K, Natori S. Molecular cloning of cDNA for sapecin and unique expression of the sapecin gene during the development of *Sarcophaga peregrina*. J Biol Chem. 1988; 263(32):17117–21. pmid:3182837.

106. Lowenberger C, Bulet P, Charlet M, Hetru C, Hodgeman B, Christensen BM, et al. Insect immunity: isolation of three novel inducible antibacterial defensins from the vector mosquito, *Aedes aegypti*. Insect Biochem Mol Biol. 1995; 25(7):867–73. pmid:7633471. doi: 10. 1016/0965-1748(95)00043-u

107. Dobson AJ, Purves J, Kamysz W, Rolff J. Comparing selection on S. aureus between antimicrobial peptides and common antibiotics. PloS one. 2013; 8(10):e76521. doi: 10. 1371/journal. pone. 0076521 pmid:24204634; PubMed Central PMCID: PMC3799789.

108. Smith C, Wang X, Yin H. Caspases come together over LPS. Trends in immunology. 2015; 36(2):59–61. doi: 10. 1016/j. it. 2014. 12. 007 pmid:25573609; PubMed Central PMCID: PMC4323722.

109. Gettins PG. Serpin structure, mechanism, and function. Chem Rev. 2002; 102(12):4751–804. pmid:12475206. doi: 10. 1021/cr010170+

110. Potempa J, Korzus E, Travis J. The serpin superfamily of proteinase inhibitors: structure, function, and regulation. J Biol Chem. 1994; 269(23):15957–60. pmid:8206889.

111. Griswold CM, Matthews AL, Bewley KE, Mahaffey JW. Molecular characterization and rescue of acatalasemic mutants of *Drosophila melanogaster*. Genetics. 1993; 134(3):781–8. pmid:8349109; PubMed Central PMCID: PMC1205515.

112. Jomori T, Natori S. Function of the lipopolysaccharide-binding protein of *Periplaneta americana* as an opsonin. FEBS letters. 1992; 296(3):283–6. pmid:1537405. doi: 10. 1016/0014-5793(92)80305-z

113. Yu XQ, Kanost MR. *Manduca sexta* lipopolysaccharide-specific immulectin-2 protects larvae from bacterial infection. Dev Comp Immunol. 2003; 27(3):189–96. pmid:12590970. doi: 10. 1016/s0145-305x(02)00099-x

114. Yu XQ, Gan H, Kanost MR. Immulectin, an inducible C-type lectin from an insect,*Manduca sexta*, stimulates activation of plasma prophenol oxidase. Insect Biochem Mol Biol. 1999; 29(7):585–97. pmid:10436935. doi: 10. 1016/s0965-1748(99)00036-3

115. Yu XQ, Kanost MR. Immulectin-2, a lipopolysaccharide-specific lectin from an insect,*Manduca sexta*, is induced in response to *gram-negative bacteria*. J Biol Chem. 2000; 275(48):37373–81. doi: 10. 1074/jbc. M003021200 pmid:10954704.

116. Chen C, Durrant HJ, Newton RP, Ratcliffe NA. A study of novel lectins and their involvement in the activation of the prophenoloxidase system in *Blaberus discoidalis*. Biochem J. 1995; 310 (Pt 1):23–31. pmid:7646450; PubMed Central PMCID: PMC1135849. doi: 10. 1042/ bj3100023

117. Pace KE, Baum LG. Insect galectins: roles in immunity and development. Glycoconjugate journal. 2004; 19(7–9):607–14. doi: 10. 1023/B:GLYC. 0000014092. 86763. 2f pmid:14758086.

118. Gough PJ, Gordon S. The role of scavenger receptors in the innate immune system. Microbes Infect. 2000; 2(3):305–11. pmid:10758407. doi: 10. 1016/s1286-4579(00)00297-5

119. Peiser L, Gordon S. The function of scavenger receptors expressed by macrophages and their role in the regulation of inflammation. Microbes Infect. 2001; 3(2):149–59. pmid:11251301. doi: 10. 1016/s1286-4579(00)01362-9

120. Peiser L, Mukhopadhyay S, Gordon S. Scavenger receptors in innate immunity. Curr Opin Immunol. 2002; 14(1):123–8. pmid:11790542. doi: 10. 1016/s0952-7915(01)00307-7

121. Agaisse H, Burrack LS, Philips JA, Rubin EJ, Perrimon N, Higgins DE. Genome-wide RNAi screen for host factors required for intracellular bacterial infection. Science. 2005; 309(5738):1248–51. doi: 10. 1126/science. 1116008 pmid:16020693.

122. Franc NC, Dimarcq JL, Lagueux M, Hoffmann J, Ezekowitz RA. Croquemort, a novel*Drosophila* hemocyte/macrophage receptor that recognizes apoptotic cells. Immunity. 1996; 4(5):431–43. pmid:8630729. doi: 10. 1016/s1074-7613(00)80410-0

123. Franc NC, Heitzler P, Ezekowitz RA, White K. Requirement for croquemort in phagocytosis of apoptotic cells in *Drosophila*. Science. 1999; 284(5422):1991–4. pmid:10373118. doi: 10. 1126/science. 284. 5422. 1991

124. Franc NC, White K. Innate recognition systems in insect immunity and development: new approaches in *Drosophila*. Microbes Infect. 2000; 2(3):243–50. pmid:10758400. doi: 10. 1016/s1286-4579(00)00304-x

125. Peiser L, De Winther MP, Makepeace K, Hollinshead M, Coull P, Plested J, et al. The class A macrophage scavenger receptor is a major pattern recognition receptor for*Neisseria meningitidis* which is independent of lipopolysaccharide and not required for secretory responses. Infect Immun. 2002; 70(10):5346–54. pmid:12228258; PubMed Central PMCID: PMC128305. doi: 10. 1128/iai. 70. 10. 5346-5354. 2002

126. Adema CM, Hertel LA, Miller RD, Loker ES. A family of fibrinogen-related proteins that precipitates parasite-derived molecules is produced by an invertebrate after infection. Proc Natl Acad Sci U S A. 1997; 94(16):8691–896. pmid:9238039; PubMed Central PMCID: PMC23082. doi: 10. 1073/pnas. 94. 16. 8691

127. Dimopoulos G, Casavant TL, Chang S, Scheetz T, Roberts C, Donohue M, et al. *Anopheles gambiae* pilot gene discovery project: identification of mosquito innate immunity genes from expressed sequence tags generated from immune-competent cell lines. Proc Natl Acad Sci U S A. 2000; 97(12):6619–24. pmid:10841561; PubMed Central PMCID: PMC18678. doi: 10. 1073/pnas. 97. 12. 6619

128. Dimopoulos G, Christophides GK, Meister S, Schultz J, White KP, Barillas-Mury C, et al. Genome expression analysis of *Anopheles gambiae*: responses to injury, bacterial challenge, and malaria infection. Proc Natl Acad Sci U S A. 2002; 99(13):8814–9. doi: 10. 1073/pnas. 092274999 pmid:12077297; PubMed Central PMCID: PMC124381.

129. Gokudan S, Muta T, Tsuda R, Koori K, Kawahara T, Seki N, et al. *Horseshoe crab*acetyl group-recognizing lectins involved in innate immunity are structurally related to fibrinogen. Proc Natl Acad Sci U S A. 1999; 96(18):10086–91. pmid:10468566; PubMed Central PMCID: PMC17846. doi: 10. 1073/pnas. 96. 18. 10086

130. Wang X, Fuchs JF, Infanger LC, Rocheleau TA, Hillyer JF, Chen CC, et al. Mosquito innate immunity: involvement of beta 1,3-glucan recognition protein in melanotic encapsulation immune responses in *Armigeres subalbatus*. Mol Biochem Parasitol. 2005; 139(1):65–73. doi: 10. 1016/j. molbiopara. 2004. 09. 009 pmid:15610820.

131. Zdobnov EM, von Mering C, Letunic I, Torrents D, Suyama M, Copley RR, et al. Comparative genome and proteome analysis of *Anopheles gambiae* and*Drosophila melanogaster*. Science. 2002; 298(5591):149–59. doi: 10. 1126/science. 1077061 pmid:12364792.

132. Jiang H, Kanost MR. The clip-domain family of serine proteinases in arthropods. Insect Biochem Mol Biol. 2000; 30(2):95–105. pmid:10696585. doi: 10. 1016/s0965-1748(99)00113-7

133. Gobert V, Gottar M, Matskevich AA, Rutschmann S, Royet J, Belvin M, et al. Dual activation of the *Drosophila* toll pathway by two pattern recognition receptors. Science. 2003; 302(5653):2126–30. doi: 10. 1126/science. 1085432 pmid:14684822.

134. Wang L, Weber AN, Atilano ML, Filipe SR, Gay NJ, Ligoxygakis P. Sensing of *Gram-positive bacteria* in *Drosophila*: GNBP1 is needed to process and present peptidoglycan to PGRP-SA. EMBO J. 2006; 25(20):5005–14. doi: 10. 1038/sj. emboj. 7601363 pmid:17024181; PubMed Central PMCID: PMC1618108.

135. Blandin S, Shiao SH, Moita LF, Janse CJ, Waters AP, Kafatos FC, et al. Complement-like protein TEP1 is a determinant of vectorial capacity in the malaria vector*Anopheles gambiae*. Cell. 2004; 116(5):661–70. pmid:15006349. doi: 10. 1016/s0092-8674(04)00173-4

136. Levashina EA, Moita LF, Blandin S, Vriend G, Lagueux M, Kafatos FC. Conserved role of a complement-like protein in phagocytosis revealed by dsRNA knockout in cultured cells of the mosquito, *Anopheles gambiae*. Cell. 2001; 104(5):709–18. pmid:11257225. doi: 10. 1016/s0092-8674(01)00267-7

137. Moita LF, Wang-Sattler R, Michel K, Zimmermann T, Blandin S, Levashina EA, et al. In vivo identification of novel regulators and conserved pathways of phagocytosis in*A. gambiae*. Immunity. 2005; 23(1):65–73. doi: 10. 1016/j. immuni. 2005. 05. 006 pmid:16039580.

138. Jose AM, Hunter CP. Transport of sequence-specific RNA interference information between cells. Annual review of genetics. 2007; 41(1):305–30. doi: 10. 1146/annurev. genet. 41. 110306. 130216

139. Xia X, Yu L, Xue M, Yu X, Vasseur L, Gurr GM, et al. Genome-wide characterization and expression profiling of immune genes in the diamondback moth, Plutella xylostella (L.). Scientific Reports. 2015; 5. doi: 10. 1038/srep09877

140. Mcnall RJ, Adang MJ. Identification of novel Bacillus thuringiensis Cry1Ac binding proteins in Manduca sexta midgut through proteomic analysis. Insect Biochemistry & Molecular Biology. 2003; 33(10):999–1010. doi: 10. 1016/s0965-1748(03)00114-0

141. Li X, Cui Z, Liu Y, Song C, Shi G. Transcriptome analysis and discovery of genes involved in immune pathways from hepatopancreas of microbial challenged mitten crab Eriocheir sinensis. PloS one. 2013; 8(7):374-. doi: 10. 1371/journal. pone. 0068233

142. Ai J, Yong Z, Duan J, Yu Q, Zhang G, Fei W, et al. Genome-wide analysis of cytochrome P450 monooxygenase genes in the silkworm, Bombyx mori. Gene. 2011; 480(1–2):42–50. doi: 10. 1016/j. gene. 2011. 03. 002. pmid:21440608

143. Friedman R. Genomic organization of the glutathione S-transferase family in insects. Molecular Phylogenetics & Evolution. 2011; 61(3):924–32. doi: 10. 1016/j. ympev. 2011. 08. 027

144. Susanna V, Jing-Huan W, Mika RM. The Drosophila Toll signaling pathway. Journal of Immunology. 2011; 186(2):649–56.

145. Boran A, Abdelnaser E, Nurper G, Grundler FMW, Andreas V, Heinz-Wilhelm D. Next Generation Sequencing Based Transcriptome Analysis of Septic-Injury Responsive Genes in the Beetle Tribolium castaneum. Plos One. 2013; 8(1):e52004. doi: 10. 1371/ journal. pone. 0052004. pmid:23326321

146. Krishnamoorthy M, Jurat-Fuentes JL, Mcnall RJ, Andacht T, Adang MJ. Identification of novel Cry1Ac binding proteins in midgut membranes from Heliothis virescens using proteomic analyses. Insect Biochemistry & Molecular Biology. 2007; 37(3):189–201. doi: 10. 1016/j. ibmb. 2006. 10. 004

147. Agaisse H, Perrimon N. The roles of JAK/STAT signaling in Drosophila immune responses. Immunological Reviews. 2004; 198(1):72–82. doi: 10. 1111/j. 0105-2896. 2004. 0133. x

148. Deng YY, J. Q. L, Wu SF, Zhu YP, Chen YW, He FC. Integrated nr Database in Protein Annotation System and Its Localization. Computer Engineering. 2006; 32(5):71–4.

149. Apweiler R, Bairoch A, Wu CH, Barker WC, Boeckmann B, Ferro S, et al. UniProt: the Universal Protein knowledgebase. Nucleic acids research. 2004; 32(Database issue):D115–9. doi: 10. 1093/nar/gkh131 pmid:14681372; PubMed Central PMCID: PMC308865.

150. Ashburner M, Ball CA, Blake JA, Botstein D, Butler H, Cherry JM, et al. Gene ontology: tool for the unification of biology. The Gene Ontology Consortium. Nat Genet. 2000; 25(1):25–9. pmid:10802651; PubMed Central PMCID: PMC3037419. doi: 10. 1038/75556

151. Koonin EV, Fedorova ND, Jackson JD, Jacobs AR, Krylov DM, Makarova KS, et al. A comprehensive evolutionary classification of proteins encoded in complete eukaryotic genomes. Genome Biol. 2004; 5(2):R7. doi: 10. 1186/gb-2004-5-2-r7 pmid:14759257; PubMed Central PMCID: PMC395751.

152. Kanehisa M, Goto S, Kawashima S, Okuno Y, Hattori M. The KEGG resource for deciphering the genome. Nucleic acids research. 2004; 32(Database issue):D277–80. doi: 10. 1093/nar/gkh063 pmid:14681412; PubMed Central PMCID: PMC308797.

153. Altschul SF, Madden TL, Schaffer AA, Zhang J, Zhang Z, Miller W, et al. Gapped BLAST and PSI-BLAST: a new generation of protein database search programs. Nucleic acids research. 1997; 25(17):3389–402. pmid:9254694; PubMed Central PMCID: PMC146917. doi: 10. 1093/ nar/25. 17. 3389

154. Finn RD, Bateman A, Clements J, Coggill P, Eberhardt RY, Eddy SR, et al. Pfam: the protein families database. Nucleic acids research. 2014; 42(Database issue):D222–30. doi: 10. 1093/nar/ gkt1223 pmid:24288371; PubMed Central PMCID: PMC3965110.

155. Eddy SR. Profile hidden Markov models. Bioinformatics. 1998; 14(9):755–63. pmid:9918945. doi: 10. 1093/ bioinformatics/14. 9. 755

CHAPTER 18

GENOME OF CNAPHALOCROCIS MEDINALIS GRANULOVIRUS, THE FIRST CRAMBIDAE-INFECTING BETABACULOVIRUS ISOLATED FROM RICE LEAFFOLDER TO SEQUENCED

Guangjie Han, Jian Xu, Qin Liu, Chuanming Li, Hongxing Xu, Zhongxian Lu

INTRODUCTION

The rice leaffolder, *Cnaphalocrocis medinalis* Güenée (Lepidoptera: Crambidae), is a migratory and important insect pest of rice in Asia [1, 2]. The larvae fold the leaves, feed on the photosynthetic leaf tissues in the folded leaves and such damages can result in reduction of rice yields [3]. In China frequent outbreaks have occurred in rice production regions and have caused rice yield reduction and farmers' overuse of insecticides. Insecticide control is the main measure farmers in China use and the pests have developed resistance to some insecticides [4]. The CnmeGV belonging to the family of *Baculoviridae*, was isolated from the infected caterpillars collected from fields in China recently. Bioassay showed that CnmeGV is a highly virulent baculovirus and suggested the potential of its use as an environmentally friendlier microbial agent for future rice leaffolder management [5].

Baculoviridae is a family ofrod-shaped baculoviurs with circular, covalently closed double-stranded DNA genomes, which has been successfully applied for the control of some agricultural and forest insect pests [6]. Based on phylogeny and host specificities, Baculoviridae is divided into four genera: *Alphabaculovirus*(lepidopteran-specificnucleopolyhedrovirus, NPVs), *Betabaculovirus* (lepidopteran-specific granulovirus, GVs), *Gammabaculovirus* (hymenopteran-specific NPVs) and *Deltabaculovirus* (dipteran-specific NPV) [7]. *Alphabaculovirus* is further subdivided into groups I and II according to the phylogenetic analysis of the *lef-8*, *lef-9* and *polh/gran* genes [8]. *Betabaculovirus* is classified into three types based on the tissue tropism [9]. To date, the complete genomes of more than 51 NPVs and 17 GVs are published or available in GenBank.

GVs are more specific than NPVs, which have been reported only from Lepidoptera [10]. Partly because of the difficulty of establishing cell lines that are permissive for GV infection, the molecular biology and genetics of GVs have been less well studied than those of NPVs [11]. CnmeGV is a new isolate and is an effective baculovirus pathogen but less studied and genomic information is lacking. In this paper we present the complete sequence and morphological characterization of the CnmeGV genome and compared them to other baculoviruses using genomic and phylogenetic analyses. This is the first completely sequenced betabaculovirus isolated from a crambidae host to be reported.

RESULTS AND DISCUSSION

Sequence Analysis of the CnmeGV Genome

A total of 53,359 reads from post-filter sequencing libraries were used for genome assembly by the hierarchical genome-assembly process (HGAP). The genome of CnmeGV was sequenced and was registered as the first complete sequence of

Citation: Han G, Xu J, Liu Q, Li C, Xu H, Lu Z (2016) Genome of Cnaphalocrocis medinalis Granulovirus, the First Crambidae-Infecting Betabaculovirus Isolated from Rice Leaffolder to Sequenced. PLoS ONE 11(2): e0147882. doi:10.1371/journal.pone.0147882.

a crambidae infecting betabaculovirus in GenBank (Accession number KP658210). The genome consisted of 112,060 bp, which was within the sizes of the 17 sequenced betabaculovirus genomes ranging from 99,657 bp in AdorGV [12] to 178,733 bp inXcGV [13] (Table 1). The G+C content of CnmeGV genome was 35.2%, close to the lowest one estimated for betabaculovirus members which ranged between 32.5% in CrleGV and 45.2% in CpGV. However, no correlation was found between these data and the biological properties. In the criteria for selecting ORFs there should be methionine-initiated ORFs of at least 50 codons having minimal overlap with other ORFs [14], 133 putative ORFs were identified and were numbered from the ATG start codon of the *granulin* gene in a clockwise direction. Coding sequences represented 85.1% of the genome of CnmeGV similar to CpGV [15]. Seventy (70) ORFs were in the same orientation as the *granulin* ORF and 63 were opposite, indicating that CnmeGV ORFs have no obvious preferred orientation. Helicase (ORF79) is the longest sequence gene encoding 1162 amino acids, while ORF8 is the shortest in CnmeGV genome. The circular map of the CnmeGV genome was established and shown in Fig 1.

Table 1. All species from the genus *Betabaculovirus* completely sequenced to date*.

Virus species	Host family	Size (bp)	ORFs	Accession	Tandem repeats number	Percentage of genome (%)	Authors
Adoxophyes orana granulovirus	Tortricidae	99,657	119	AF547984	22	2.31	Wormleaton et al.
Agrotis segetum granulovirus Xinjiang	Noctuidae	131,680	132	AY522332	14	0.76	Ai et al.
Choristoneura occidentalis granulovirus	Tortricidae	104,710	116	DQ333351	56	8.38	Escasa et al.
Clostera anachoreta granulovirus	Notodontidae	101,487	123	HQ116624	9	0.59	Liang et al.
Clostera anastomosis L. granulovirus	Notodontidae	101,818	123	KC179784	4	0.24	Liang et al.
Cnaphalocrocis medinalis granulovirus	Crambidae	112,060	133	KP658210	53	3.83	Han et al.
Cryptophlebia leucotreta granulovirus	Tortricidae	110,907	129	AY229987	46	4.66	Jehle et al.
Cydia pomonella granulovirus	Tortricidae	123,500	143	U53466	40	2.38	Crook et al.
Epinotia aporema granulovirus	Tortricidae	119,092	132	JN408834	22	1.14	Ferrelli et al.
Erinnyis ello granulovirus	Sphingidae	102,759	135	KJ406702	24	1.66	Ardisson-Araújo et al.
Helicoverpa armigera granulovirus	Noctuidae	169,794	179	EU255577	26	2.53	Harrison et al.
Phthorimaea operculella granulovirus	Gelechiidae	119,217	130	AF499596	134	7.65	Taha et al.
Pieris rapae granulovirus China	Pieridae	108,592	120	GQ884143	42	2.92	Wang et al.
Plutella xylostella granulovirus	Plutellidae	100,999	120	AF270937	32	9.38	Hashimoto et al.
Pseudaletia unipuncta granulovirus	Noctuidae	176,677	183	EU678671	20	1.40	Li et al.
Spodoptera litura granulovirus	Noctuidae	124,121	136	DQ288858	25	2.21	Wang et al.
Xestia c-nigrum granulovirus	Noctuidae	178,733	181	AF162221	27	1.82	Hayakawa et al.

*The same field of different types were not shown.

doi:10.1371/journal.pone.0147882.t001

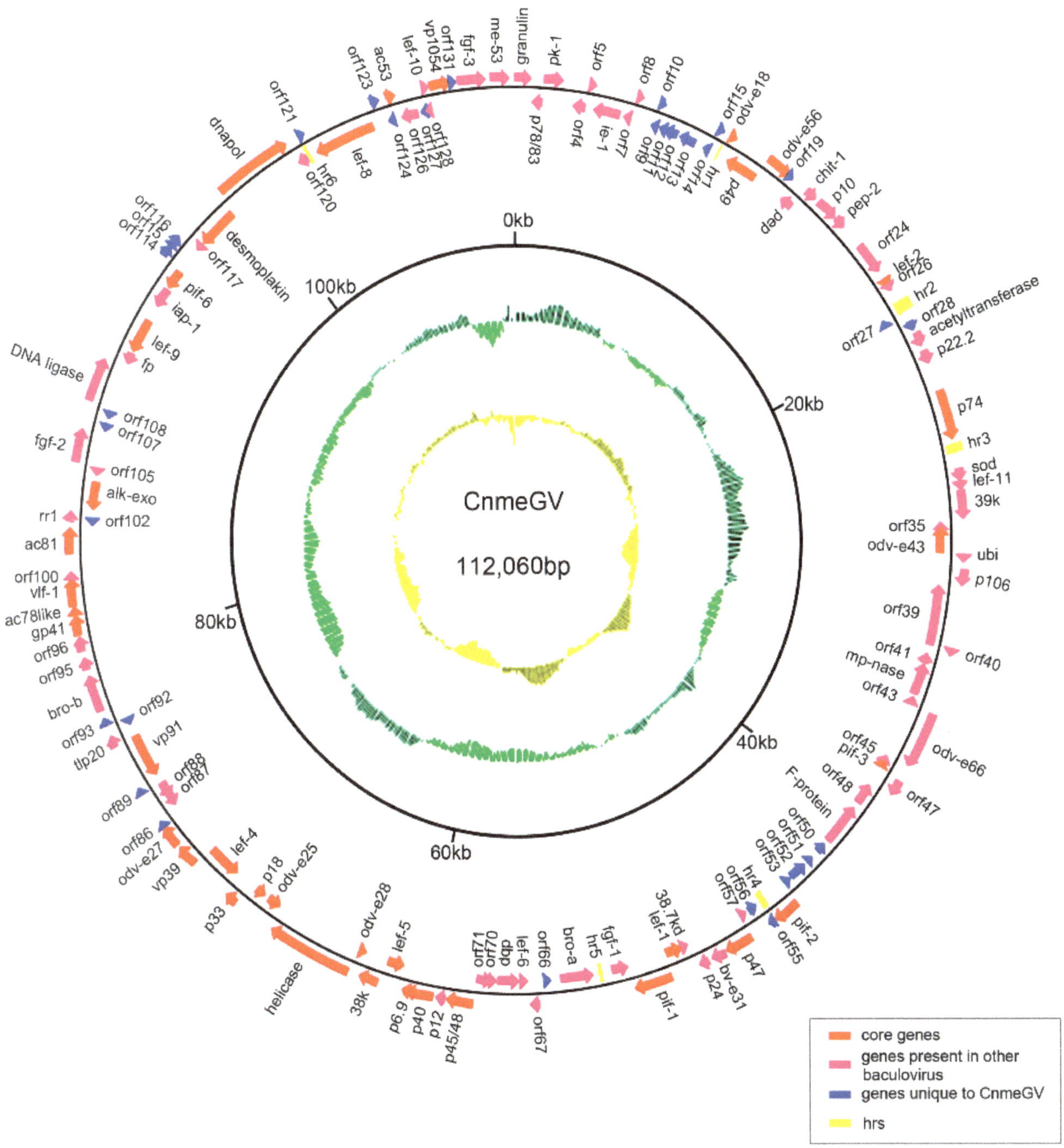

Figure 1. Circular map of the CnmeGV genome. ORFs and transcription direction are indicated as arrows. Core genes were indicated by red arrows, genes present in other baculovirus were indicated by pink arrows, unique genes were indicated by blue arrows and *hrs* were indicated by yellow squares. The innermost circle shows GC skew, which indicates possible locations of the DNA leading strand, lagging strand, replication origin, and replication terminal during DNA replication. Below average GC skew is light orange and above average dark orange. The next innermost circle is a GC plot, with light green representing below average GC content, and dark green indicating above average GC content.

The putative proteins of those ORFs were predicted by BlastX search which had an E-value of less than 10^{-6} in NCBI. In total, 101 of the 133 putative ORFs encoding similar proteins are found in other organisms, while 32 of these were shown to be unique. Core Genes were a set of factors strongly conserved in the *Baculoviridae* family for they provide the essentials roles needed to complete the virus cycle [16]. When compared to the ORFs encoding the 37 described core proteins for *Betabaculovirus* genus [17], the 37 core genes were found in CnmeGV genome, representing the essential functions for replication and transcription; cell cycle interaction and/or arrest with host proteins; packaging and assembly; viral release; and oral infectivity. Baculovirus repeated ORFs (*bro* genes) were striking features of many baculovirus genomes. Two repeated *bro* genes were identified in the CnmeGV genome (ORF65, 94) and were designated as *bro-a*, *bro-b* respectively based on their order in the genome. This highly repetitive and conserved family might have functioned as DNA binding proteins that influenced host DNA replication or transcription and improve the infection capability of virus [18, 19].

Replication Genes

The core genes of CnmeGV involved in DNA replication, *alk-exo* (ORF104), *dnapol*(ORF119), *lef-1* (ORF62), *lef-2*

(ORF25), *helicase* (ORF79), were detected. Other replication genes that belonged to lepidoptera baculovirus conserved genes discovered in CnmeGV were *dbp* (ORF69), *ie-1* (ORF6) and *me-53* (ORF133). Similarly, the *lef-7* gene was not found in CnmeGV while present in most of NPVs and only 4 GVs (HearGV, XcGV, PsunGV, SpfrGV). In the EpapGV genome, a protein, epap36, was found to have a match with PsunGV *lef-7*, but revealed a lower e-value (E = 0.54) [20].

The non-conserved baculoviurs gene *rr1* (ORF103) was also identified in CnmeGV. This putative protein had about 133 aa present and had lower identity than that of PhopGV rr1 (E = 7e-07, 33% amino acid identity) and CpGV rr1 (E = 0.088, 26% amino acid identity). Proteins of rr1 present in most NPVs generally have higher identity [21, 22]. In other GVs genomes, genes of *rr1* encoded proteins usually have about 609–782 aa and higher identity with NPVs (among 25%-53%). So, ORF103 in the CnmeGV genome might be a truncated sequence of the *rr1* gene.

Transcriptional Genes

Transcriptional genes presented as core genes in the *Baculoviridae* family, includes *lef-4*(ORF83), *lef-5* (ORF76), *lef-8* (ORF122), *lef-9* (ORF111), *p47* (ORF58) and *vlf-1* (ORF99) were detected in the CnmeGV DNA. Other genes, *39k* (ORF34), *lef-6* (ORF68), *lef-11*(ORF33) and *pk-1* (ORF3), related to the transcription process in all lepidopteran baculovirus were also identified in CnmeGV genome. The codes of CnmeGV ORF33, which was named *lef-11*, similar to ClanGV ORF47 (64% amino acid identity), might be necessary for efficient transcription as a gene for viral late gene expression [23]. Another late transcription gene *lef-10*, presented in most alphabaculoviruses (except SujuNPV, ClbiNPV and OrleNPV [24]) and betabaculoviruses, was also detected in the CnmeGV genome (ORF129). ORF129 also showed a 61% (E = 3e-18) and 39% (E = 0.003) amino acid identity to PrGV and AdorGV, respectively. *Lef-10* might be possibly the components of multi-subunit RNA polymerase that might be involved in late and very late transcription [25].

Structural Genes

All the core genes associated with structure were found in CnmeGV genome. The *pifs*, including *pif-1* (ORF63), *pif-2* (ORF54), *pif-3* (ORF46), *pif-4/odv-28* (ORF78), *pif-5/odv-e56*(ORF18), *pif-6* (ORF113) were found. *Pif* genes encode an essential structural protein of the occlusion-derived virus envelope [26]. In the early stages of virus infection, *pif-1*, *pif-2* and*pif-3* perform an essential function in association with *p74* [27]. In addition, 7 other conserved genes related to structure were also found in the CnmeGV genome: *fp* (ORF110),*p12* (ORF73), *p24* (ORF60), *tlp20* (ORF71), *F-protein* (ORF49), *granulin* (ORF1), and *p10*(ORF22). ORF1 was identified to reveal 86% amino acid identity to AnbiGV *granulin*, the major component of occlusion bodies as conserved baculovirus structural protein [28]. ORF22 in CnmeGV genome revealed 52% homologous to p10 of AdorGV. It shared a high conserve region at N- terminal (3–102 aa) and C- terminal (180–271 aa). P10 protein formed fibrillar structures, advantageous to occlusion body morphogenesis that result in the disseminating of occlusion bodies (OBs) [29]. In addition, this protein includes various common structural and functional domains: a coiled-coil region followed by a proline-rich domain, a variable region and finally a basic region at the C-terminal, which is characterized from different baculoviruses [13]. ORF22 also showed similarity to the*calyx/pep/pp34* genes of GVs. Therefore, more studies should be done to understand the role of the *p10* homologues in CnmeGV.

Unique ORFs

Thirty two (32) ORFs appeared to be unique to CnmeGV compared to the rest of the members of Baculoviridae (ORF9, 10, 11, 12, 13, 14, 15, 19, 27, 28, 50, 51, 52, 53, 55, 56, 66, 86, 89, 92, 93, 102, 107, 108, 114, 115, 116, 121, 123, 124, 127, 131). The predicated proteins were peptides with no significant similarities to any other sequences in GenBank. Among these unique ORFs in the CnmeGV genome, 14 ORFs were in the same orientation as the *granulin* ORF and 18 in the opposite. Three (3) ORFs (ORF15, 27 and 131) presented a late promoter motif (GATA), suggesting expression at a late stage of viral infection. Early promoter motifs, including CAKT and TATAWAW, were also detected at upstream of the start codon in other ORFs. ORF14 was the longest sequence of the unique ORFs encoded for a putative protein of 271 aa. It had no significant BlastP hits, and had early promoter elements upstream of the first ATG (TATAAAT). ORF55 encoded the shortest hypothetical protein of 52 aa in unique proteins. An early promoter motif (ATTTATA) was also found 57 nt upstream ORF55. The proteins encoded with others ORFs also showed no significant BlastP hits. It seemed apparent that the CnmeGV shared much more unique genes. Whether these are functional ORFs of CnmeGV would require further experimentation.

The SMART program detected 15 unique ORFs that contained at least one region which encoded a limited set of amino acids of special domains. Three (3) ORFs (ORF107, 116, 124) were found with trans-membrane helix regions by the TMHMM v2.0 program. Thirteen (13) ORFs (ORF13, 14, 28, 55, 56, 66, 86, 92, 107, 108, 114, 123, 131) were detected with low-complexity regions (LCRs) by the SMART program. ORF56 encoded three LCRs with the longest one containing 61 aa within 62–122 aa. ORF28 was found to be a coiled-coil region by the COILS program. The coiled-coil segments lie in areas that are possibly playing a functionally important role [30]. Interestingly, the predicated protein of ORF28 also contained a domain of zinc finger-like of 4–50 residues specialized by two C2C2 zinc finger motifs that putatively bound two atoms of zinc. The function of this domain was hypothesized to involve protein dimerization [31], or suggested as an ubiquitin ligase [32], or necessary for DNA binding and zinc-dependent repression [33]. In addition, zinc fingers are typical motifs distributed in DNA/RNA regulatory proteins whereas the coordination of heavy metals is often a characteristic of different metallothioneins in some cases [17]. These assumptions would need further experimentation.

Tandem Repeats (TRs) and Homologous Regions (*HRS*)

Tandem repeats (TRs) are DNA repeat sequences of each repeat unit located right next to each other, reflecting their origin in local duplications. These ubiquitous, unstable elements were found to combine characteristics of genetic and epigenetic changes that might facilitate organismal evolvability [34]. In the genome of PhopGV, 134 TRs were detected in a frequency of 7.65% in the genome. It was the highest TRs composed in the genomes of all betabaculovirus to date. The least TRs were detected in the genome of CalGV, which has 4 TRs with a frequency of 0.24% in genome (Table 1). In a screening of the CnmeGV genome for repeated sequences with TRs Finder [35], 53 TRs were found with sequence lengths from 25 to 551 nt. Fifteen (15) TRs were located in the coding region, 22 TRs were in the non-coding region and 16 TRs were in both the regions. All the 53 TRs contained 3.83% of genome of CnmeGV. These TRs in baculovirus genomes enhance the transcription of early gene in promoters and act as mediators for rapid phenotypic changes in coding sequences [34, 36].

Mutations in these repeats often have fascinating phenotypic consequences [36]. The number of the repeating unit changes, recombination and replication slippage will bring about mutation in TRs [37]. *Tr51*, the least repeat unit of TRs in CnmeGV genome, contained 3 repeat units. The secondary structure of hairpin-loops was predicted by DNAMAN 8 (Minimum free energy of the structure is -14.01kcal/mol) (Fig 2B). This structure was a part of variability [36]. In addition to inherent instability, TR mutation can also be affected by external factors [38]. For example, CAG repeat stability is modulated by the chaperone protein hsp90 in the human cell. Hsp90 function can be overwhelmed by severe environmental stresses, resulting in a role of mediating an influence by the environment on TR mutation rates [39]. In the genome of CnmeGV, a CAG repeat unit in the TRs of *Tr6*contained 12.3 repeat units was found in the coding regions (Fig 2C). It coded a hypothetical protein of 333 amino acids. This might be possible that the correlation between CAG repeat units and the stability of the hypothetical protein would response to the environment. This assumption would need further verification.

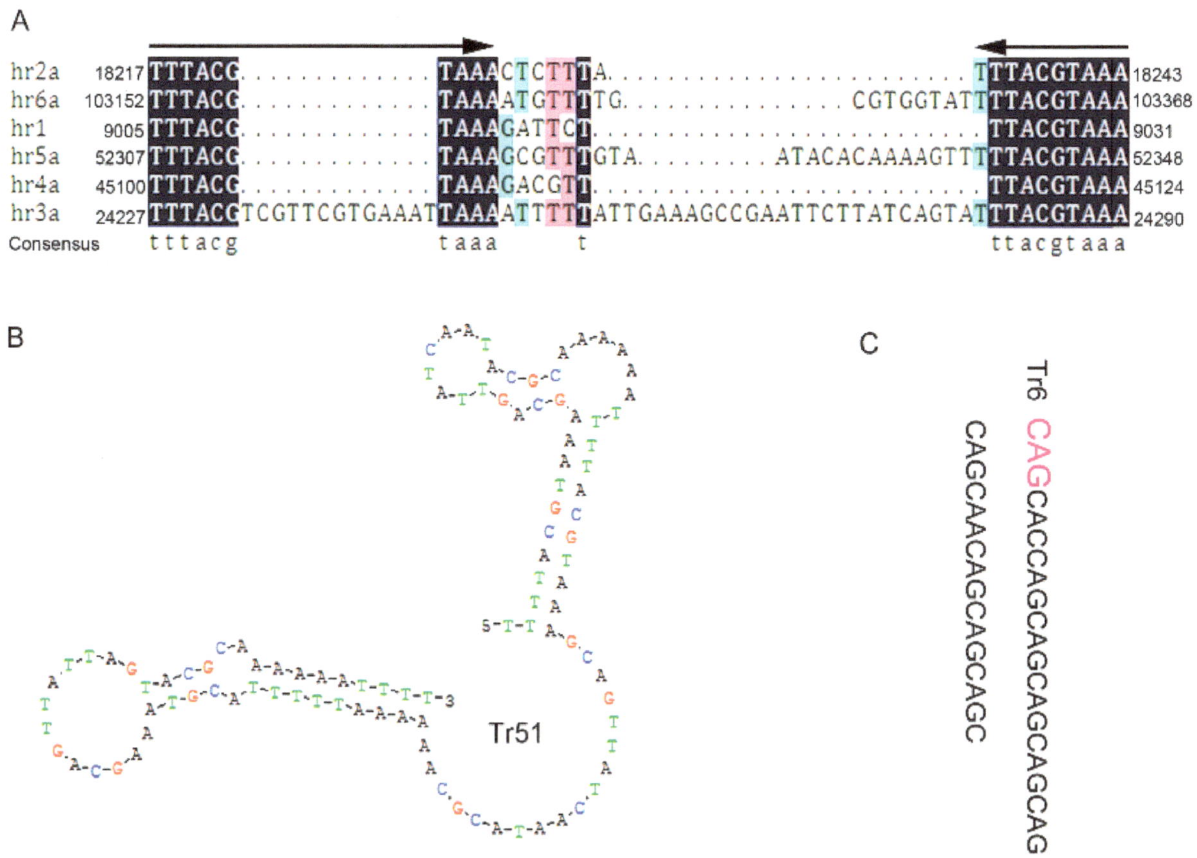

Figure 2. TRs and *hrs* analysis. (A) Alignment of the part of *hrs* in the CnmeGV genome. Palindromes within the repeats are indicated by arrows on the alignment. (B) Predicted secondary structure of the Tr51. (C) The repetitive sequence in coding regions of Tr6, CAG is repeating unit.

In the baculovirus genomes sequenced so far, it is common to find 1 to 16 homologous regions (*hrs*) present. [40]. Generally, the *hrs* in baculoviruses are the intergenic repeats that play putative or demonstrated roles as enhancers of transcription and origins of replication [41]. Twenty seven (27) (within 6 *hrs*) imperfect palindromes were identified in CnmeGV genome. However, only 4 *hrs* were identified in ClanGV genome sequence [42]. The alignment of these sequences revealed a typical structure of palindrome. The alignment of these shorter palindromes shows that they have a 10 bp conserved inverted repeats (Fig 2A). Similarly, these palindrome structures of *hrs* were found in numerous GVs (EpapGV, CrleGV, AdorGV, and ChocGV) [43, 44, 15].

The largest intergenic region, which contains imperfect palindromes, was found between ORF26 and ORF27 in CnmeGV. It contained 757 bp in size and a high content of AT sequence (about 73%). Eleven (11) imperfect palindromes were identified in this region, which was assigned in *hr2*, but it revealed no significant homology in tBlastx searches.

ORF21, with Double Chitin-Binding Domains

Chitin is an important component of the insect cuticle and the peritrophic matrix (PM) lining the gut epithelium. The *chitinase* gene of baculovirus is usually expressed in the late phase of virus replication in insects that can hydrolyze chitin in the body of the insect that promotes terminal host liquefaction [45]. The CnmeGV ORF21 encoding a predicted protein of 173 aa with a size of 19.85 kDa is homologous to *chit-1* gene. A baculovirus consensus late promoter motif TTAAG was found at 8 nt upstream of the start codon ATG, indicating that ORF21 may express in the later stages of the infection cycle. ORF21 protein contains a trans-membrane helix region as detected by the TMHAMM V2.0 program (Fig 3B). The region started at position 7 aa and ended at position 24 aa. Moreover, two special domains, CBM_14A and CBM_14B, belonging to the carbohydrate-binding module family 14 (CBM_14) were found by the SMART to be located at the sites of 40–95 aa and 99–154 aa of the protein (Fig 3A).

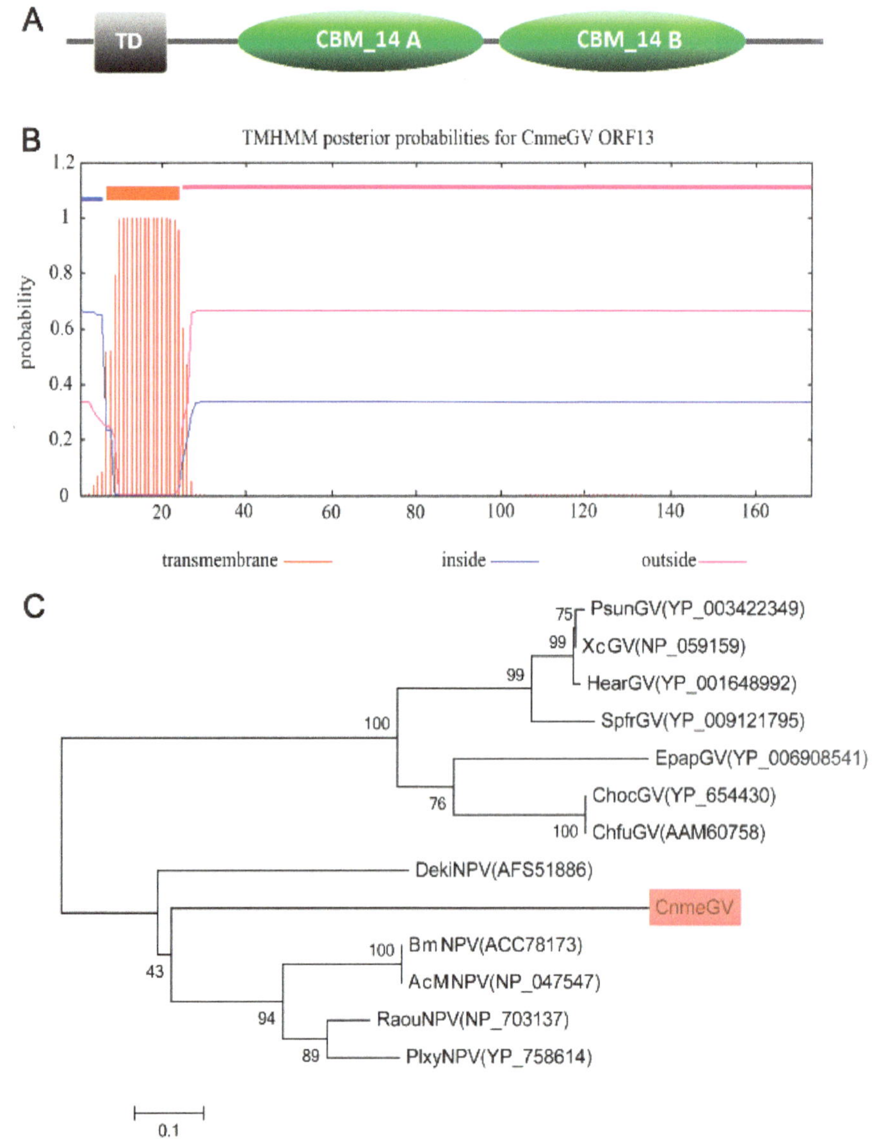

Figure 3. Structural domain and phylogenetic analysis of ORF21. (A) The structural organization of ORF21 protein. It contains a transmembrane helix domain (TD) and two carbohydrate-binding modules family 14 (CBM_14A and CBM_14B). (B) Transmembrane helix region was detected by the TMHAMM V2.0 program. (C)The NJ tree was inferred using the conserved amino-terminal region alignment of ORF21 gene for 13 baculoviruses. The postulated horizontal gene transfer (HGT) events are highlighted for CnmeGV.

The family of CBM_14 was known as the peritrophin-A domain found in chitin binding proteins, particularly, the PM proteins of insects and animal chitinases [46–48]. Homologous genes were also found in some other betabaculoviruses, but these genes only contained one chit-binding region (Fig 4). All the chitin-binding domains were characterized by processing a six-cysteine-containing motif: C-x(13,20)-C-x(5,6)-C-x(9,19)-C-x(10,14)-C-x(4,14)-C [49]. Comparing the homologous genes among sequenced betabaculoviruses using the BlastP and subsequently aligning by the ClustalX program, we found high similarity shared among HearGV, XcGV, SpfrGV, ChfuGV, ChocGV, EpapGV, except CnmeGV and PsunGV (Fig 4). Chitin-binding proteins encoded by baculoviruses might be involved in the virus-host interactions during the infection cycle [50]. The protein GP37 binding to the chitin of *Spodoptera litura* PM was found to facilitate virus infection by targeting the chitin component of PM [51]. The double chitin-binding domains of CnmeGV OFR21 might be more beneficial to virus infection and host liquefaction.

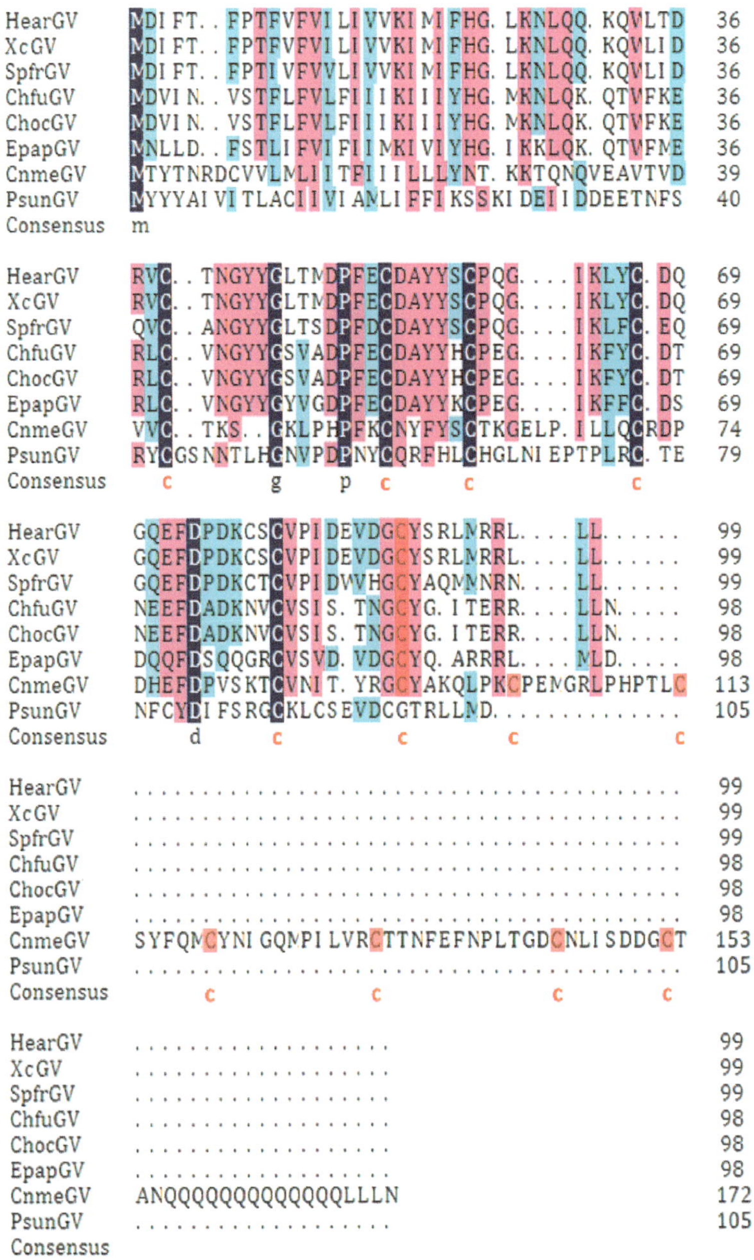

Figure 4. Multiple sequence alignment of proteins with the chit-binding domain. This domain contains one or two six-cysteine-containing motif that are indicated with red font. The alignment was generated by using ClustalX and edited using DNAMAN software. The sequence used were: HearGV gp010 (YP_001648992); XcGV ORF11 (NP_059159); SpfrGVchit-1 (YP_009121795); ChfuGV ORF9 (AAM60758); ChocGV gp009 (YP_654430); EpapGV chit binding protein (YP_006908541); PsunGV gp010 (YP_003422349).

Homologous proteins of ORF21 were also found in 5 alphabaculoviruses. There were DekiNPV, BmNPV, AcMNPV, RaouNPV, and PlxyNPV. A phylogenetic tree was reconstructed based on these conserved domains (Fig 3C). The virus samples were divided into the two major groups of GV and NPV. The *Betabaculovirus* genus of CnmeGV was classified into the NPV group rather than the group of GV, which suggested possible horizontal gene transfer (HGT) occurring in ORF21 of CnmeGV. Two hypotheses are proposed for ORF21 introduction in CnmeGV: 1, the CnmeGV acquired the ORF21 gene from NPV during co-infection of *C. medinalis*; and 2, the CnmeGV acquired from the host itself and the viruses might acted as vectors of HGT between insects or animals [52]. Although the transposable elements (TEs) had not been detected in the ORF21 protein, a poly-glutamine residue of trinucleotide repeat ($[CAA]_{13}$) at 156–168 aa was found. This polyglutamine-containing protein appeared to be over-represented in spliceosome components [53].

Relationships with Other Baculoviruses

Gene colinearity was analyzed by comparing CnmeGV to all the other sequenced GVs and type species of the *Alphabaculovirus* genus, AcMNPV, using Artemis Comparison Tool. Syntenic maps of CnmeGV and other baculovirus genomes were constructed through tBlastX comparison between genomes with blue stripes indicating inversions and colour intensity to reveal the different percentages of identities [20]. The conserved gene colinearity of all 17 GV genomes and the poorly conserved synteny between GVs and AcMNPV are listed in Fig 5. It was apparent that the synteny maps were conserved among betabaculovirus species differing from that of alphabaculoviruses with greater gene order correlation among the CnmeGV and other GVs and some inversions and drifts. Nevertheless, CnmeGV is different from the rest of the GVs by two main gene block inversions about 19.8 kb and 16.8 kb. Inversion of large portions of the genomes was observed in a region between nt 17931–37024 and 81922–98743, that contained eight major ORFs: *p74*, *ubi*, *p106*, *odv-e66*, *alk-exo*, *dna ligase*, *lef-9* and *dnapol*. Most of the differences in the organization among GVs genomes could be explained by insertions and deletions that contributed to the plasticity of the viral population [54, 55].

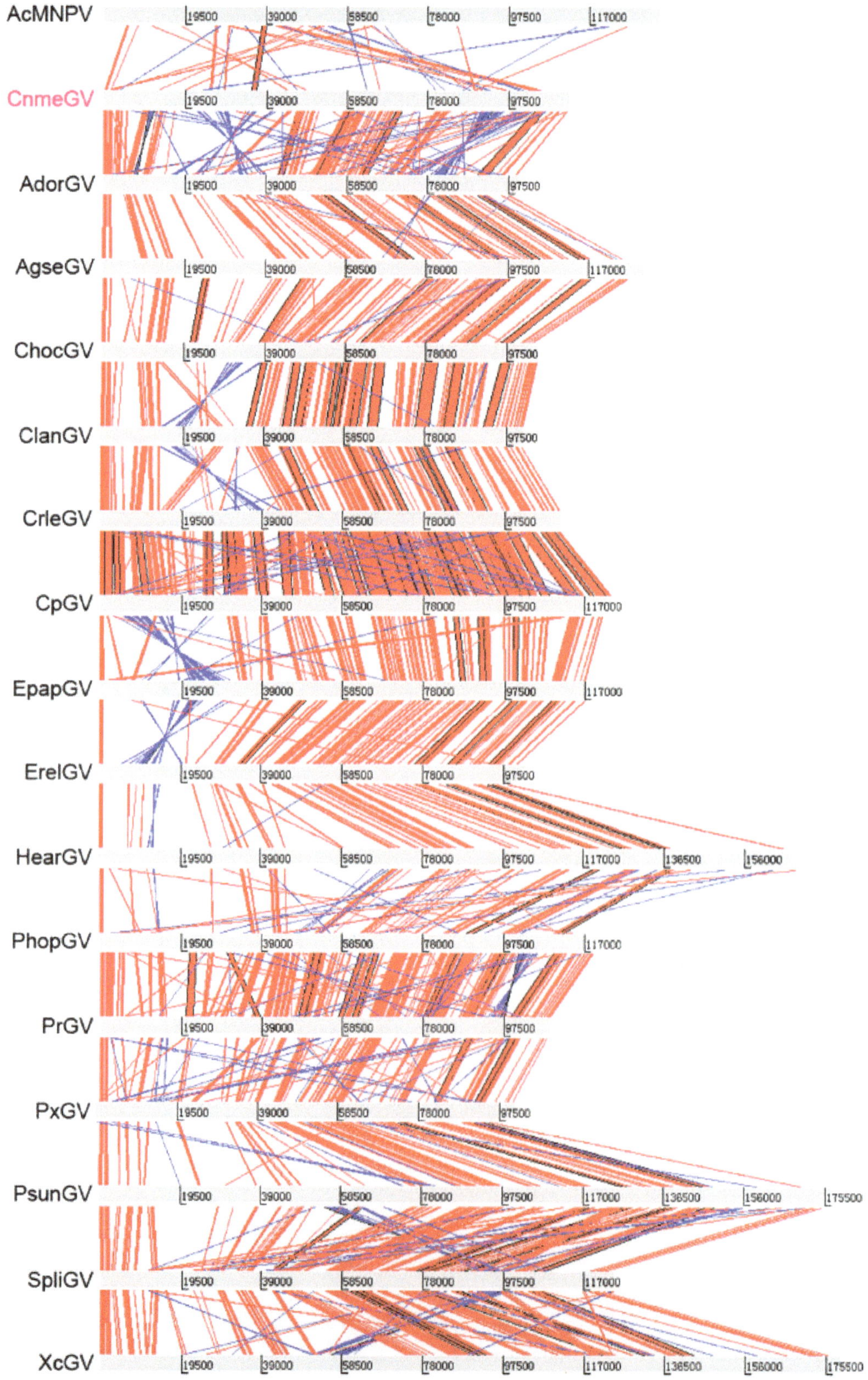

Figure 5. Syntenic map of CnmeGV and other baculovirus genomes. The illustration shows the comparison of gene colinearity based on genome physical positions and protein similarities among 17 GVs and AcMNPV. Each genome is represented by a grey line where nucleotide positions are indicated (kb). Red stripes connecting the genomes indicate syntenic regions in the same strand, whereas blue stripes indicate syntenic regions in opposite strands (inversions). Color intensity is proportional to %identity (darker is more conserved).

The neighbor-joining (NJ) and unweighted pair-group method with arithmetic means (UPMGA) trees were generated using the concatenated amino acid sequences of the partial*polh/gran*, *lef-8* and *lef-9* from 68 baculovirus genomes. The UPMGA tree revealed higher bootstrap values (Fig 6). The obtained cladogram reproduced the grouping of four genera reflecting the current systematic assignment of the virus family [56]. As expected, CnmeGV was grouped in the *Betabaculovirus* genus. A close relationship between CnmeGV with AdorGV was supported by high bootstrap value. In previous reports betabaculoviruswas mainly divided in two well separated monophyletic clades, Clade "a" and Clade "b". GVs of Clade "a" were isolated mainly from noctuidae hosts, while those of Clade "b" isolated from other hosts [57]. But the cladogram obtained in this work based on 17 complete GV genomes did not support the division of betabaculovirus in two separated monophyletic clades. The same result was also shown by other authors who constructed the tree using all core genes or *polh/gran*, *lef-8* and *lef-9* genes [58, 17]. Additionally, compared to the evolution and phylogenetic utility in lepidoptera, there are no direct correlation between the classification of insect and host's virus [59].

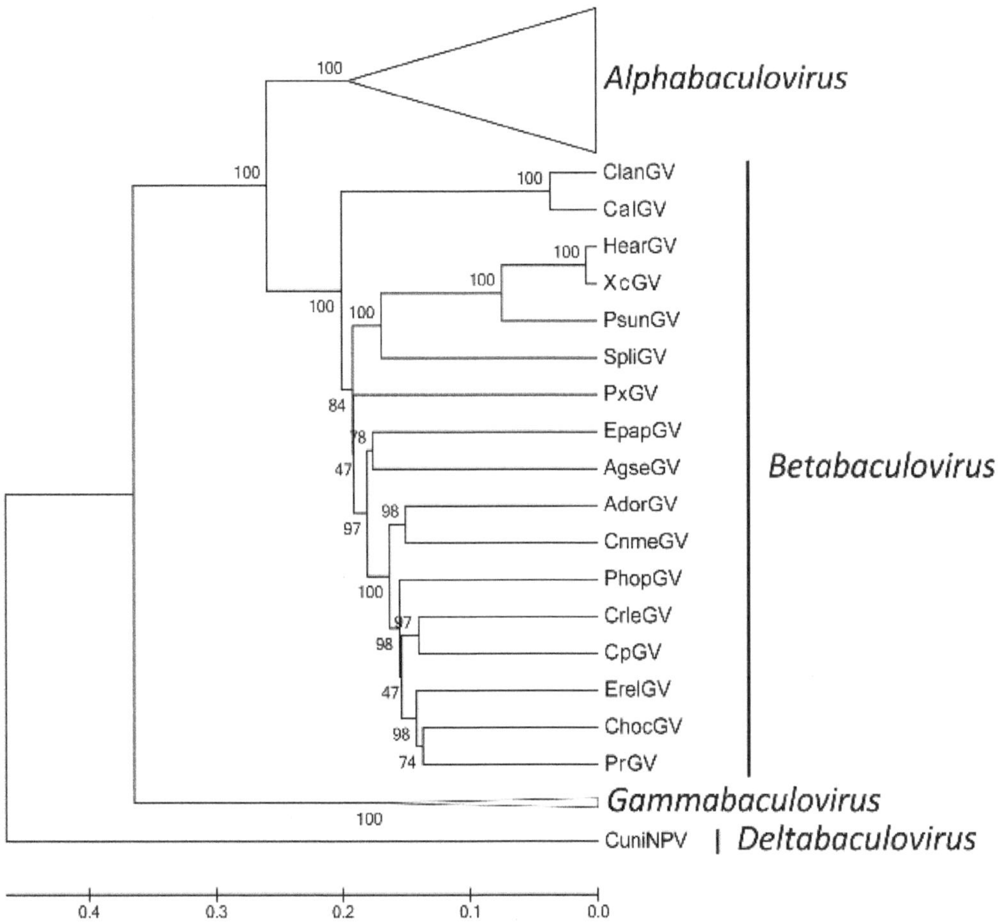

Figure 6. UPMGA tree for all baculovirus. Cladogram based on amino acid sequences of the partial *polh/gran*, *lef-8* and *lef-9*genes in all complete baculovirus genome sequences. We collapsed all the Gammabaculovirus and Alphabaculovirus. The phylogenetic tree was inferred using MEGA 5.1 program.

CONCLUSION

In this study, the first crambidae host-isolated betabaculovirus CnmeGV was sequenced and characterized. Its genome encodes 133 putative ORFs including 37 core genes from baculoviurs. In addition, it contained 32 unique genes that were not shared with the rest of the family with unknown functions. The unique ORF28 protein contained a specialized zinc finger-like domain and a coiled-coil region hypothesized to involve special functions. Fifty one (51) TRs and 6 *hrs* were identified interspersed throughout the genome. ORF21 presented two peritrophin-A domains of CBM_14 that were beneficial to virus infection and host liquefaction. There was also evidence of HGT events from *Alphabaculovirus* to*Betabaculovirus*. Phylogenetic analysis revealed that the CnmeGV is a new*Betabaculovirus* species closely related to AdorGV. The cladogram obtained in this work grouped the 17 complete GV genomes into one monophyletic clade.

MATERIALS AND METHODS

Virus and Viral DNA Separation

This study was carried out on private land (E: 119.388888, N: 32.479142), the owner permitted us to conduct the study on this site. CnmeGV was isolated from a larva of the rice leaffolder *C. medinalis* collected in 2008 and stored in the lab. The virus was not an endangered or protected species. It was multiplied in the laboratory by feeding the second instar larvae with CnmeGV OBs. Infected larvae were homogenized with ddH_2O, filtered through four layers of gauze, and centrifuged at 7727 x g for 10min. The pellet was suspended in 0.5% (w/v)

SDS and centrifugation steps were repeated 5 times until the liquid became clear. Then, the pellet was suspended in 40–65% sucrose gradient and centrifuged at 400 x g for 10 min to remove the debris of larvae tissue. Finally, the OBs were collected in ddH_2O. Viral DNA was extracted according to Wang et al [60]. Its integrity and identity was analyzed by Nano Drop and Agilent 2100 Bioanalyzer.

DNA Sequencing and Analysis

The CnmeGV genomic DNA was sequenced with PacBio RS II at Nextomics inWuhan of China and assembled *de novo* using HGAP2.2.0 [61, 62]. The annotation was performed using RAST [63] to identify the ORF that started with a methionine codon (ATG). The criterion for defining an ORF was the size of at least 150 nt (50 aa) with minimal overlap. Homology searches were done using BlastP in database of NCBI. The complete genome was compared with other betabaculovirus genomes using the Artemis Comparative Tool (ACT) [64] (http://www.sanger.ac.uk/resources/software/act/) and the tBlastX program. The Tandem Repeats Finder [35] (http://tandem.bu.edu/trf/trf.html) was used to locate and analyze tandem repeats. The REPuter program [65] (http://bibiserv.techfak.uni-bielefeld.de/reputer) was applied to analyze homologous repeat regions. The secondary DNA structure and alignment of these sequences were predicted with the DNAMAN 8 and ClustalX programs [66].

he phylogenetic analysis was based on amino acid sequences of 3 core genes (*polh/gran,lef-8*, *lef-9*) from CnmeGV and the other 67 baculoviruses listed in NCBI genome database. NJ and UPGMA phylogenetic trees (1000 bootstrap replicates) were inferred from the amino acid sequence alignments by using MEGA, version 5.1.

ACKNOWLEDGMENTS

We thank an anonymous peer reviewer for constructive comments on drafts of this paper. We also thank the 5 services from the Jiangsu Lixiahe Institute of Agricultural Sciences.

AUTHOR CONTRIBUTIONS

Conceived and designed the experiments: GJH JX ZXL. Performed the experiments: QL CML GJH. Analyzed the data: GJH JX HXX. Contributed reagents/materials/analysis tools: QL CML HXX ZXL. Wrote the paper: JX GJH ZXL.

REFERENCES

1. Kawazu K, Setokuchi O, Kohno K, Takahashi K, Yoshiyasu Y, Tatsuki S. Sex pheromone of the rice leaffolder moth, Cnaphalocrocis medinalis (Lepidoptera: Crambidae): Synthetic Indian and Philippine blends are not attractive to male C. medinalis, but are attractive to C. pilosa in the South-Western islands in Japan. Appl Entomol Zool. 2001; 36(4):471–474. doi: 10. 1303/aez. 2001. 471
2. Inoue H, Kamiwada H, Fukamachi S. Seasonal changes in adult density and female mating status of the rice leaf roller Cnaphalocrocis medinalis (Lepidoptera: Pyralidae) in paddy fields of Southern Kyushu, Japan. Appl Entomol Zool. 2004; 48:177–183. doi: 10. 1303/jjaez. 2004. 177
3. Nathan SS, Chung PG, Murugan K. Effect of botanical insecticides and bacterial toxins on the gut enzyme of the rice leaffolder *Cnaphalocrocis medinalis*. Phytoparasitica. 2004; 32:433–443. doi: 10. 1007/bf02980437
4. Zheng X, Ren X, Su J. Insecticide susceptibility of Cnaphalocrocis medinalis (Lepidoptera: Pyralidae) in China. J Econ Entomol. 2011; 104(2):653–658. pmid:21510218 doi: 10. 1603/ec10419
5. Liu Q, Xu J, Wang Y, Li CM, Han GJ, Qi JH, et al. Symergism of CmGV and *Bacillus thuringiensis* against larvae of *Cnaphalocrocis medinalis* Guenee. Journal of Yangzhou University (Agricultural and Life Science Edition). 2013; 34(4):89–93.
6. Vega FE, Kaya HK. Insect pathology. California: Academic Press; 2012. pp. 17–19.
7. Jehle JA, Lange M, Wang HL, Hu ZH, Wang YJ, Hauschild W. Molecular identification and phylogenetic analysis of baculoviruses from lepidoptera. Virology. 2006; 346:180–193. pmid:16313938 doi: 10. 1016/j. virol. 2005. 10. 032
8. Paolo MA, Kessing BD, Maruniak JE. Phylogenetic interrelationships among baculoviruses: evolutionary rates and host associations. J Invertebr Pathol. 1993; 62(2):147–164. pmid:8228320 doi: 10. 1006/jipa. 1993. 1090
9. Federici BA. Baculovirus pathogenesis. New York: Plenum Press; 1997. pp. 33–59.
10. Theilmann DA, Blissard GW, Bonning B, Jehle JA, O'Reilly DR, Rohrmann GF, et al. Virus Taxonomy-Classification and Nomenclature of Viruses. 8th Report of the International Committee on the Taxonomy of Viruses. Fauquet CX, Mayo MA, Maniloff J, Desselberger U, Ball LA editor. Elsevier Academic Press; 2005. pp. 177–185.
11. Winstanley D, Crook N E. Replication of *Cydia pomonella* granulosis virus in cell cultures. J Gen Virol. 1993; 74:1599–1609. pmid:8345351 doi: 10. 1099/0022-1317-74-8-1599
12. Wormleaton S, Kuzio J, Winstanley D. The complete sequence of the *Adoxophyes orana* granulovirus genome. Virology. 2003; 311(2):350–365. pmid:12842624 doi: 10. 1016/s0042-6822(03)00149-1
13. Hayakawa T, Ko R, Okano K, Seong SI, Goto C, Maeda S. Sequence analysis of the*Xestia c-nigrum* granulovirus genome. Virology. 1999; 262(2):277–297. pmid:10502508 doi: 10. 1006/viro. 1999. 9894
14. Possee RD, Rohrmann GF. Baculovirus genome organization and evolution. *The Baculoviruses*. New York: Plenum; 1997. pp. 109–140.
15. Escasa SR, Lauzon HAM, Mathur AC, Krell P J, Arif BM. Sequence analysis of the*Choristoneura occidentalis* granulovirus genome. J Gen Virol. 2006; 87(7):1917–1933. pmid:16760394 doi: 10. 1099/vir. 0. 81792-0
16. Wang Y, Choi JY, Roh JY, Woo SD, Jin BR, Je YH. Molecular and phylogenetic characterization of *Spodoptera litura* granulovirus. J Microbiol. 2008; 46(6):704–708. doi: 10. 1007/s12275-008-0133-z. pmid:19107401

17. Cuartas PE, Barrera GP, Belaich MN, Barreto E, Ghiringhelli PD, Villamizar LF. The complete sequence of the first *Spodoptera frugiperda* betabaculovirus genome: a natural multiple recombinant virus. Viruses. 2015; 7(1):394–421. doi: 10. 3390/v7010394. pmid:25609309
18. Kang W, Suzuki M, Zemskov E, Okana K, Maeda S. Characterization of baculovirus repeated open reading frames (bro) in Bombyx mori nucleopolyhedrovirus. J Virol. 1999; 73(12):10339–10345. pmid:10559352
19. Zemskov EA, Kang WK, Maeda S. Evidence for nucleic acid binding ability and nucleosome association of *Bombyx mori* nucleopolyhedrovirus BRO proteins. J Virol. 2000; 74(15):6784–6789. pmid:10888617 doi: 10. 1128/jvi. 74. 15. 6784-6789. 2000
20. Ferrelli ML, Salvador R, Biedma ME, Berretta MF, Haase S, Sciocco-Cap A, et al. Genome of Epinotia aporema granulovirus (EpapGV), a polyorganotropic fast killing betabaculovirus with a novel thymidylate kinase gene. BMC Genomics. 2012; 13(1):548. doi: 10. 1186/1471-2164-13-548. pmid:23051685
21. Wolff JLC, Valicente FH, Martins R, de Castro Oliveira JV, de Andrade Zanotto PM. Analysis of the genome of Spodoptera frugiperda nucleopolyhedrovirus (SfMNPV-19) and of the high genomic heterogeneity in group II nucleopolyhedroviruses. J Gen Virol. 2008, 89(5):1202–1211. doi: 10. 1099/vir. 0. 83581-0. pmid:18420798
22. Nai YS, Wu CY, Wang TC, Chen YR, Lau WH, Lo CF, et al. Genomic sequencing and analyses of *Lymantria xylina* multiple nucleopolyhedrovirus. BMC Genomics. 2010; 11(1):116. doi: 10. 1186/1471-2164-11-116. pmid:20167051
23. Todd JW, Passarelli AL, Miller LK. Eighteen baculovirus genes, including lef-11, p35, 39K, and p47, support late gene expression. J Virol. 1995; 69(2):968–974. pmid:7815564
24. Liu X, Yin F, Zhu Z, Hou D, Wang J, Zhang L, et al. Genomic sequencing and analysis of Sucra jujuba nucleopolyhedrovirus. PLoS One. 2014; 9(10):e110023. doi: 10. 1371/journal. pone. 0110023. pmid:25329074
25. Lu A, Miller LK. The roles of eighteen baculovirus late expression factor genes in transcription and DNA replication. J Virol. 1995; 69(2):975–982. pmid:7815565
26. Pijlman GP, Pruijssers AJ, Vlak JM. Identification of pif-2, a third conserved baculovirus gene required for per os infection of insects. J Gen Virol. 2003; 84(8):2041–2049. pmid:12867634 doi: 10. 1099/vir. 0. 19133-0
27. Peng K, van Oers MM, Hu ZH, van Lent JWM, Vlak JM. Baculovirus per os infectivity factors form a complex on the surface of occlusion-derived virus. J Virol. 2010; 84:9497–9504. doi: 10. 1128/JVI. 00812-10. pmid:20610731
28. Rohrmann GF. Baculovirus structural proteins. J Gen Virol. 1992; 73:749–761. doi: 10. 1099/0022-1317-73-4-749. pmid:1634870
29. Van OMM, Vlak JM. The baculovirus 10-kDa protein. J Inverteb Pathol. 1997; 70(1):1–17. doi: 10. 1006/jipa. 1997. 4675.
30. Lupas A, Van Dyke M, Stock J. Predicting coiled coils from protein sequences. Science. 1991; 252(5009):1162–1164. pmid:2031185 doi: 10. 1126/science. 252. 5009. 1162
31. Feuerstein R, Wang X, Song D, Cooke NE, Liebhaber SA. The LIM/double zinc-finger motif functions as a protein dimerization domain. Proc Natl Acad Sci U S A. 1994; 91(22):10655–10659. pmid:7938009 doi: 10. 1073/pnas. 91. 22. 10655
32. Araki K, Kawamura M, Suzuki T, Matsuda N, Kanbe D, Ishii K, et al. A palmitoylated RING finger ubiquitin ligase and its homologue in the brain membranes. J Neurochem. 2003; 86(3):749–762. pmid:12859687 doi: 10. 1046/j. 1471-4159. 2003. 01875. x
33. Ehrensberger KM, Corkins ME, Choi S, Bird AJ. The double zinc finger domain and adjacent accessory domain from the transcription factor loss of zinc sensing 1 (loz1) are necessary for DNA binding and zinc sensing. J Biol Chem. 2014; 289(26):18087–18096. doi: 10. 1074/jbc. M114. 551333. pmid:24831008
34. Buard J, Bourdet A, Yardley J, Dubrova Y, Jeffreys AJ. Influences of array size and homogeneity on minisatellite mutation. EMBO J. 1998; 17(12):3495–3502. pmid:9628885 doi: 10. 1093/emboj/17. 12. 3495
35. Benson G. Tandem repeats finder: a program to analyze DNA sequences. Nucleic Acids Res. 1999; 27(2):573. pmid:9862982 doi: 10. 1093/nar/27. 2. 573
36. Gemayel R, Vinces MD, Legendre M, Verstrepen KJ. Variable tandem repeats accelerate evolution of coding and regulatory sequences. Annu Rev Genet. 2010; 44:445–477. doi: 10. 1146/annurev-genet-072610-155046. pmid:20809801
37. Richard GF, Kerrest A, Dujon B. Comparative genomics and molecular dynamics of DNA repeats in eukaryotes. Microbiol Mol Biol Rev. 2008; 72(4):686–727. doi: 10. 1128/MMBR. 00011-08. pmid:19052325
38. Wierdl M, Greene CN, Datta A, Jinks-Robertson S, Petes TD. Destabilization of simple repetitive DNA sequences by transcription in yeast. Genetics. 1996; 143(2):713–721. pmid:8725221
39. Mittelman D, Sykoudis K, Hersh M, Lin Y, Wilson JH. *Hsp90* modulates CAG repeat instability in human cells. Cell Stress Chaperon. 2010; 15(5), 753–759. doi: 10. 1007/s12192-010-0191-0. pmid:20373063
40. Thumbi DK, Eveleigh RJM, Lucarotti CJ, Lapointe R, Graham RI, Pavlik L, et al. Complete sequence, analysis and organization of the *Orgyia leucostigma*nucleopolyhedrovirus genome. Viruses. 2011; 3(11):2301–2327. doi: 10. 3390/v3112301. pmid:22163346
41. Hilton S, Winstanley D. Identification and functional analysis of the origins of DNA replication in the *Cydia pomonella* granulovirus genome. J Gen Virol. 2007; 88(5):496–1504. pmid:17412979 doi: 10. 1099/vir. 0. 82760-0
42. Liang Z, Zhang X, Yin X, Cao S, Xu F. Genomic sequencing and analysis of *Clostera anachoreta* granulovirus. Arch Virol. 2011; 156(7):1185–1198. doi: 10. 1007/s00705-011-0977-0. pmid:21442228

43. Lange M, Jehle JA. The genome of the *Cryptophlebia leucotreta* granulovirus. Virology. 2003; 317(2):220–236. pmid:14698662 doi: 10. 1016/s0042-6822(03)00515-4

44. Garcia-Maruniak A, Maruniak JE, Zanotto PM, Doumbouya AE, Liu JC, Merritt TM, et al. Sequence analysis of the genome of the *Neodiprion sertifer* nucleopolyhedrovirus. J Virol. 2004; 78(13):7036–7051. pmid:15194780 doi: 10. 1128/jvi. 78. 13. 7036-7051. 2004

45. Oh S, Kim DH, Patnaik BB, Jo YH, Noh MY, Lee HJ, et al. Molecular and immunohistochemical characterization of the *chitinase* gene from *Pieris rapae* granulovirus. Arch Virol. 2013; 158(8):1701–1718. doi: 10. 1007/ s00705-013-1649-z. pmid:23512574

46. Shen Z, Jacobs-Lorena M. A type I peritrophic matrix protein from the malaria vector *Anopheles gambiae* binds to chitin. Cloning, expression, and characterization. J Biol Chem. 1998; 273(28):17665–17670. pmid:9651363 doi: 10. 1074/jbc. 273. 28. 17665

47. Elvin CM, Vuocolo T, Pearson RD, East IJ, Riding GA, Eisemann CH, et al. Characterization of a major peritrophic membrane protein, peritrophin-44, from the larvae of *Lucilia cuprina* cDNA and deduced amino acid sequences. J Biol Chem. 1996; 271(15):8925–8935. pmid:8621536 doi: 10. 1074/jbc. 271. 15. 8925

48. Casu R, Eisemann C, Pearson R, Riding G, East I, Donaldson A, et al. Antibody-mediated inhibition of the growth of larvae from an insect causing cutaneous myiasis in a mammalian host. Proc Natl Acad Sci U S A. 1997; 94(17):8939–8944. pmid:9256413 doi: 10. 1073/ pnas. 94. 17. 8939

49. Gaines PJ, Walmsley SJ, Wisnewski N. Cloning and characterization of five cDNAs encoding peritrophin-A domains from the cat flea, *Ctenocephalides felis*. Insect Biochem Mol Biol. 2003; 33:1061–1073. pmid:14563358 doi: 10. 1016/s0965-1748(03)00096-1

50. Wang D, Zhang CX. HearSNPV orf83 encodes a late, nonstructural protein with an active chitin-binding domain. Virus Res. 2006; 117(2):237–243. pmid:16313991 doi: 10. 1016/j. virusres. 2005. 10. 019

51. Li Z, Li C, Yang K, Wang L, Yin C, Gong Y, Pang Y: Characterization of a chitin-binding protein GP37 of *Spodoptera litura* multicapsid nucleopolyhedrovirus. *Virus Res* 2003, 96(1):113–122. doi: 10. 1016/s0168-1702(03)00179-5

52. Gilbert C, Chateigner A, Ernenwein L, Barbe V, Bézier A, Herniou EA, et al. Population genomics supports baculoviruses as vectors of horizontal transfer of insect transposons. Nat Commun. 2014; 5:3348. doi: 10. 1038/ ncomms4348. pmid:24556639

53. Eichinger L, Pachebat JA, Glockner G, Rajandream MA, Sucgang R, Berriman M, et al. The genome of the social amoeba *Dictyostelium discoideum*. Nature. 2005; 435(7038):43–57. pmid:15875012 doi: 10. 1038/ nature03481

54. de Castro Oliveira JV, Wolff JLC, Garcia-Maruniak A, Ribeiro BM, de Castro MEB, de Souza ML, et al. Genome of the most widely used viral biopesticide: *Anticarsia gemmatalis* multiple nucleopolyhedrovirus. J Gen Virol. 2006; 87(11):3233–3250. pmid:17030857 doi: 10. 1099/vir. 0. 82161-0

55. Li Q, Donly C, Li L, Willis LG, Theilmann DA, Erlandson M. Sequence and organization of the *Mamestra configurata* nucleopolyhedrovirus genome. Virology. 2002; 294(1):106–121. pmid:11886270 doi: 10. 1006/viro. 2001. 1313

56. Herniou EA, Arif BM, Becnel JJ, Blissard GW, Bonning B, Harrison R, et al. Baculoviridae. In Virus taxonomy: classification and nomenclature of viruses: Ninth Report of the International Committee on Taxonomy of Viruses. Edited by King AMQ, Adams MJ, Carstens EB, Lefkowitz EJ. Elsevier Academic Press; 2011. pp. 163–173.

57. Miele SAB, Garavaglia MJ, Belaich MN, Ghiringhelli PD. Baculovirus: molecular insights on their diversity and conservation. Int J Evol Biol. 2011; 2011:379424. doi: 10. 4061/2011/379424. pmid:21716740

58. Ardisson-Araújo DM, de Melo FL, de Andrade M, Sihler W, Báo SN, Ribeiro BM, et al. Genome sequence of Erinnyis ello granulovirus (ErelGV), a natural cassava hornworm pesticide and the first sequenced sphingid-infecting betabaculovirus. BMC Genomics. 2014; 15(1):856. doi: 10. 1186/1471-2164-15-856. pmid:25280947

59. Regier JC, Fang QQ, Mitter C, Peigler RS, Friedlander TP, Solis MA. Evolution and phylogenetic utility of the period gene in Lepidoptera. Mol Bio Evol. 1998; 15(9):1172–1182. pmid:9729881 doi: 10. 1093/ oxfordjournals. molbev. a026024

60. Wang Y, Choi JY, Roh JY, Liu Q, Tao XY, Park JB, et al. Genomic sequence analysis of granulovirus isolated from the tobacco cutworm, Spodoptera litura. PLoS One. 2011; 6(11):e28163. doi: 10. 1371/journal. pone. 0028163. pmid:22132235

61. Roberts RJ, Carneiro MO, Schatz MC. The advantages of SMRT sequencing. Genome Biol. 2013; 14(7):405. pmid:23822731 doi: 10. 1186/gb-2013-14-6-405

62. Chin C-S, Alexander DH, Marks P, Klammer AA, Drake J, Heiner C, et al. Nonhybrid, finished microbial genome assemblies from long-read SMRT sequencing data. Nat Methods. 2013;10(6):563–569. doi: 10. 1038/nmeth. 2474. pmid:23644548

63. Aziz RK, Bartels D, Best AA, Dejongh M, Disz T, Edwards RA, et al. The RAST Server: rapid annotations using subsystems technology. BMC Genomics. 2008; 9(1):75. doi: 10. 1186/1471-2164-9-75. pmid:18261238

64. Carver T J, Rutherford K M, Berriman M, Rajandream MA, Barrell BG, Parkhill J. ACT: the Artemis comparison tool. Bioinformatics. 2005; 21(16):3422–3423. pmid:15976072 doi: 10. 1093/bioinformatics/ bti553

65. Kurtz S, Choudhuri JV, Ohlebusch E, Schleiermacher C, Stoye J, Giegerich R. REPuter: the manifold applications of repeat analysis on a genomic scale. Nucleic Acids

Res. 2001; 29(22):4633–4642. pmid:11713313 doi: 10. 1093/nar/29. 22. 4633

66. Thompson JD, Gibson TJ, Plewniak F, Jeanmougin F, Higgins DG. The CLUSTAL_X windows interface: flexible strategies for multiple sequence alignment aided by quality analysis tools. Nucleic Acids Res. 1997; 25(24):4876–4882. pmid:9396791 doi: 10. 1093/nar/25. 24. 4876

CHAPTER 19

INVESTIGATION OF POTENTIAL BIOMARKERS FOR THE EARLY DIAGNOSIS OF CELLULAR STABILITY AFTER THE EXPOSURE OF AGRICULTURAL WORKERS TO PESTICIDES

Jodel S. Alves, Fernanda R. Da Silva, Gabrieli F. Da Silva, Mirian Salvador, Kátia Kvitko, Paula Rohr, Carla E. I. Dos Santos, Johnny F. Dias, João A. P. Henriques, Juliana Da Silva

Keywords: Pesticides, Comet assay, micronucleus test, superoxide dismutase activity, PON1 gene

INTRODUCTION

Globally, the use of pesticides has increased in recent years. Among the groups involved in the preparation and the final distribution of pesticide mixtures, farmers and agricultural workers are commonly the most exposed individuals (Bolognesi 2003). Many of these pesticides are classified as "carcinogenic" by the International Agency for Research on Cancer (IARC). Various pesticides interacting with the DNA have been identified to induce miscarriages, degenerative diseases, immune disorders, and cancer (Bolognesi 2003, Au et al. 1999). Although it is difficult to establish a correlation between the exposure to pesticides and the prevalence of cancer, especially due to the large number of compounds involved, some studies suggested a higher prevalence of certain types of cancer in population groups exposed to pesticides. According to Buckley et al. (2000) and Meinert et al. (2000), increased incidences of leukemia and non-Hodgkin lymphomas can be observed for individuals exposed to pesticides. A meta-analysis showed that farmers were at risk of specific tumors including leukemia and multiple myeloma (Bolognesi 2003, Bolognesi et al. 2011). Bull et al. (2006) and Bolognesi et al. (2011) argued that the assessment of cytogenetic effects in exposed subjects might act as an early indicator for an increased risk to develop cancer, although the evidence presented was somewhat contradictory.

Numerous reports on chromosomal aberrations (CA; Au et al. 1999, Zeljesic and Garaj-Vrhovac 2001), sister chromatid exchange (SCE; Shaham et al. 2001, Zeljesic and Garaj-Vrhovac 2002), micronucleous (MN; Falck et al. 1999, Da Silva et al. 2008) and Comet assay (Zeljesic and Garaj-Vrhovac 2002, Da Silva et al. 2012a, Da Silva et al. 2008, Grover et al. 2003) observed significant increases in these biomarkers posterior to exposure to pesticides, providing suggestive evidence for the genotoxic effects induced by these chemicals. However, the conclusions drawn with respect to the genotoxic damage attributed to these pesticides remain conflicting. Some investigations indicated a significant increase in MN, SCE, and CA frequencies, while others did not observe any significant differences (Gauthier et al. 2001, Bolognesi 2003, Bull et al. 2006, Da Silva et al. 2012a, b). The responses reported in these studies were found to depend predominantly upon the type of pesticide, the exposure period, and the use of personal protective equipment.

Genotypes responsible for interindividual differences in the ability to activate or detoxify genotoxic substances are recognized as biomarkers of susceptibility towards mutations, cancer, and other diseases (Zeljesic and Garaj-Vrhovac 2001, Bolognesi 2003, Bolognesi et al. 2011). Paraoxonases (PON) are responsible for the metabolism of organophosphate-based insecticides (Bolognesi 2003), and many enzymatic isoforms, such as e.g. *PON1* were suggested to contribute to the individual cancer susceptibility, as they act as genetic risk modifiers in the development of cancer after the exposure to pesticides (Bolognesi 2003).

Citation: Jodel S. Alves, Fernanda R. Da Silva, Gabrieli F. Da Silva, Mirian Salvador, Kátia Kvitko, Paula Rohr, Carla E. I. Dos Santos, Johnny F. Dias, João A. P. Henriques, Juliana Da Silva. Investigation of potential biomarkers for the early diagnosis of cellular stability after the exposure of agricultural workers to pesticides, Anais da Academia Brasileira de Ciências (2016); 88(1): 349-360. http://dx.doi.org/10.1590/0001-3765201520150181.

In order to evaluate potential biomarkers of exposure to pesticides for the early diagnosis of cellular stability, agricultural workers exposed to pesticides were investigated by the MN frequency test in exfoliated buccal cells, and the Comet assay in peripheral leukocytes. In order to evaluate, if genetically determined individual variations in xenobiotic metabolizing capacity might modify the individual susceptibility to the potential genotoxic effects induced by pesticides, subjects were genotyped for *PON1*. Moreover, SOD activity, hematological parameters, and content levels of inorganic compounds were examined in the blood samples of these workers.

MATERIALS AND METHODS

Study Population and Sample Collection

With an annual production in excess of 687,180 tons, cultivation of tobacco (*Nicotiana tabacum*) generates employment for over 223,000 farmers in Brazil (AFUBRA 2011). The municipality of Santa Cruz do Sul (Rio Grande do Sul, Brazil) is an important production area for tobacco leaves. In order to improve crop characteristics and yield, tobacco farmers usually distribute several pesticides (insecticides, herbicides, fungicides, and plant growth regulators) over tobacco plants by hand pumps.

Between 2007 and 2008 blood and buccal samples were collected from 77 agricultural workers on a tobacco plantation (collected during the peak periods of pesticide application), and 60 unexposed individuals to pesticides from the same region (control group). Individuals who had smoked more than five cigarettes/day for at least one year were considered smokers.

Selected population groups were contacted by community health agents, in order to set an appointment with a general practitioner, where a medical interview was conducted and the individual occupational history was recorded. Clinical data were collected and evaluated by medical personnel. Workers applying pesticides also completed an occupational questionnaire, which included questions related to work activity, duration of the pesticide application, types of pesticides applied, and potentially implemented protective measures. All individuals gave informed consent in writing, and the study was approved by the *Brazilian National Committee on Research Ethics* (*Comissão Nacional de Ética em Pesquisa - CONEP*).

The individuals examined in this study were asked to fill out a Portuguese version of a questionnaire of the*International Commission for Protection against Environmental Mutagens and Carcinogens* (Carrano and Natarajan 1988), before being asked to participate in a face-to-face interview, which included questions related to age, gender, medical issues (they should not have diabetes or cancer or receiving medication treatment), life style (e.g. smoking), and occupation (e.g. working hours per day, exposure times to organic solvents, use of personal protective equipment).

Blood samples were obtained by venopuncture using vacutainers and were processed as quickly as possible, in order to prevent storage-related damage. Blood-cell samples were transported to the laboratory facilities at or below 8 ºC, and processed within 8 h of collection.

Evaluation of Genotoxicity

Micronucleus Test (MN) in Exfoliated Buccal Cells

Buccal cell samples were obtained by gently rubbing the inside of the subjects' cheeks (right and left side) with a cytobrush, which was subsequently immersed in cold, aqueous saline solution (5 mL; 0.9% NaCl w/v) in a conical tube, before being transported under refrigeration to the laboratory. The suspension was centrifuged (1,500 rpm; 8 min), the supernatant decanted, and the deposited sediment of buccal cells was washed twice with saline (5 mL), and once with Carnoy's fixative (methanol and glacial acetic acid; 3:1 v/v) using the same centrifugation conditions. The cell suspension was dropped onto a slide and allowed to dry in air. The slides were stained with 2% Giemsa solution for 10 min, rinsed with distilled water and dried in air. Thus, 2,000 cells were scored per subject and used to determine the MN frequency. The criteria used for MN analysis were those defined by Tolbert et al. (1992) and Titenko-Holland et al. (1998).

For each subject, 2,000 buccal cells (1,000 on each of the duplicate slides) were analyzed by two different scorers using identical microscopes (Axiophot, Carl Zeiss, Oberkochem, Germany), whereby no significant variability was observed between scorers. All slides were coded for blind analysis, and exposed/unexposed groups were analyzed simultaneously.

Comet Assay

The alkaline Comet assay was carried out as described by Singh et al. (1988), whereby the modifications suggested by Tice et al. (2000), Da Silva et al. (2000) and Collins et al. (2008) were implemented. For that purpose, blood cells (5 µL) were embedded in 95 µL of 0.75% low-melting point agarose. After solidification of the agarose, slides were placed for a minimum of 1 h and a maximum of 2 weeks in lysis buffer (2.5 M NaCl, 10 mM EDTA, 10 mM Tris; pH 10.0-10.5), containing 1% (v/v) of freshly added Triton X-100 and 10% (v/v) dimethyl sulfoxide (DMSO). After treatment with lysis buffer, slides were incubated for 20 min in a freshly prepared alkaline buffer (300 mM NaOH, 1 mM EDTA; pH >13), before DNA was submitted to electrophoresis (20 min; 25 V; 0.90 V/cm; 300 mA). Following electrophoresis, slides were neutralized (400 mM Tris; pH 7.5), fixed (15% w/v trichloroacetic acid, 5% w/v zinc sulfate, 5% glycerol), washed in distilled water, and dried overnight at room temperature. Gels were rehydrated for 5 min in distilled water, and subsequently stained (15 min; 37 ºC) with a solution containing 34 mL of solution B (0.2% w/v NH_4NO_3, 0.2% w/v $AgNO_3$, 0.5% w/v tungstosilicic acid, 0.15% v/v formaldehyde, 5% w/v Na_2CO_3) and 66 mL of solution A (5% Na_2CO_3). The staining was stopped with 1% acetic acid and the gels were dried in air (Nadin et al. 2001).

Images of 100 randomly selected cells (50 cells from each of two replicate slides) were analyzed for each test subject using a fluorescence microscope equipped with a 12 nm BP546 excitation filter and a 590 nm barrier filter. Two parameters were evaluated: (1) damage index (DI), in which each cell was categorized according to tail size and shape (Heuser et al. 2007) into one of five classes 0-4 (no damage = 0; maximum damage = 4). DI values thus obtained for each individual ranged from 0 (0×200) to 400 (4×100); and (2) damage frequency (DF), calculated as the percentage of damaged cells. All slides were coded for blind analysis, whereby exposed and unexposed groups were analyzed simultaneously by two different scorers.

Enzymatic Activity of Superoxide Dismutase (SOD)

The activity of superoxide dismutase was determined spectrophotometrically in serum samples by measuring the inhibition of the autocatalytic formation rate of adrenochrome at 480 nm. The reaction medium of these samples was adjusted to a pH value of 10.0 and contained a final concentration of 1 mM/L adrenaline (pH 2.0) and 50 mM/L glycine (pH 10.2). Both reagents were purchased from E. Merck. The reaction was allowed to proceed for 3 min at a constant temperature of 30 °C. The enzymatic activity is expressed in SOD units per g of protein. One unit is defined as the amount of enzyme that inhibits the rate of adrenochrome formation by 50%. Total protein levels were analyzed spectrophotometrically at 545 nm according to the Biureto method (Total Protein Kit-labtest Diagnostica S.A., Brazil).

Hematological Evaluations

The hematological evaluation was carried out in a commercial laboratory (ENZILAB Laboratory, Santa Cruz do Sul-RS, Brazil) according to standard methods.

DNA Extraction and Genotyping

Genomic DNA was isolated from whole blood samples (collected using vacutainers with EDTA) by the salting out method (Lahiri and Nurnberger 1991). The genotyping of five polymorphic markers was accomplished by using the polymerase chain reaction-restriction fragment length polymorphism (PCR-RFLP) method.

The polymorphism of *PON1* in amino acid 192 (*Gln192Arg*) was examined by PCR/RFLP according to Humbert et al. (1993). For that purpose, an aliquot of the PCR product was digested with *Alw I*, and the genotypes were resolved in agarose gel. The *Arg* allele (G or B type) is present in the high activity allozyme, and the *Gln* allele (A type) is present in the low activity allozyme at position 192 of the protein (*Gln* = 99 bp fragment; *Arg* = 69 bp and 30 bp fragments).

Chemical Analysis

The content of inorganic elements in the blood samples was analyzed by particle-induced X-ray emission (PIXE) (Johansson et al. 1995). Blood samples were briefly dried (t = 6 h; T = 60 °C), ground into a fine powder, homogenized, and pressed into approximately 3 mm thick pellets, before being placed in the target holder inside the reaction chamber. During the experiments, the pressure inside the reaction chamber was about 10^{-5} mbar. The experiments were carried out at the Ion Implantation Laboratory of the Physics Institute of the Federal University of Rio Grande do Sul (IF-UFRGS). A 3 MV Tandetron accelerator provided a 2.0 MeV proton beam with an average current of 5 nA at the target. The X-rays produced in the samples were detected by a Si(Li) detector with an energy resolution of ca. 160 eV at 5.9 keV. The spectra were analyzed with the GUPIX software package and the data are expressed in parts per million (ppm) (Campbell et al. 2000).

Statistical Analysis

The normality of the variables was evaluated using the Kolmogorov-Smirnov test. To compare the parameters of the study population, c^2Student *t*-, and Mann-Whitney U non-parametric tests were used. The critical level for rejection of the null hypothesis was considered to be at $P < 0.05$.

RESULTS

The subjects were classified by gender and age, and divided into smokers and non-smokers. The unexposed group consisted of 60 subjects (38 males and 22 females; mean age: 45.34 ± 13.19 years), whereby 15 subjects were smokers; the exposed group consisted of 77 subjects (49 males and 28 females; mean age: 44.27 ± 12.44 years), whereby 20 subjects were smokers. The mean age did not differ significantly between groups, and about 30% of the male agricultural workers in each group were smokers (no female smokers).

In the exposed group, all agricultural workers were regularly exposed (ca. 2-3 times per week) to complex mixtures of pesticides, the composition of which depends on weather conditions (~600 h/year). The list of pesticides used is shown in Table I. Agricultural activities were conducted by the subjects mainly in open fields, and the main crop was tobacco. As a rule, pesticides were applied onto the tobacco plants at a level above the workers› heads. All pesticide-exposed workers (100%) confirmed that they had not used any kind of personal protection equipment (e.g. gloves, breathing masks, protective goggles, impermeable boots) during the preparation or application of the pesticides. During the exposure period, all individuals presented symptoms related to pesticide poisoning and green tobacco sickness (GTS), such as headaches, abdominal pain, nausea, and vomiting.

Table I. List of pesticides used by exposed group in tobacco cultivation.

Pesticides	**Compounds**	**Chemical Class**	**Hazard Classifica-tiona**
Fungicides	**Mancozeb**	**Dithiocarbamate**	**Moderately haz-ardous**
Propinebe	Carbamate	Moderately hazardous	
Iprodione	Dicarboximide	Slightly hazardous	

Insecticides	*O,S*- Dimethyl acethylphosphoramidothiate	Organophosphate	Moderately hazardous
Chlorpyrifos	Organophosphate	Highly hazardous	
Imidachlopride	Neonicotinoid	Moderately hazardous	
Carbofuran	Carbamate	Extremely hazardous	
Magnesium aluminum phosphide	Inorganic	Highly hazardous	
Deltamethrin	Pyrethoid	Moderately hazardous	
Herbicides	Glyphosate	Organophosphate	Slightly hazardous
Sethoxydim	Organophosphate	Moderately hazardous	
Clomazone	Isoxazolidinone	Moderately hazardous	
Pendimethalin	Dinitroaniline	Moderately hazardous	

[a]Hazard classification from National Agency of Sanitary Surveillance (ANVISA: Agência Nacional de Vigilância Sanitária).

The results of the cytogenetic parameter analysis are presented in Table II. No significant differences were found between males and females in the unexposed or exposed groups, both in the Comet assay [unexposed: DI = 9.72 ± 7.50 (males: 9.34 ± 7.79; females: 10.36 ± 7.08); exposed: DI = 28.01 ± 21.43 (males: 30.37 ± 21.83; females: 23.74 ± 20.39)] and MN test [unexposed: MN = 1.33 ± 1.86 (males: 1.42 ± 1.72; females: 1.17 ± 2.09); exposed: MN = 7.14 ± 6.49 (males: 6.3 ± 5.4; females: 8.6 ± 7.9)]. Therefore, samples were combined and divided into exposed and unexposed workers. The analysis of the Comet assay exhibited a significant increase (ca. threefold) in DI and DF in the exposed group compared to the unexposed group (Table II). The exposed group also presented a ca. fivefold increased number of MN in exfoliated buccal cells relative to the control group. No significant difference was observed between smokers and non-smokers.

Table II. Mean values (mean ± standard deviation) obtained in the cytogenetic analysis in unexposed and exposed groups.

Group (n)	Comet assay (100 leukocytes/subject)		MN in Exfoliated buccal cells (2000 cell/subject)
	Damage Index	Damage Frequency	
Unexposed (60)	**9.72 ± 7.50**	**6.75 ± 4.73**	**1.33 ± 1.86**
Smokers (15)	11.45 ± 10.01	6.82 ± 5.40	1.31 ± 2.50
Non-smokers (45)	9.33 ± 6.88	6.73 ± 4.63	1.32 ± 1.65
Exposed (77)	28.01 ± 21.43*	19.54 ± 13.03*	7.14 ± 6.49*
Smokers (20)	32.24 ± 21.24*	22.76 ± 13.05*	5.22 ± 5.64*
Non-smokers (57)	24.36 ± 20.12*	17.29 ± 12.47*	7.73 ± 6.66*

*Significant difference as compared to the respective control group (unexposed) at P < 0.001 (Student *t*-test).

Figure 1 shows the effect of the *PON1* (*Gln/Gln*, or *Gln/Arg* or *Arg/Arg*) genotype on the level of different biomarkers evaluated in the exposed individuals. No deviations from Hardy-Weinberg expectations were detected, either relative to the polymorphism or the sample group. Similar genotype and allele distributions (*PON1 192Arg/-* = 36.6%) were observed in the control and the exposed group. A significant increase in MN frequencies was observed for *PON1 Gln/Gln* individuals in the exposed group compared to *PON1 Arg/-* individuals (P < 0.01; Student *t*-test).

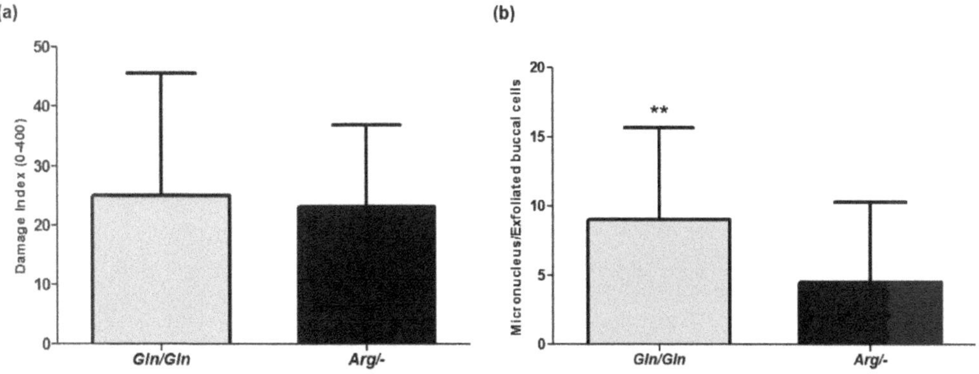

Figure 1. Effect of individual genotype (*PON1*) on the level of different biomarkers evaluated in exposed group (mean ± S.D.): (a) Comet assay and (b) Micronucleus test. **Significant at P<0.01 in relation to *PON1 Arg/-* genotype; *t*-Student Test.

The SOD activity is shown in Figure 2. After exposure to pesticides, a drastic increase in SOD activity was observed for the exposed group relative to the unexposed control group (P < 0.001; Student *t*-test).

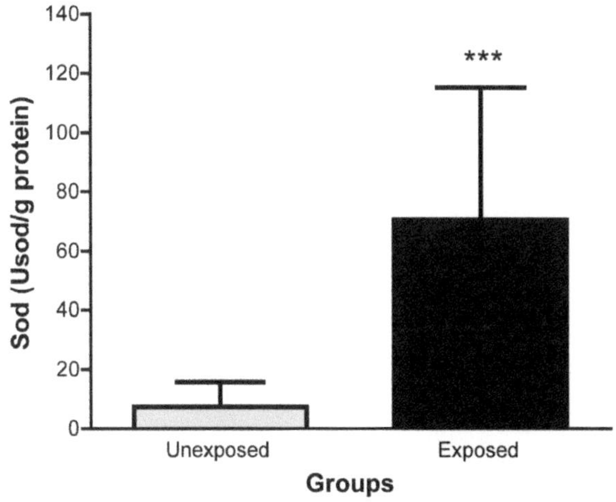

Figure 2. Superoxide dismutase activity in the unexposed and exposed groups. ***Significant at P<0.001 in relation to unexposed group; Student *t*-Test.

Mean values for the hematological parameters of unexposed and exposed groups, and reference values are presented in Table III. In this analysis, the values obtained were within the normal limits. In the first sample (unexposed group), the number of eosinophils of two females and one male was higher than the reference values. With the exception of one subject, all volunteers presented normal eosinophil counts in the second sample (exposed group). Nevertheless, significant differences were observed with respect to band neutrophils and monocyte counts between the exposed and the unexposed group (P < 0.05; Student *t*-test).

Table III. Mean values of hematological parameters (± standard deviation) analyzes of unexposed and exposed groups to pesticides.

Parameters	Unexposed group	Exposed group	Reference values
Hematocrit	44.17 ± 3.50	42.71 ± 3.95	39.00-53.00 g/dl
Hemoglobin	14.22 ± 1.18	14.51 ± 1.43	12.80-17.80 %
White blood cells	6872.72 ± 1794.74	6621.05 ± 1780.94	3.600-11.500/mm^3
Band neutrophils	2.68 ± 1.08	1.78 ± 1.31*	0-5%
Segmented neutrophils	62.7 ± 6.57	60.57 ± 7.52	45-70%
Eosinophils	5.40 ± 11.40	2.94 ± 3.36	1-7%
Basophils	0.00 ± 0.00	0.05 ± 0.22	0-3%
Lymphocytes	30.22 ± 6.61	29.84 ± 6.90	20-50%
Monocytes	1.31 ± 0.64	4.78 ± 1.35*	2-10%
Red blood cells	4.87 ± 0.31	4.74 ± 0.32	4.50-6.10 million/mm^3

Mean Corpuscular Volume	90.51 ± 3.22	89.81 ± 3.15	80.00-98.00m^3
Red cell Distribution Width	12.65 ± 0.52	12.70 ± 0.51	Max. 15.00%
Platelets	206,545.50 ± 47,336.41	233,894.70 ± 50,094.24	150,000-450,000/mm^3

*Significant difference as compared to the control group (unexposed) at $P < 0.05$ (Student *t*-test).

The content of inorganic elements in blood samples of unexposed and exposed groups was analyzed by PIXE, and the presence of eight elements was detected (Table IV). The inorganic elements that appear in significantly increased concentrations in the blood samples of exposed subjects are: zinc (Zn), magnesium (Mg), and aluminum (Al). Moreover, higher levels of chlorine (Cl) were detected relative to the unexposed group.

Table IV. Inorganic element detection by PIXE analyses (ppm; mean values± standard deviation) in the blood samples of unexposed and exposed groups to pesticides.

Inorganic element	Unexposed	Exposed
Zinc	99 ± 28	207 ± 67*
Iron	2077 ± 165	2033 ± 189
Magnesium	562 ± 134	1019 ± 271*
Aluminum	76 ± 4	648 ± 83*
Phosphorus	1599 ± 341	1510 ± 221
Sulfur	3705 ± 155	3664 ± 161
Chlorine	9081 ± 2192	9670 ± 1242
Bromide	12 ± 4	13 ± 8

*$P<0.05$ in relation to unexposed group. Mann-Whitney *U*-test.

DISCUSSION

Pesticides are a heterogeneous category of chemicals, especially designed to control pests, weeds, and plant diseases. The abuse or misuse of pesticides can lead to significant levels of exposure, particularly among those occupationally exposed. The potential cytogenetic damage caused by these compounds has been investigated by researchers in several countries, and literature reviews on this subject (Bolognesi 200[3]) report that most studies found an increase in genotoxicity-biomonitoring end points in subjects using or applying pesticides.

Working environment, personal protective equipment, exposure times and conditions are described in the literature as factors that may affect cytogenetic damage levels (Bolognesi 2003). The agricultural workers in our study did not use any protective measures. In the present study, Comet assay and MN test results showed significantly higher values for the exposed group compared to the unexposed group. Accordingly, this study demonstrates a significant genotoxic and mutagenic effect of exposure to pesticides in tobacco farmers, which is consistent with previous reports. The most frequently exposed agricultural workers are those distributing pesticides by spraying. For these subjects, positive findings were obtained in 18 of 27 studies, representing a 1.12-7.67 times higher exposure rate compared to other workers (Bolognesi 2003). The agricultural workers in this study were exposed to a large number of pesticides, some of which are classified as "hazardous" by the World Health Organization (WHO), e.g. glyphosate, sethoxydim, and chlorpyrifos (Table I). In another study from our group involving tobacco workers from same region in Rio Grande do Sul (Brazil), an increase in DNA damage was observed by various early biological biomarkers of exposure (Da Silva et al. 2012a, b). It is noteworthy that no difference between males and females was observed by Comet assay or MN test in the unexposed and exposed groups. This result is supported by a recent study of Costa et al. (2006), where no significant differences between males and females regarding chromosome aberrations and SCE was observed.

The Comet assay detects recent repairable lesions, such as breaks and alkali-labile sites, while the MN test detects non-repairable damage, such as clastogenic and aneugenic lesions (Buckley et al. 2000, Collins 2004,Iarmarcovai et al. 2008). In this study, we observed predominantly damage classes related to single strand breaks (classes 1, 2, and 3; data not shown), i.e. some reminiscent lesions or incorrect rejoining of the DNA molecules, which could potentially lead to MN formation (Videla et al. 2003, Iarmarcovai et al. 2008). Some evidence has been presented, suggesting that metal ions contained in some pesticides may interfere with distinct steps of various DNA repair systems and produce reactive oxygen species (ROS), leading to oxidative DNA damage (Videla et al. 2003). It is also noteworthy that in our study, the MN frequency was observed to be increased fivefold between exposed and unexposed individuals. In a review about biomonitoring DNA damage by MN in human buccal cells, studies about pesticide exposure have shown a ca. onefold increase in MN frequency (Holland et al. 2008).

In addition, many pesticides used extensively on tobacco leaves have metal-based formulations, some of which are associated with DNA damage. Pesticides classified as "moderately hazardous", such as Mancozeb and Propinebe contain zinc (Zn) and manganese (Mn). Furthermore, the insecticide magnesium aluminum phosphide that is classified as "highly hazardous" (Table I) contains magnesium (Mg) and aluminum (Al). In the whole blood samples of tobacco farmers analyzed in this study, significantly increased levels of Zn, Mg, and Al, and chlorine (Cl) were detected by PIXE in the exposed group (Table IV). However, this only confirms exposure to these chemicals. In general, the genotoxicity of metals proceeds via indirect mechanisms. Recently, Beyersmann and Hartwig (2008) described three predominant mechanisms for the genotoxicity of metals: (1) interference with cellular redox regulation and induction of oxidative stress, which may cause oxidative DNA damage or trigger signaling cascades leading to stimulation of cell growth; (2) inhibition of major DNA repair systems, resulting in genomic instability and accumulation of critical mutations; (3) deregulation of cell proliferation by induction of signaling pathways or inactivation of growth controls such as tumor suppressor genes. In addition, several other studies in the literature relate exposure of individuals to inorganic elements with cancer, including those observed in this study (Beyersmann and Hartwig 2008, Khlifi and Hamza-Chaffai 2010, Tokar et al. 2011, Zhao et al. 2014). A previous study from our group investigating tobacco farmers was able to associate DNA damage with increased blood levels of Mg, Al, Zn, and Cl (Da Silva et al. 2012a).

We also noticed a significantly increased SOD activity (Figure 2) in the agricultural workers exposed to pesticides. SOD is an important antioxidant enzyme that metabolizes the superoxide anion into peroxide. The increased activity of this enzyme suggests an increased generation of the superoxide anion by exposure to pesticides.Shadnia et al. (200[5]) previously demonstrated that chronic exposure to pesticides is associated with increased SOD activity. Other studies have reported that exposure to different categories of pesticides, e.g. organophosphates, carbamates, or pyrethroids leads to oxidative stress in individuals spraying pesticides (Prakasam et al. 2001, Da Silva et al. 2012a).

A significant increase in MN frequency was observed for the *PON1 Gln/Gln* genotype (Figure 1). Enzymes from the *PON1* genes have been reported to be responsible for the metabolism of pesticides, suggesting that the 'unfavorable' alleles could have increased the body load of reactive genotoxic agents in the exposed subjects (Bolognesi 2003, Da Silva et al. 2008). The hydrolysis of paraoxon is catalyzed by serum paraoxonase (*PON1*) / arylesterase, an enzyme associated with the lipoprotein fraction of the serum. Our study revealed that the MN frequency increased in *PON1 Gln/Gln* individuals compared to *PON1 Arg/-* individuals. It has been suggested that individuals with low enzyme levels may be more susceptible to the toxic effects of organophosphates (Bolognesi 2003). This was also observed in our study, where the exposed *Gln/Gln* homozygote individuals presented higher levels of genotoxicity, i.e. higher MN frequencies as a result of exposure to these pesticides. Da Silva et al. (2008)showed that the mean MN frequency in a group of farmers from Caxias do Sul (Rio Grande do Sul, Brazil) exposed to organophosphates was significantly higher for *Gln/Gln* homozygote individuals relative to other genotypes. No increase in DI or DF (repairable damage) was observed for *PON1 Gln/Gln*, which allows the conclusion that the micronuclei observed may be due to a deficiency in the DNA repair mechanisms.

In conclusion, our study demonstrates the presence of genotoxic effects in blood cells and exfoliated buccal cells of agricultural workers exposed to pesticides. DNA damage may be a consequence of oxidative damage, resulting from exposure to complex pesticide mixtures containing metals. These genotoxic effects were observed during the high-exposure period, when pesticides are applied on a daily basis. However, it is important to emphasize that protective measures are essential to avoid potential health risks. In addition, some effects of genetic polymorphism in the *PON1* gene were observed in the modulation of MN results in the exposed group. Application of new techniques to the occupational exposure to pesticides may facilitate a better understanding of the mechanisms involved in toxicity and carcinogenesis of inorganic elements. Such studies may eventually result in the development of molecular markers for pesticide exposure, toxicity, and carcinogenesis. However, a risk assessment of pesticide users is crucial to prevent long-term health hazards leading to the development of cancer and other degenerative diseases.

ACKNOWLEDGMENTS

The authors would like to thank the agricultural workers who participated in this study. Moreover, the Technical and Agricultural Assistance Extension Institute (EMATER, RS), the Laboratório de Santa Cruz do Sul ENZILAB (RS, Brazil), as well as Silvane de Melllo and Clarice Maria de Araújo Hochscheidt (communitary health agents) are gratefully acknowledged for their collaboration. This work was financially supported by Conselho Nacional de Desenvolvimento Científico e Tecnológico (CNPq), and the Fundação de Amparo à Pesquisa do Estado do Rio Grande do Sul (FAPERGS).

REFERENCES

1. AFUBRA - Associação dos Fumicultores do Brasil. 2011. http://www.afubra.com.br (September 9, 2011).
2. Au WW, Sierra-Torres CH, Cajas-Salazar N, Shipp BK and Legator MS. 1999. Cytogenetic effects from exposure to mixed pesticides and the influence from genetic susceptibility. Environ Health Perspect 107: 501-505.
3. Beyersmann D and Hartwig A. 2008. Carcinogenic metal compounds: recent insight into molecular and cellular mechanisms. Arch Toxicol 82: 493-512.
4. Bolognesi C. 2003. Genotoxicity of pesticides: a review of human biomonitoring studies. Mutat Res 543: 251-272.
5. Bolognesi C, Creus A, Ostrosky-Wegman P and Marcos R. 2011. Micronuclei and pesticide exposure. Mutagenesis 26: 19-26.
6. Buckley JD, Meadows AT, Kadin ME, Le Beau MM, Siegel S and Robison LL. 2000. Pesticide exposures in children with non-Hodgkin lymphoma. Cancer 89: 2315-2321.
7. Bull S, Fletcher K, Boobis AR and Battershill JM. 2006. Evidence for genotoxicity of pesticides in pesticide applicators: a review. Mutagenesis21: 93-103.
8. Campbell JL, Hopman TL, Maxwell JA and Nejedly Z. 2000. The Guelph PIXE software package III: alternative proton database. Nucl Instrum Methods Phys Res B 170: 193-204.
9. Carrano AV and Natarajan AT. 1988. International Commission for Protection Against Environmental

Mutagens and Carcinogens. ICPEMC publication no. 14. Considerations for population monitoring using cytogenetic techniques. Mutat Res204: 379-406.

10. Collins AR. 2004. The comet assay for DNA damage and repair: principles, applications, and limitations. Mol Biotechnol 26: 249-261.

11. Collins AR, Oscoz AA, Brunborg G, Gaivao I, Giovannelli L, Kruszewski M, Smith CC and Stetina R. 2008. The comet assay: topical issues. Mutagenesis23: 143-151.

12. Costa C, Teixeira JP, Silva S, Roma-Torres J, Coelho P, Gaspar J, Alves M, Laffon B, Rueff J and Mayan O. 2006. Cytogenetic and molecular biomonitoring of a Portuguese population exposed to pesticides. Mutagenesis21: 343-350.

13. Da Silva FR et al. 2012a. Genotoxic biomonitoring of tobacco farmers: Biomarkers of exposure, of early biological effects and of susceptibility. J Hazard Mater 225-226: 81-90.

14. Da Silva FR, Da Silva J, Nunes E, Benedetti D, Kahl V, Rohr P, Abreu MB, Thiesen FV and Kvitko K. 2012b. Application of the buccal micronucleus cytome assay and analysis of PON1Gln192Arg and CYP2A6*9(-48T>G) polymorphisms in tobacco farmers. Environ Mol Mutagen 53: 525-534.

15. Da Silva J, Freitas TRO, Marinho JR, Speit G and Erdtmann B. 2000. An alkaline single-cell gel electrophoresis (comet) assay for environmental biomonitoring with native rodents. Genet Mol Biol 23: 241-245.

16. Da Silva J et al. 2008. Evaluation of genetic damage in a Brazilian population occupationally exposed to pesticides and its correlation with polymorphisms in metabolizing genes. Mutagenesis23: 415-422.

17. Falck GC, Hirvonen A, Scarpato R, Saarikoski ST, Migliore L and Norppa H. 1999. Micronuclei in blood lymphocytes and genetic polymorphism for GSTM1, GSTT1 and NAT2 in pesticide-exposed greenhouse workers. Mutat Res441: 225-237.

18. Gauthier E, Fortier I, Courchesne F, Pepin P, Mortimer J and Gauvreau D. 2001. Environmental pesticide exposure as a risk factor for Alzheimer's disease: a case-control study. Environ Res 86: 37-45.

19. Grover P, Danadevi K, Mahboob M, Rozati R, Banu BS and Rahman MF. 2003. Evaluation of genetic damage in workers employed in pesticide production utilizing the Comet assay. Mutagenesis18: 201-205.

20. Heuser VD, Erdtmann B, Kvitko K, Rohr P and Da Silva J. 2007. Evaluation of genetic damage in Brazilian footwear-workers: biomarkers of exposure, effect, and susceptibility. Toxicology 232: 235-247.

21. Holland N, Bolognesi C, Kirsch-Volders M, Bonassi S, Zeiger E, Knasmueller S and Fenech M. 2008. The micronucleus assay in human buccal cells as a tool for biomonitoring DNA damage: the HUMN project perspective on current status and knowledge gaps. Mutat Res 659(1-2): 93-108.

22. Humbert R, Adler DA, Disteche CC, Hassett C, Omiecinski CJ and Furlong CE. 1993. The molecular basis of the human serum paraoxonase activity polymorphism. Nat Genet 3: 73-76.

23. Iarmarcovai G, Bonassi S, Botta A, Baan RA and Orsiere T. 2008. Genetic polymorphisms and micronucleus formation: a review of the literature. Mutat Res658: 215-233.

24. Johansson SAE, Campbell JL and Malmqvist KG. 1995. Particle-induced X-ray emission spectrometry (PIXE), Chemical Analysis, Vol. 133, Wiley, New York, 415 p.

25. Khlifi R and Hamza-Chaffai A. 2010. Head and neck cancer due to heavy metal exposure via tobacco smoking and professional exposure: a review. Toxicol Appl Pharmacol 248(2): 71-88.

26. Lahiri DK and Nurnberger Jr JI. 1991. A rapid non-enzymatic method for the preparation of HMW DNA from blood for RFLP studies. Nucleic Acids Res 19: 5444.

27. Meinert R, Schuz J, Kaletsch U, Kaatsch P and Michaelis J. 2000. Leukemia and non-Hodgkin's lymphoma in childhood and exposure to pesticides: results of a register-based case-control study in Germany. Am J Epidemiol 151: 639-646.

28. Nadin SB, Vargas-Roig LM and Ciocca DR. 2001. A silver staining method for single-cell gel assay. J Histochem Cytochem 49: 1183-1186.

29. Prakasam A, Sethupathy S and Lalitha S. 2001. Plasma and RBCs antioxidant status in occupational male pesticide sprayers. Clin Chim Acta 310: 107-112.

30. Shadnia S, Azizi E, Hosseini R, Khoei S, Fouladdel S, Pajoumand A, Jalali N and Abdollahi M. 2005. Evaluation of oxidative stress and genotoxicity in organophosphorus insecticide formulators. Hum Exp Toxicol 24: 439-445.

31. Shaham J, Kaufman Z, Gurvich R and Levi Z. 2001. Frequency of sister-chromatid exchange among greenhouse farmers exposed to pesticides. Mutat Res491: 71-80.

32. Singh NP, McCoy MT, Tice RR and Schneider EL. 1988. A simple technique for quantitation of low levels of DNA damage in individual cells. Exp Cell Res 175: 184-191.

33. Tice RR, Agurell E, Anderson D, Burlinson B, Hartmann A, Kobayashi H, Miyamae Y, Rojas E, Ryu JC and Sasaki YF. 2000. Single cell gel/comet assay: guidelines for in vitro and in vivo genetic toxicology testing. Environ Mol Mutagen35: 206-221.

34. Titenko-Holland N, Jacob RA, Shang N, Balaraman A and Smith MT. 1998. Micronuclei in lymphocytes and exfoliated buccal cells of postmenopausal women with dietary changes in folate. Mutat Res417: 101-114.

35. Tokar EJ, Benbrahim-Tallaa L and Waalkes MP. 2011. Metal ions in human cancer development. Met Ions Life Sci 8: 375-401.

36. Tolbert PE, Shy CM and Allen JW. 1992. Micronuclei and other nuclear anomalies in buccal smears: methods development. Mutat Res271: 69-77.

37. Videla LA, Fernandez V, Tapia G and Varela P. 2003. Oxidative stress-mediated hepatotoxicity of iron and copper: role of Kupffer cells. Biometals 16: 103-111.

38. Zeljezic D and Garaj-Vrhovac V. 2001. Chromosomal aberration and single cell gel electrophoresis (Comet) assay in the longitudinal risk assessment of occupational exposure to pesticides. Mutagenesis16: 359-363.

39. Zeljezic D and Garaj-Vrhovac V. 2002. Sister chromatid exchange and proliferative rate index in the longitudinal risk assessment of occupational exposure to pesticides. Chemosphere 46: 295-303.

40. Zhao Q et al. 2014. Potential health risks of heavy metals in cultivated topsoil and grain, including correlations with human primary liver, lung and gastric cancer, in Anhui province, Eastern China. Sci Total Environ 470-471: 340-347.

INDEX

I

L

M

N

O

P

R

S

T

V

W